Liquid Crystals

Liquid Crystals

Second Edition

IAM-CHOON KHOO

WILEY-INTERSCIENCE
A JOHN WILEY & SONS, INC., PUBLICATION

Published by John Wiley & Sons, Inc., Hoboken, New Jersey
Published simultaneously in Canada

For general information on our other products and services or for technical support, please contact our Customer Care Department within the United States at 877-762-2974, outside the United States at 317-572-3993 or fax 317- 572-4002.

Wiley also publishes its books in a variety of electronic formats. Some content that appears in print may not be available in electronic formats. For more information about Wiley products, visit our web site at www.wiley.com.

Wiley Bicentennial Logo : Richard J. Pacifico

Library of Congress Cataloging-in-Publication Data:

Khoo, Iam-Choon.
  Liquid crystals / Iam-Choon Khoo. —2nd ed.
      p. cm.
  Includes bibliographical references and index.
  ISBN 978-0-471-75153-3
  1. Liquid crystals. I. Title.
  QD923. K49   2007
  530.4'29—dc22   2006048260

Printed in the United States of America

10 9 8 7 6 5 4 3 2 1

Contents

Chapter 10.   Electronic Optical Nonlinearities   253

Chapter 11.   Introduction to Nonlinear Optics   273

Preface

Liquid crystal science and applications now permeate almost all segments of the society—from large industrial displays to individual homes and offices. Nondisplay applications in nonlinear optics, optical communication, and data/signal/image processing are receiving increasing attention and are growing at a rapid pace. Since the last edition (1995), tremendous progress has been made in the study of optics of liquid crystals, and advances are measured by orders of magnitude rather than small increments. Feature sizes have shrunk from microns to nanometers; optical nonlinearities, such as the refractive index coefficient, have "grown" a millionfold from 10^{-3} to 10^3 cm^2/W. This book is intended to capture the essentials of these fundamental breakthroughs and point out new exciting possibilities, while providing the reader a basic overall grounding in liquid crystal optics.

This edition of *Liquid Crystals*, consisting of ten original book chapters that have been completely revised with the addition of the latest concepts, devices, applications, literature, and two new chapters, provides a comprehensive coverage of the fundamentals of liquid crystal physical, optical, electro- and nonlinear optical properties, and related optical phenomena. It is intended for students in their final years of undergraduate studies in the sciences and beginning graduate students and researchers. It will also serve as a general useful introductory guide to all newcomers to the field of liquid crystals and contemporary optics and photonics materials and devices.

Studies of liquid crystals are highly inter- and multidisciplinary, encompassing physics, materials science, optics, and engineering. Liquid crystal science and technologies are rapidly advancing. As such, details of actual devices and applications may likely change or become obsolete. In order to present a treatment that is useful and more readily understandable to the intended readers, I have limited the discussion to only the fundamentals that can withstand the passage of time. Wherever possible and without loss of the physics, I have replaced vigorous theoretical formalisms with their simplified versions, for the sake of clarity.

Chapters 1–5 cover the basic physics and optical properties of liquid crystals intended for beginning workers in liquid crystal related areas. Although the major focus is on nematics, we have included sufficient discussions on other mesophases of liquid crystals such as the smectics, ferroelectrics, and cholesterics to enable the readers to proceed to more advanced or specialized topics elsewhere. New sections have also been added. For example, in Chapter 4, a particularly important addition is a quantitative discussion of the optical properties and fundamentals of one-dimensional photonic crystal band structures. Dispersion is added to fill in an important gap in most treatments of cholesteric liquid crystals.

In Chapter 6, we explore the fundamentals of liquid crystals for electro-optics and display, and nondisplay related applications such as sensing, switching and specialized nanostructured tunable photonic crystals and frequency selective surfaces. In Chapter 7 we provide a thorough account of the various theoretical and computational techniques used to describe optical propagation through liquid crystals and anisotropic materials.

Chapters 8–12 provide the most comprehensive self-contained treatment of nonlinear optics of liquid crystals available anywhere, and have greatly expanded on the coverage of the same subject matter in the previous edition of the book with updated literature reviews and fundamental discussions. In particular, readers will find quantitative and complete theories and analysis of important nonlinear optical processes such as photorefractivity, various all-optical image/beam processing, stimulated scattering and optical phase conjugation, nonlinear multiphoton absorptions, and optical limiting of short laser pulses and continuous-wave lasers.

During the course of writing this book, as in my other work, I have enjoyed valuable encouragement and support from a wide spectrum of people. First and foremost is my wife, Chor San. Her patience, understanding, and unqualified support are important sources of strength and motivation. I would also like to express my gratitude to my present and former students and co-authors for valuable contributions to the advances we have made together and the rewarding life experiences. Support from the National Science Foundation, Army Research Office, Air Force Office of Scientific Research, Defense Advanced Research Projects Agency, and Naval Air Development Center over the years is also gratefully acknowledged.

IAM-CHOON KHOO

University Park, Pennsylvania

1

Introduction to Liquid Crystals

1.1. MOLECULAR STRUCTURES AND CHEMICAL COMPOSITIONS

Liquid crystals are wonderful materials. In addition to the solid crystalline and liquid phases, liquid crystals exhibit intermediate phases where they flow like liquids, yet possess some physical properties characteristic of crystals. Materials that exhibit such unusual phases are often called mesogens (i.e., they are mesogenic), and the various phases in which they could exist are termed mesophases.[1,2] The well-known and widely studied ones are thermotropics, polymerics,[3] and lyotropics. As a function of temperature, or depending on the constituents, concentration, substituents, and so on, these liquid crystals exist in many so-called mesophases—nematic, cholesteric, smectic, and ferroelectric. To understand the physical and optical properties of these materials, we will begin by looking into their constituent molecules.[4]

1.1.1. Chemical Structures

Figure 1.1 shows the basic structures of the most commonly occurring liquid crystal molecules. They are aromatic, and, if they contain benzene rings, they are often referred to as benzene derivatives. In general, aromatic liquid crystal molecules such as those shown in Figure 1.1 comprise a side chain R, two or more aromatic rings A and A', connected by linkage groups X and Y, and at the other end connected to a terminal group R'.

Examples of side-chain and terminal groups are alkyl (C_nH_{2n+1}), alkoxy ($C_nH_{2n+1}O$), and others such as acyloxyl, alkylcarbonate, alkoxycarbonyl, and the nitro and cyano groups. The Xs of the linkage groups are simple bonds or groups such as stilbene

$(-CH=CH-)$, ester ($-\overset{\displaystyle O}{\underset{\displaystyle C}{\|}}-O-$), tolane ($-C\equiv C-$), azoxy ($-N=N-$), Schiff base

$(-CH=N-)$, acetylene ($-C\equiv C-$), and diacetylene ($-C\equiv C-C\equiv C-$). The names of liquid crystals are often fashioned after the linkage group (e.g., Schiff-base liquid crystal).

Liquid Crystals, Second Edition   By Iam-Choon Khoo
Copyright © 2007 John Wiley & Sons, Inc.

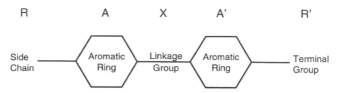

Figure 1.1. Molecular structure of a typical liquid crystal.

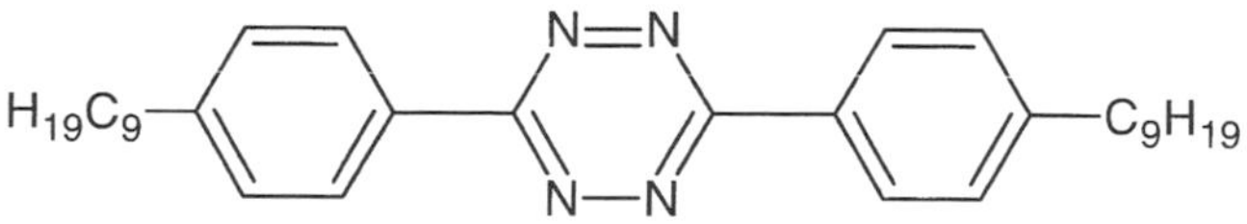

Figure 1.2. Molecular structure of a heterocyclic liquid crystal.

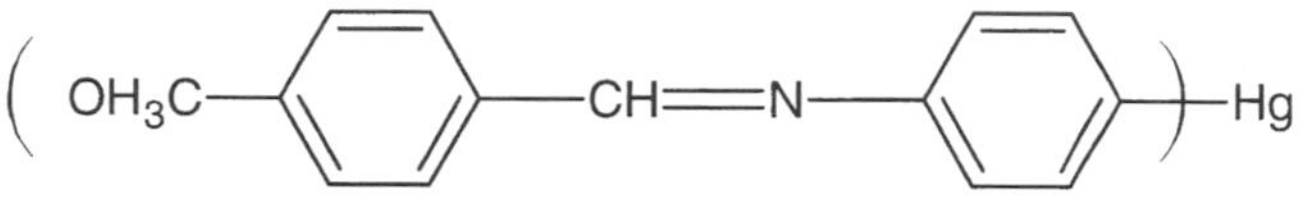

Figure 1.3. Molecular structure of an organometallic liquid crystal.

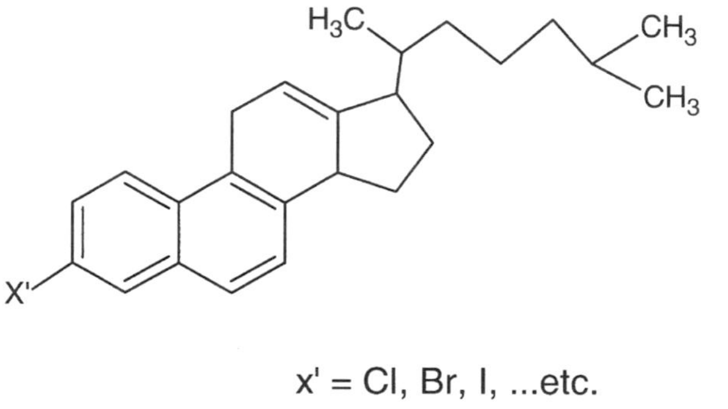

x' = Cl, Br, I, ...etc.

Figure 1.4. Molecular structure of a sterol.

There are quite a number of aromatic rings. These include saturated cyclohexane or unsaturated phenyl, biphenyl, and terphenyl in various combinations.

The majority of liquid crystals are benzene derivatives mentioned previously. The rest include heterocyclics, organometallics, sterols, and some organic salts or fatty acids. Their typical structures are shown in Figures 1.2–1.4.

Heterocyclic liquid crystals are similar in structure to benzene derivatives, with one or more of the benzene rings replaced by a pyridine, pyrimidine, or other similar groups. Cholesterol derivatives are the most common chemical compounds that exhibit the cholesteric (or chiral nematic) phase of liquid crystals. Organometallic

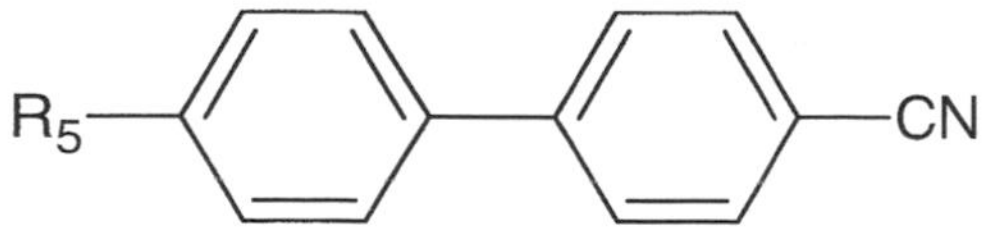

Figure 1.5. Molecular structure of pentylcyanobiphenyl (5CB).

compounds are special in that they contain metallic atoms and possess interesting dynamical and magneto-optical properties.[4]

All the physical and optical properties of liquid crystals are governed by the properties of these constituent groups and how they are chemically synthesized together. Dielectric constants, elastic constants, viscosities, absorption spectra, transition temperatures, existence of mesophases, anisotropies, and optical nonlinearities are all consequences of how these molecules are engineered. Since these molecules are quite large and anisotropic, and therefore very complex, it is practically impossible to treat all the possible variations in the molecular architecture and the resulting changes in the physical properties. Nevertheless, there are some generally applicable observations on the dependence of the physical properties on the molecular constituents. These will be highlighted in the appropriate sections.

The chemical stability of liquid crystals depends very much on the central linkage group. Schiff-base liquid crystals are usually quite unstable. Ester, azo, and azoxy compounds are more stable, but are also quite susceptible to moisture, temperature change, and ultraviolet (UV) radiation. Compounds without a central linkage group are among the most stable liquid crystals ever synthesized. The most widely studied one is pentylcyanobiphenyl (5CB), whose structure is shown in Figure 1.5. Other compounds such as pyrimide and phenylcyclohexane are also quite stable.

1.2. ELECTRONIC PROPERTIES

1.2.1. Electronic Transitions and Ultraviolet Absorption

The electronic properties and processes occurring in liquid crystals are decided largely by the electronic properties of the constituent molecules. Since liquid crystal constituent molecules are quite large, their energy level structures are rather complex. As a matter of fact, just the process of writing down the Hamiltonian for an isolated molecule itself can be a very tedious undertaking. To also take into account interactions among the molecular groups and to account for the difference between individual molecules' electronic properties and the actual liquid crystals' responses will be a monumental task. It is fair to say that existing theories are still not sufficiently precise in relating the molecular structures and the liquid crystal responses. We shall limit ourselves here to stating some of the well-established results, mainly from molecular theory and experimental observations.

In essence, the basic framework of molecular theory is similar to that described in Chapter 10, except that much more energy levels, or bands, are involved. In general,

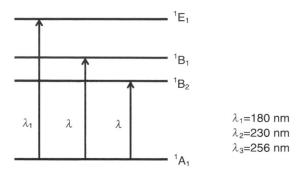

Figure 1.6. $\pi \rightarrow \pi^*$ electronic transitions in a benzene molecule.

the energy levels are referred to as orbitals. There are π, n, and σ orbitals, with their excited counterparts labeled as π^*, n^*, and σ^*, respectively. The energy differences between these electronic states which are connected by dipole transitions give the so-called resonant frequencies (or, if the levels are so large that bands are formed, give rise to absorption bands) of the molecule; the dependence of the molecular suscepti-bility on the frequency of the probing light gives the dispersion of the optical dielec-tric constant (see Chapter 10).

Since most liquid crystals are aromatic compounds, containing one or more aro-matic rings, the energy levels or orbitals of aromatic rings play a major role. In par-ticular, the $\pi \rightarrow \pi^*$ transitions in a benzene molecule have been extensively studied. Figure 1.6 shows three possible $\pi \rightarrow \pi^*$ transitions in a benzene molecule.

In general, these transitions correspond to the absorption of light in the near-UV spectral region (≤ 200 nm). These results for a benzene molecule can also be used to interpret the absorption of liquid crystals containing phenyl rings. On the other hand, in a saturated cyclohexane ring or band, usually only σ electrons are involved. The $\sigma \rightarrow \sigma^*$ transitions correspond to the absorption of light of shorter wavelength (≤ 180 nm) in comparison to the $\pi \rightarrow \pi^*$ transition mentioned previously.

These electronic properties are also often viewed in terms of the presence or absence of conjugation (i.e., alternations of single and double bonds, as in the case of a benzene ring). In such conjugated molecules the π electron's wave function is delo-calized along the conjugation length, resulting in the absorption of light in a longer wavelength region compared to, for example, that associated with the σ electron in compounds that do not possess conjugation. Absorption data and spectral depend-ence for a variety of molecular constituents, including phenyl rings, biphenyls, terphenyls, tolanes, and diphenyl-diacetylenes, may be found in Khoo and Wu.[5]

1.2.2.   Visible and Infrared Absorption

From the preceding discussion, one can see that, in general, liquid crystals are quite absorptive in the UV region, as are most organic molecules. In the visible and near-infrared regimes (i.e., from 0.4 to 5 μm), there are relatively fewer absorption bands, and thus liquid crystals are quite transparent in these regimes.

As the wavelength is increased toward the infrared (e.g., ≥ 9 μm), rovibrational transitions begin to dominate. Since rovibrational energy levels are omnipresent in all large molecules, in general, liquid crystals are quite absorptive in the infrared regime.

The spectral transmission dependence of two typical liquid crystals is shown in Figures 1.7a and 1.7b. The absorption coefficient α in the ultraviolet (~ 0.2 μm) regime is on the order of 10^3 cm^{-1}; in the visible (~ 0.5 μm) regime, $\alpha \approx 10^0$ cm^{-1}; in the near-infrared (~ 10 μm) regime, $\alpha \leq 10^2$ cm^{-1}; and in the infrared (~ 10 μm) regime, $\alpha \leq 10^2$ cm^{-1}. There are, of course, large variations among the thousands of liquid crystals "discovered" or engineered so far, hence it is possible to identify liquid crystals with the desired absorption/transparency for a particular wavelength of interest.

Outside the far-infrared regime, e.g., in the microwave region, there have also been active studies.[6] At the 20–60 GHz region, for example, liquid crystals continue to exhibit sizable birefringence. Studies have shown that for a typical liquid crystal such as E7, the dielectric permittivities for extraordinary and ordinary waves are

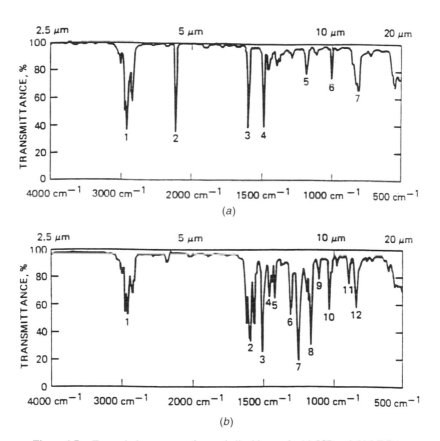

Figure 1.7. Transmission spectra of nematic liquid crystals: (a) 5CB and (b) MBBA.

ε_e = 3.17 (refractive index n_e = 1.78) and ε_0 = 2.72 (refractive index n_0 = 1.65), respectively, i.e., a birefringence of $\Delta n \sim 0.13$.

1.3. LYOTROPIC, POLYMERIC, AND THERMOTROPIC LIQUID CRYSTALS

One can classify liquid crystals in accordance with the physical parameters controlling the existence of the liquid crystalline phases. There are three distinct types of liquid crystals: lyotropic, polymeric, and thermotropic. These materials exhibit liquid crystalline properties as a function of different physical parameters and environments.

1.3.1. Lyotropic Liquid Crystals

Lyotropic liquid crystals are obtained when an appropriate concentration of a material is dissolved in some solvent. The most common systems are those formed by water and amphiphilic molecules (molecules that possess a hydrophilic part that interacts strongly with water and a hydrophobic part that is water insoluble) such as soaps, detergents, and lipids. Here the most important variable controlling the existence of the liquid crystalline phase is the amount of solvent (or concentration). There are quite a number of phases observed in such water-amphiphilic systems, as the composition and temperature are varied; some appear as spherical micelles, and others possess ordered structures with one-, two-, or three-dimensional positional order. Examples of these kinds of molecules are soaps (Fig. 1.8) and various phospholipids like those present in cell membranes. Lyotropic liquid crystals are of interest in biological studies.[2]

1.3.2. Polymeric Liquid Crystals

Polymeric liquid crystals are basically the polymer versions of the monomers discussed in Section 1.1. There are three common types of polymers, as shown in Figures 1.9a–1.9c, which are characterized by the degree of flexibility. The vinyl type (Fig. 1.9a) is the most flexible; the Dupont Kevlar polymer (Fig. 1.9b) is semirigid; and the polypeptide chain (Fig. 1.9c) is the most rigid. Mesogenic (or liquid

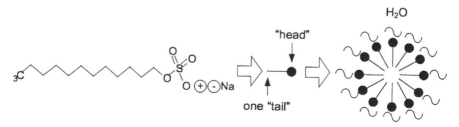

Figure 1.8. Chemical structure and cartoon representation of sodium dodecylsulfate (soap) forming micelles.

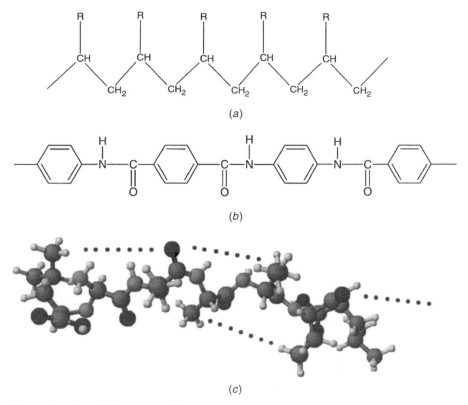

Figure 1.9. Three different types of polymeric liquid crystals. (a) Vinyl type; (b) Kevlar polymer; (c) polypeptide chain.

crystalline) polymers are classified in accordance with the molecular architectural arrangement of the mesogenic monomer. Main-chain polymers are built up by joining together the rigid mesogenic groups in a manner depicted schematically in Figure 1.10a; the link may be a direct bond or some flexible spacer. Liquid crystal side-chain polymers are formed by the pendant side attachment of mesogenic monomers to a conventional polymeric chain, as depicted in Figure 1.10b. A good account of polymeric liquid crystals may be found in Ciferri et al.[3] In general, polymeric liquid crystals are characterized by much higher viscosity than that of monomers, and they appear to be useful for optical storage applications.

1.3.3. Thermotropic Liquid Crystals: Nematics, Cholesterics, and Smectics

The most widely used liquid crystals, and extensively studied for their linear as well as nonlinear optical properties, are thermotropic liquid crystals. They exhibit various liquid crystalline phases as a function of temperature. Although their molecular structures, as discussed in Section 1.1, are, in general, quite complicated, they are

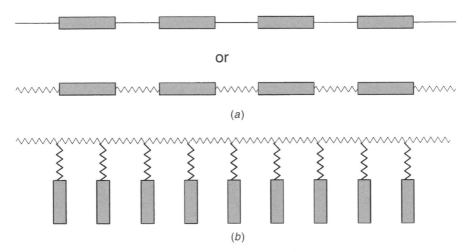

Figure 1.10. Polymeric liquid crystals: (a) main chain and (b) side chain.

often represented as "rigid rods." These rigid rods interact with one another and form distinctive ordered structures. There are three main classes of thermotropic liquid crystals: nematic, cholesteric, and smectic. There are several subclassifications of smectic liquid crystals in accordance with the positional and directional arrangements of the molecules.

These mesophases are defined and characterized by many physical parameters such as long- and short-range order, orientational distribution functions, and so on. They are explained in greater detail in the following chapters. Here we continue to use the rigid-rod model and pictorially describe these phases in terms of their molecular arrangement.

Figure 1.11a depicts schematically the collective arrangement of the rodlike liquid crystal molecules in the nematic phase. The molecules are positionally random, very much like liquids; x-ray diffraction from nematics does not exhibit any diffraction peak. These molecules are, however, directionally correlated; they are aligned in a general direction defined by a unit vector $\tilde{n}$, the so-called director axis.

In general, nematic molecules are centrosymmetric; their physical properties are the same in the $+\hat{n}$ and the $-\hat{n}$ directions. In other words, if the individual molecules carry a permanent electric dipole (such a polar nature is typically the case), they will assemble in such a way that the bulk dipole moment vanishes.

Cholesterics, now often called chiral nematic liquid crystals, resemble nematic liquid crystals in all physical properties except that the molecules tend to align in a helical manner as depicted in Figure 1.11b. This property results from the synthesis of cholesteric liquid crystals; they are obtained by adding a chiral molecule to a nematic liquid crystal. Some materials, such as cholesterol esters, are naturally chiral.

Smectic liquid crystals, unlike nematics, possess positional order; that is, the position of the molecules is correlated in some ordered pattern. Several subphases of smectics have been "discovered," in accordance with the arrangement or ordering of the molecules and their structural symmetry properties.[1,2] We discuss here three representative ones: smectic-A, smectic-C, and smectic-C* (ferroelectrics).

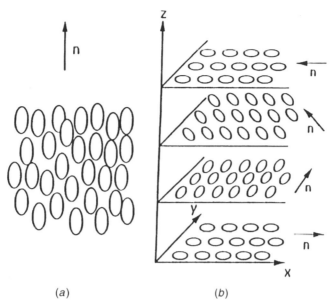

(a)   (b)

Figure 1.11. Molecular alignments of liquid crystals: (a) nematic and (b) cholesteric or chiral nematic.

Figure 1.12a depicts the layered structure of a smectic-A liquid crystal. In each layer the molecules are positionally random, but directionally ordered with their long axis normal to the plane of the layer. Similar to nematics, smectic-A liquid crystals are optically uniaxial, that is, there is a rotational symmetry around the director axis.

The smectic-C phase is different from the smectic-A phase in that the material is optically biaxial, and the molecular arrangement is such that the long axis is tilted away from the layer normal $\hat{z}$ (see Fig. 1.12b).

In smectic-C* liquid crystals, as depicted in Figure 1.12c, the director axis $\hat{n}$ is tilted away from the layer normal $\hat{z}$ and "precesses" around the $\hat{z}$ axis in successive layers. This is analogous to cholesterics and is due to the introduction of optical-active or chiral molecules to the smectic-C liquid crystals.

Smectic-C* liquid crystals are interesting in one important respect—namely, that they comprise a system that permits, by the symmetry principle, the existence of a spontaneous electric polarization. This can be explained simply in the following way.

The spontaneous electric polarization $\hat{p}$ is a vector and represents a breakdown of symmetry; that is, there is a directional preference. If the liquid crystal properties are independent of the director axis $\hat{n}$ direction (i.e., $+\hat{n}$ is the same as $-\hat{n}$), $\hat{p}$, if it exists, must be locally perpendicular to $\hat{n}$. In the case of smectic-A, which possesses rotational symmetry around $\hat{n}$, $\hat{p}$ must therefore be vanishing. In the case of smectic-C, there is a reflection symmetry (mirror symmetry) about the plane defined by the $\hat{n}$ and $\hat{z}$ axes, so $\hat{p}$ is also vanishing.

This reflection symmetry is broken if a chiral center is introduced to the molecule, resulting in a smectic-C* system. By convention, $\hat{p}$ is defined as positive if it is along the direction of $\hat{z} \times \hat{n}$, and as negative otherwise. Figure 1.12c shows that since $\hat{n}$ precesses around $\hat{z}$, $\hat{p}$ also precesses around $\hat{z}$. If, by some external field, the helical

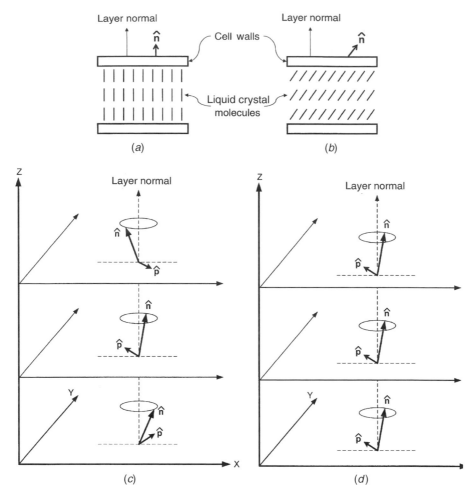

Figure 1.12. Molecular arrangements of liquid crystals: (a) smectic-A, (b) smectic-C, (c) smectic-C* or ferroelectric, and (d) unwound smectic-C*.

structure is unwound and $\hat{n}$ points in a fixed direction, as in Figure 1.12d, then $\hat{p}$ will point in one direction. Clearly, this and other director axis reorientation processes are accompanied by considerable change in the optical refractive index and other properties of the system, and they can be utilized in practical electro- and opto-optical modulation devices. A detailed discussion of smectic liquid crystals is given in Chapter 4.

1.3.4. Other Liquid Crystalline Phases and Molecular Engineered Structures

Besides those phases mentioned above, many other phases of liquid crystals such as smectic G, H, I, F,...,Q,..., and cholesteric blue phase have been identified,[2,7,8] to name a few. Numerous new molecular engineered liquid crystalline

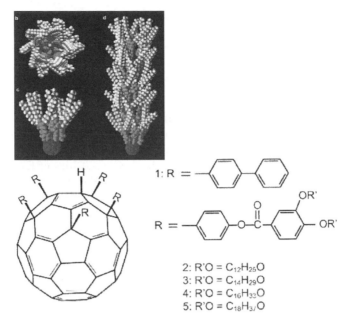

Figure 1.13. Shuttlecock-shaped liquid crystal formed by incorporating fullerene C60 to various liquid crystals reported.[9]

compounds/structures have also emerged.[9,10] Figure 1.13 shows, for example, the shuttlecock-shaped liquid crystal formed by incorporating fullerene C60 to various crystals and liquid crystals reported by Sawamura et al.[9]

1.4. MIXTURES AND COMPOSITES

In general, temperature ranges for the various mesophases of pure liquid crystals are quite limited. This and other physical limitations impose severe shortcomings on the practical usage of these materials. Accordingly, while much fundamental research is still performed with pure liquid crystals, industrial applications employ mostly mixtures, composites, or specially doped liquid crystals with tailor-made physical and optical properties. Current progress and large-scale application of liquid crystals in optical technology are largely the result of tremendous advances in such new-material development efforts.

There are many ways and means of modifying a liquid crystal's physical properties. At the most fundamental level, various chemical groups such as bonds or atoms can be substituted into a particular class of liquid crystals. A good example is the cyanobiphenyl homologous series nCB ($n=1$, 2, 3,...). As n is increased through synthesis, the viscosities, anisotropies, molecular sizes, and many other parameters are greatly modified. Some of these physical properties can also be modified by substitution. For example, the hydrogen in the 2, 3, and 4 positions of the phenyl ring may be substituted by some fluoro (F) or chloro (Cl) group.[11]

Besides these molecular synthesis techniques, there are other physical processes that can be employed to dramatically improve the performance characteristics of liquid crystals. In the following sections we describe three well-developed ones, focusing our discussion on nematic liquid crystals.

1.4.1. Mixtures

A large majority of liquid crystals in current device usage are eutectic mixtures of two or more mesogenic substances. A good example is E7 (from EM Chemicals), which is a mixture of four liquid crystals (see Fig. 1.14).

The optical properties, dielectric anisotropies, and viscosities of E7 are very different from those of the individual mixture constituents. Creating mixtures is an art, guided of course by some scientific principles.[11]

One of the guiding principles for making the right mixture can be illustrated by the exemplary phase diagram of two materials with different melting (i.e., crystal → nematic) and clearing (i.e., nematic → isotropic) points, as shown in Figure 1.15. Both substances have small nematic ranges ($T_i - T_n$ and $T_i' - T_n'$). When mixed at the right concentration,[4] however, the nematic range ($T_i^m - T_n^m$) of the mixture can be several magnitudes larger.

If the mixture components do not react chemically with one another, clearly their bulk physical properties, such as dielectric constant, viscosity, and anisotropy, are some weighted sum of the individual responses; that is, the physical parameter α_m of the mixture is related to the individual responses' α_i's by $\alpha_m = \Sigma c_i \alpha_i$, where c_i is the corresponding molar fraction. However, because of molecular correlation effects and the critical dependence of the constituents on their widely varying transition temperatures and other collective effects, the simple linear additive representation of the mixture's response is at best a rough approximation. In general, one would expect that optical and other parameters (e.g., absorption lines or bands), which depend

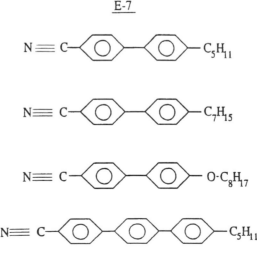

Figure 1.14.   Molecular structures of the four constituents making up the liquid crystal E7 (from EM Chemicals).

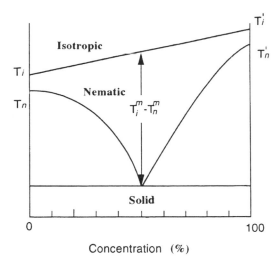

Figure 1.15. Phase diagram of the mixture of two liquid crystals.

largely on the electronic responses of individual molecules, will follow the simple additive rule more closely than physical parameters (e.g., viscosities), which are highly dependent on intermolecular forces.

In accordance with the foregoing discussion, liquid crystal mixtures formed by different concentrations of the same set of constituents should be regarded as physically and optically different materials.

1.4.2. Dye-Doped Liquid Crystals

From the standpoint of optical properties, the doping of liquid crystals by appropriately dissolved concentrations and types of dyes clearly deserves special attention. The most important effect of dye molecules on liquid crystals is the modification of their well-known linear, and more recently observed nonlinear, optical properties (see Chapters 8 and 12).

An obvious effect of dissolved dye is to increase the absorption of a particular liquid crystal at some specified wavelength region. If the dye molecules undergo some physical or orientational changes following photon absorption, they could also affect the orientation of the host liquid crystal, giving rise to nonlinear or storage-type optical effects[12] (see Chapter 8).

In linear optical and electro-optical applications, another frequently employed effect is the so-called guest–host effect. This utilizes the fact that the absorption coefficients of the dissolved dichroic dyes are different for optical fields polarized parallel or perpendicular to the long (optical) axis of the dye molecule. In general, a dichroic dye molecule absorbs much more for optical field polarization parallel to its long axis than for optical field polarization perpendicular to its long axis. These molecules are generally elongated in shape and can be oriented and reoriented by the host nematic liquid crystals. Accordingly, the transmission of the cell can be switched with the application of an external field (see Fig. 1.16).

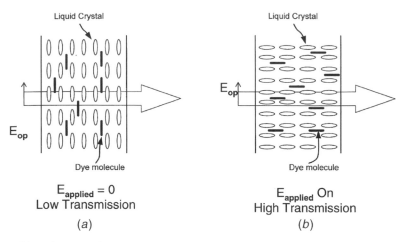

Figure 1.16. Alignment of a dichroic dye-doped nematic liquid crystal: (a) before application of switching electric field; (b) switching field on.

1.4.3. Polymer-Dispersed Liquid Crystals

Just as the presence of dye molecules modifies the absorption characteristics of liquid crystals, the presence of a material interdispersed in the liquid crystals of a different refractive index modifies the scattering properties of the resulting "mixed" system. Polymer-dispersed liquid crystals are formed by introducing liquid crystals as micron- or sub-micron-sized droplets into a polymer matrix. The optical indices of these randomly oriented liquid crystal droplets, in the absence of an external alignment field, depend on the liquid crystal–polymer interaction at the boundary, and therefore assume a random distribution (see Fig. 1.17a). This causes large scattering. Upon the application of an external field, the droplets will be aligned (Fig. 1.17b), and the system will become clear as the refractive index of the liquid crystal droplets matches the isotropic polymer backgrounds.

Polymer-dispersed liquid crystals were introduced many years ago.[13] There are now several techniques for preparing such composite liquid crystalline materials, including the phase separation and the encapsulation methods.[14] More recently, optical holographic interference methods[15–17] have been employed successfully in making polymer-dispersed liquid crystal photonic crystals (regular array of materials of different refractive indices). Caputo et al.[18] and Strangi et al.[19] have also demonstrated one-dimensional (1D) polymer/liquid crystal layered structures that exhibit high diffraction efficiency as well as laser emission capabilities.

1.5. LIQUID CRYSTAL CELLS AND SAMPLE PREPARATION

Liquid crystals, particularly nematics which are commonly employed in many electro-optical devices, behave physically very much like liquids. Milk is often a good analogy

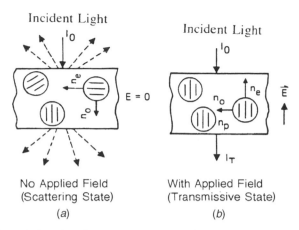

Figure 1.17. Schematic depiction of a polymer-dispersed liquid crystal material: (a) in the absence of an external alignment field (highly scattered state); (b) when an external alignment field is on (transparent state).

to liquid crystals in such bulk, "unaligned" states. Its crystalline properties become apparent when such milky liquids are contained in (usually) flat thin cells. The alignment of the liquid crystal axis in such cells is essentially controlled by the cell walls, whose surfaces are treated in a variety of ways to achieve various director axis alignments.

1.5.1. Bulk Thin Film

For nematics, two commonly used alignments are the so-called homogeneous (or planar) and homeotropic alignments, as shown in Figures 1.18a and 1.18b, respectively. To create homeotropic alignment, the cell walls are treated with a surfactant such as hexadecyl-trimethyl-ammoniumbromide (HTAB).[20] These surfactants are basically soaps, whose molecules tend to align themselves perpendicular to the wall and thus impart the homeotropic alignment to the liquid crystal.

In the laboratory, a quick and effective way to make a homeotropic nematic liquid crystal sample is as follows: Dissolve 1 part of HTAB in 50 parts of distilled deionized water by volume. Clean two glass slides (or other optical flats appropriate for the spectral region of interest). Dip the slides in the HTAB solution and slowly withdraw them. This effectively introduces a coating of HTAB molecules on the glass slides. The glass slides should then be dried in an oven or by other means. To prepare the nematic liquid crystal sample, prepare a spacer (Mylar or some nonreactive plastic) of desirable dimension and thinness and place the spacer on one of the slides. Fill the inner spacer with the nematic liquid crystal under study (it helps to first warm it to the isotropic phase). Place the second slide on top of this and clamp the two slides together. Once assembled, the sample should be left alone, and it will slowly (in a few minutes) settle into a clear homeotropically aligned state.

Planar alignment can be achieved in many ways. A commonly employed method is to first coat the cell wall with some polymer such as polyvinyl alcohol (PVA) and then rub it unidirectionally with a lens tissue. This process creates elongated stress/strain

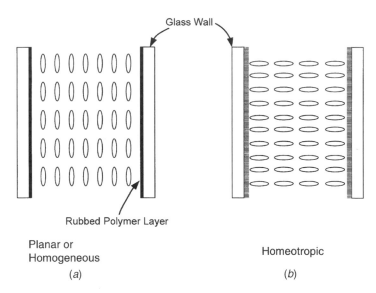

Figure 1.18. Nematic liquid crystal cells: (a) homeogeneous (or planar) aligned and (b) homeotropic aligned.

on the polymer and facilitates the alignment of the long axis of the liquid crystal molecules along the rubbed direction (i.e., on the plane of the cell wall). Another method is to deposit silicon oxide obliquely onto the cell wall.

In preparing a PVA-coated planar sample in the laboratory, the following technique has been proven to be quite reliable. Dissolve chemically pure PVA (which is solid at room temperature) in distilled deionized water at an elevated temperature (near the boiling point) at a concentration of about 0.2%. Dip the cleaned glass slide into the PVA solution at room temperature and slowly withdraw it, thus leaving a film of the solution on the slide. (Alternatively, one could place a small amount of the PVA solution on the slide and spread it into a thin coating.) The coated slide is then dried in an oven, followed by unidirectional rubbing of its surfaces with a lens tissue. The rest of the procedure for cell assembly is the same as that for homeotropic alignment.

Ideally, of course, these cell preparation processes should be performed in a clean room and preferably in an enclosure free of humidity or other chemicals (e.g., a nitrogen-filled enclosure) in order to prolong the lifetime of the sample. Nevertheless, the liquid crystal cells prepared with the techniques outlined previously have been shown to last several months and can withstand many temperature cyclings through the nematic–isotropic phase transition point, provided the liquid crystals used are chemically stable. In general, nematics such as 5CB and E7 are quite stable, whereas p-methoxybenzylidene-p'-n-butylaniline (MBAA) tends to degrade in a few days.

Besides these two standard cell alignments, there are many other variations such as hybrid, twisted, supertwisted, fingerprint, multidomain vertically aligned, etc. Industrial processing of these nematic cells, as well as the transparent conductive coating of the cell windows for electro-optical device applications, is understandably more elaborate.

For chiral nematic liquid crystals, the method outlined previously for a planar nematic cell has been shown to be quite effective. For smectic-A the preparation method is similar to that for a homeotropic nematic cell. In this case, however, it helps to have an externally applied field to help maintain the homeotropic alignment as the sample (slowly) cools down from the nematic to the smectic phase. The cell preparation methods for a ferroelectric liquid crystal (FLC), smectic-C* for surface stabilized FLC (SSFLC) operation, is more complicated as it involves surface stabilization.[21,22] On the other hand, smectic-A* (Sm-A*) cells for soft-mode FLC (SMFLC) operation are easier to prepare using the methods described above.[23]

1.5.2. Liquid Crystal Optical Slab Waveguide, Fiber, and Nanostructured Photonic Crystals

Besides the bulk thin film structures discussed in the preceding section, liquid crystals could also be fabricated into optical waveguides[24–30] or nanostructured photonic crystals.[31]

Both slab and cylindrical (fiber) waveguide structures have been investigated. A typical liquid crystal slab waveguide[24,25] is shown in Figure 1.19. A thin film (approximately 1 μm) of liquid crystal is sandwiched between two glass slides (of lower refractive index than the liquid crystal), one of which has been deposited with an organic film into which an input laser is introduced via the coupling prism. The laser excites the transverse electric (TE) and/or transverse magnetic (TM) modes in the organic film, which are then guided into the nematic liquid crystal region. Using such optical waveguides, Whinnery et al.[24] and Giallorenzi et al.[25] have measured the scattering losses in nematic and smectic liquid crystals and introduced electro-optical and integrated optical switching devices. However, the large losses in nematics (about 20 dB/cm) and their relatively slow responses impose serious limitations in practical integrated electro-optical applications. The scattering losses in smectic waveguides are generally much lower, and they may be useful in nonlinear optical applications (see Chapter 10).

Liquid crystal "fibers" are usually made by filling hollow fibers (microcapillaries) made of material of lower indices of refraction.[26,27] The microcapillaries are usually

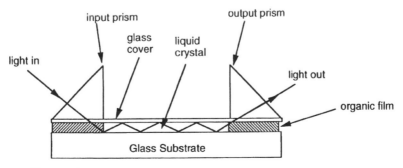

Figure 1.19. Schematic depiction of a liquid crystal slab waveguide structure.

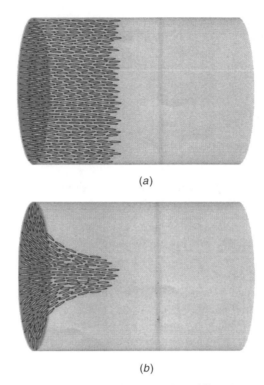

(a)

(b)

Figure 1.20. (a) Axial alignment of a nematic liquid crystal cored fiber; (b) mixed radial and axial alignments of a nematic liquid crystal cored fiber.

made of Pyrex or silica glass, whose refractive indices are 1.47 and 1.45, respectively. It was reported[26] that the scattering losses of the nematic liquid crystal fiber core are considerably reduced for a core diameter smaller than 10 μm; typically, the loss is about 3 dB/cm (compared to 20 dB/cm for a slab waveguide or bulk thin film). Also, the director axis alignment within the core is highly dependent on the liquid crystals–capillary interface interaction (i.e., the capillary material). In silica or Pyrex capillaries the nematic director tends to align along the axis of the fiber (Fig. 1.20a), whereas in borosilicate capillaries the nematic director tends to align in a radial direction, occasionally mixed in with a thread of axially aligned material running down the axis of the fiber (Fig. 1.20b).

Fabrications of such fibers with isotropic phase liquid crystals are much easier.[27,29] Because of the fluid property and much lower scattering loss, liquid crystal fibers of much longer dimension have been fabricated and shown to exhibit interesting nonlinear optical properties; high quality image transmitting fiber arrays[28,29] have also been fabricated for passive pulsed laser limiting applications. Other optical devices based on liquid crystal filled photonic crystal (holey) fibers have also been reported.[30]

Recently, photonic crystals[31] in one-, two-, and three-dimensional forms have received intense research interest owing to the rich variety of possibilities in terms of

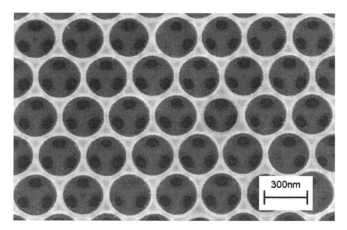

Figure 1.21. TiO$_2$ inverse opal structure for liquid crystal infiltration.

material compositions, lattice structures, and their electronic as well as optical properties. By using an active tunable material as a constituent, photonic crystals can function as tunable filters, switches, and lasing devices. In particular, liquid crystals have been employed in many studies involving opals and inverse opal structures (see Fig. 1.21). In particular, Graugnard et al.[31] has reported non-close-packed inverse opals, consisting of overlapping air spheres in a TiO$_2$ matrix, which were infiltrated with liquid crystal. Because of the higher volume fraction for nematic liquid crystal (NLC) infiltration, a larger electrical tuning range ($>$ 20 nm) of the Bragg reflection peak can be achieved.

REFERENCES

1. deGennes, P. G. 1974. *The Physics of Liquid Crystals*. Oxford: Clarendon Press.

2. Chandrasekhar, S. 1992. *Liquid Crystals*. 2nd ed. Cambridge: Cambridge University Press.

3. Ciferri, A., W. R. Krigbaum, and R. B. Meyer, eds. 1982. *Polymer Liquid Crystals*. New York: Academic Press.

4. Blinov, L. M., and V. G. Chigrinov. 1994. *Electrooptic Effects in Liquid Crystal Materials*. New York: Springer-Verlag.

5. Khoo, I. C., and S. T. Wu. 1992. *Optics and Nonlinear Optics of Liquid Crystals*. Singapore: World Scientific.

6. Yang, F., and J. Roy Sambles. 2003. Determination of the permittivity of nematic liquid crystals in the microwave region. *Liq. Cryst.* 30: 599–602.

7. Wright, D. C., and N. D. Mermin. 1989. *Rev. Mod. Phys.* 61:385.

8. Etchegoin, P., 2000. Blue phases of cholesteric liquid crystals as thermotropic photonic crystals. *Phys. Rev. E.* 62: 1435–1437.

9. Sawamura, M., K. Kawai, Y. Matsuo, K. Kanie, T. Kato, and E. Nakamura. 2002. Stacking of conical molecules with a fullerene apex into polar columns in crystals and liquid crystals. *Nature.* 419: 702–705.

10. See, for example, Nishiyama, Isa, Jun Yamamoto, John W. Goodby, and Hiroshi Yokoyama. 2004. Chirality-induced liquid crystalline nanostructures and their properties. In *Liquid Crystals VIII. SPIE Proceedings*, Vol. 5518, 201–220 and references therein. I. C. Khoo, Bellingham, WA: SPIE; see also Nishiyama, Isa, Jun Yamamoto, John W. Goodby, and Hinshi Yokoyama. 2002. *J. Mater. Chem.* 12: 1709–1716.

11. Gray, G. W., M. Hird, and K. J. Toyne. 1991. The synthesis of several lateral difluorosubstituted 4,4″-dialkyl- and 4,4″-alkoxyalkyl-terphenyls. *Mol. Cryst. Liq. Cryst.* 204:43; see also Wu, S. T., D. Coates, and E. Bartmann. 1991. Physical properties of chlorinated liquid crystals. *Liq. Cryst.* 10:635.

12. See, for example, Khoo, I. C., Min-Yi Shih, M. V. Wood, B. D. Guenther, P. H. Chen, F. Simoni, S. Slussarenko, O. Francescangeli, and L. Lucchetti. 1999. Dye-doped photorefractive liquid crystals for dynamic and storage holographic grating formation and spatial light modulation. In *IEEE Proceedings Special Issue on Photorefractive Optics: Materials, Devices and Applications*. Vol. 87 (11): 1897–1911.

13. Doane, J. W., N. A. Vaz, B. G. Wu, and S. Zumer. 1986. Field controlled light scattering from nematic microdroplets. *Appl. Phys. Lett.* 48:269.

14. West, J. L. 1988. Phase separation of liquid crystals in polymers. *Mol. Cryst. Liq. Cryst.* 157:428; Urzaic, P. 1986. Polymer dispersed nematic liquid crystal for large area displays and light valves. *J. Appl. Phys.* 60:2142.

15. Khoo, I. C., Yana Zhang Williams, B. Lewis, and T. Mallouk, 2005. Photorefractive CdSe and gold nanowire-doped liquid crystals and polymer-dispersed-liquid-crystal photonic crystals. *Mol. Cryst. Liq. Cryst.* 446:233–244.

16. Tondiglia, V. P., L. V. Natarajan, R. L. Sutherland, D. Tomlin, and T. J. Bunning. 2002. Holographic formation of electro-optical polymer-liquid crystal photonic crystals. *Adv. Mater.* 14:187–191.

17. Vita, F., A. Marino, V. Tkachenko, G. Abbate, D.E. Lucchetta, L. Criante, and F. Simoni. 2005. Visible and near infrared characterization and modeling of nanosized holographic-polymer dispersed liquid crystals gratings. *Phys. Rev. E.* 72:011702 and references therein.

18. Caputo, R., L. De Sio, A.V. Sukhov, A. Veltri, and C. Umeton. 2004. Development of a new kind of holographic grating made of liquid crystal films separated by slices of polymeric material. *Opt. Lett.* 29:1261.

19. Strangi, G., V. Barna, R. Caputo, A. de Luca, C. Versace, N. Scaramuzza, C. Umeton, and R. Bartolino. 2005. Color tunable distributed feedback organic micro-cavity laser. *Phys. Rev. Lett.* 94:63903.

20. Jen, S., N. A. Clark, P. S. Pershan, and E. B. Priestley. 1977. Polarized Raman-scattering studies of orientational order in uniaxial liquid-crystalline phases. *J. Chem Phys.* 66:4635–4661.

21. Clark, N. A., and S. T. Lagerwall. 1980. Submicrosecond bistable electro-optic switching in liquid crystals. *Appl. Phys. Lett.* 36:899.

22. Macdonald, R., J. Schwartz, and H. I. Eichler. Laser-induced optical switching of a ferroelectric liquid crystal. *Int. J. Nonlinear Opt. Phys.* 1:103; Ouchi, Y., H. Takezoe, and A. Fukuda. 1987. Switching process in ferroelectric liquid crystals: Disclination dynamics of the surface stabilized states. *Jpn. J. Appl. Phys.* 26:1.

23. Anderson, G., I. Dahl, L. Komitov, S. T. Lagerwall, K. Skarp, and B. Stebler. 1989. Device physics of the soft-mode electro-optic effect. *J. Appl. Phys.* 66:4983.

24. Whinnery, J. R., C. Hu, and Y. S. Kwon. 1977. Liquid crystal waveguides for integrated optics. *IEEE J. Quantum Electron.* QE13:262.

25. Giallorenzi, G., J. A. Weiss, and J. P. Sheridan. 1976. Light scattering from smectic liquid-crystal waveguides. *J. Appl. Phys.* 47:1820.

26. Geren, M., and S. J. Madden. 1989. Low loss nematic liquid crystal cored fiber waveguide. *Appl. Opt.* 28:5202.

27. Khoo, I. C., H. Li, P. G. LoPresti, and Yu Liang. 1994. Observation of optical limiting and backscattering of nanosecond laser pulses in liquid crystal fibers. *Opt. Lett.* 19:530.

28. Khoo, I. C., M. V. Wood, B. D. Guenther, Min-Yi Shih, P. H. Chen, Zhaogen Chen, and Xumu Zhang. 1998. Liquid crystal film and nonlinear optical liquid cored fiber array for ps-cw frequency agile laser optical limiting application. *Opt. Express.* 2:471–82.

29. Khoo, I. C., Andres Diaz, and J. Ding. 2004. Nonlinear-absorbing fiber array for large dynamic range optical limiting application against intense short laser pulses. *J. Opt. Soc. Am. B.* 21:1234–1240.

30. Larsen, Thomas Tanggaard, Anders Bjarklev, David Sparre Hermann, and Jes Broeng. 2003. Optical devices based on liquid crystal photonic bandgap fibers. *Opt. Express.* 11:2589–2596.

31. Graugnard, E., J. S. King, S. Jain, C. J. Summers, Y. Zhang-Williams, and I. C. Khoo. 2005. Electric field tuning of the Bragg peak in large-pore TiO_2 inverse shell opals. *Phys. Rev. B.* 72:233105; see the references quoted therein for other similar studies.

2

Order Parameter, Phase Transition, and Free Energies

2.1. BASIC CONCEPTS

2.1.1. Introduction

Generally speaking, we can divide liquid crystalline phases into two distinctly different types: the ordered and the disordered. For the ordered phase, the theoretical framework invoked for describing the physical properties of liquid crystals is closer in form to that pertaining to solids; it is often called elastic continuum theory. In this case various terms and definitions typical of solid materials (e.g., elastic constant, distortion energy, torque, etc.) are commonly used. Nevertheless, the interesting fact about liquid crystals is that in such an ordered phase they still possess many properties typical of liquids. In particular, they flow like liquids and thus require hydrodynamical theories for their complete description. These are explained in further detail in the next chapter.

Liquid crystals in the disordered or isotropic phase behave very much like ordinary fluids of anisotropic molecules. They can thus be described by theories pertaining to anisotropic fluids. There is, however, one important difference.

Near the isotropic $\rightarrow$ nematic phase transition temperature, liquid crystals exhibit some highly correlated pretransitional effects. In general, the molecules become highly susceptible to external fields, and their responses tend to slow down considerably.

In the next few sections we introduce some basic concepts and definitions, such as order parameter, short- and long-range order, phase transition, and so on, which form the basis for describing the ordered and disordered phases of liquid crystals.

2.1.2. Scalar and Tensor Order Parameters

The physics of liquid crystals is best described in terms of the so-called order parameters[1,2]. If we use the long axis of the molecule as a reference and denote it as $\hat{k}$, the microscopic scalar order parameter S is defined[1,2] as follows:

$$S = \frac{1}{2}\langle 3(\hat{k}\cdot\hat{n})(\hat{k}\cdot\hat{n}) - 1\rangle$$
$$= \frac{1}{2}\langle 3\cos^2\theta - 1\rangle. \tag{2.1}$$

With reference to Figure 2.1, θ is the angle made by the molecular axis with the director axis. The average $\langle\ \rangle$ is taken over the whole ensemble; this kind of order is usually termed long-range order. It is called microscopic because it describes the average response of a molecule. The scalar order parameter defined previously is sufficient to describe liquid crystalline systems composed of molecules that possess cylindrical or rotational symmetry around the long axis $\hat{k}$.

On the other hand, for molecules lacking such symmetry, or in cases where such rotational symmetry is "destroyed" by the presence of asymmetric dopants or intramolecular material interactions, a more general tensor order parameter S_{ij} is needed. S_{ij} is defined as

$$S_{ij} = \frac{1}{2}\langle 3(\hat{n}\cdot\hat{i})(\hat{n}\cdot\hat{j}) - 1\rangle, \tag{2.2}$$

where $\hat{i}, \hat{j}$, and $\hat{k}$ are unit vectors along the molecular axes. With reference to Figure 2.1, the three diagonal components S_{ii}, S_{jj}, and S_{kk} are given by

$$S_{ii} = \frac{1}{2}\langle 3\sin^2\theta\cos^2\phi - 1\rangle, \tag{2.3a}$$

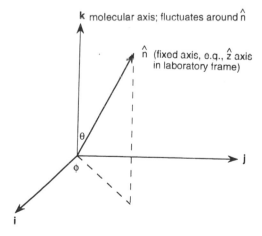

Figure 2.1. Coordinate system defining the microscopic order parameter of a nematic liquid crystal molecule. i, j, and k are the molecular axes, whereas $\hat{n}$ is the laboratory axis that denotes the average direction of liquid crystal alignment.

$$S_{jj} = \frac{1}{2}\left\langle 3\sin^2\theta\sin^2\phi - 1\right\rangle, \tag{2.3b}$$

$$S_{ii} = \frac{1}{2}\left\langle 3\cos^2\theta - 1\right\rangle. \tag{2.3c}$$

Note that $S_{ii}+S_{jj}+S_{kk}=0$. Put another way, S is a traceless tensor because its diagonal elements add up to zero.

For a more complete description of the statistical properties of the liquid crystal orientation, functions involving higher powers of $\cos^2\theta$ are needed. The most natural functions to use are the Legendre polynomials $P_l(\cos\theta)$ ($l = 0, 1, 2,\ldots$), in terms of which we can write Equation (2.1) as $S =\langle P_2\rangle$, which measures the average of $\cos^2\theta$. The next nonvanishing term is $\langle P_4\rangle$, which provides a measure of the dispersion of $\langle\cos^2\theta\rangle$.

The order parameters defined previously in terms of the directional averages can be translated into expressions in terms of the anisotropies in the physical parameters such as magnetic, electric, and optical susceptibilities. For example, in terms of the optical dielectric anisotropies $\Delta\varepsilon = \varepsilon_\parallel-\varepsilon_\perp$, one can define a so-called macroscopic order parameter which characterizes the bulk response

$$Q_{\alpha\beta} \equiv \varepsilon_{\alpha\beta} - \frac{1}{3}\delta_{\alpha\beta}\sum_\gamma \varepsilon_{\gamma\gamma} \equiv \delta\varepsilon_{\alpha\beta}. \tag{2.4}$$

It is called macroscopic because it describes the bulk property of the material. To be more explicit, consider a uniaxial material such that in the molecular axis system is of the form

$$\varepsilon_{\alpha\beta} = \begin{pmatrix} \varepsilon_\perp & 0 & 0 \\ 0 & \varepsilon_\perp & 0 \\ 0 & 0 & \varepsilon_\parallel \end{pmatrix}. \tag{2.5}$$

Writing $Q_{\alpha\beta}$ explicitly in terms of their diagonal components, we thus have

$$Q_{xx} = Q_{yy} = -\frac{1}{3}\Delta\varepsilon \tag{2.6}$$

and

$$Q_{zz} = \frac{2}{3}\Delta\varepsilon. \tag{2.7}$$

It is useful to note here that, in tensor form, $\varepsilon_{\alpha\beta}$ can be expressed as

$$\varepsilon_{\alpha\beta} \equiv \varepsilon_\perp\delta_{\alpha\beta} + \Delta\varepsilon n_\alpha n_\beta \tag{2.8}$$

Note that this form shows that $\varepsilon = \varepsilon_\parallel$ for an optical field parallel to $\hat{n}$ and $\varepsilon = \varepsilon_\perp$ for an optical field perpendicular to $\hat{n}$.

Similarly, other parameters such as the magnetic (χ^m) and electric (χ) susceptibilities may be expressed as

$$\chi^m_{\alpha\beta} = \chi^m_{\perp}\delta_{\alpha\beta} + \Delta\chi^m n_\alpha n_\beta \qquad (2.9a)$$

and

$$\chi_{\alpha\beta} = \chi_{\perp}\delta_{\alpha\beta} + \Delta\chi n_\alpha n_\beta, \qquad (2.9b)$$

respectively, in terms of their respective anisotropies $\Delta\chi^m$ and $\Delta\chi$.

In general, however, optical dielectric anisotropy and its dc or low-frequency counterpart (the dielectric anisotropy) provide a less reliable measure of the order parameter because they involve electric fields. This is because of the so-called local field effect: the effective electric field acting on a molecule is a superposition of the electric field from the externally applied source and the field created by the induced dipoles surrounding the molecules. For systems where the molecules are not correlated, the effective field can be fairly accurately approximated by some local field correction factor;[3] in liquid crystalline systems these correction factors are much less accurate. For a more reliable determination of the order parameter, one usually employs non-electric-field-related parameters, such as the magnetic susceptibility anisotropy:

$$Q_{\alpha\beta} = \chi^m_{\alpha\beta} - \tfrac{1}{3}\delta_{\alpha\beta}\sum_\gamma \chi^m_{\gamma\gamma}. \qquad (2.10)$$

2.1.3. Long- and Short-Range Order

The order parameter, defined by Equation (2.2) and its variants such as Equations (2.4) and (2.8), is an average over the whole system and therefore provides a measure of the long-range orientation order. The smaller the fluctuation of the molecular axis from the director axis orientation direction, the closer the magnitude of S is to unity. In a perfectly aligned liquid crystal, as in other crystalline materials, $\langle\cos^2\theta\rangle = 1$ and $S = 1$; on the other hand, in a perfectly random system, such as ordinary liquids or the isotropic phase of liquid crystals, $\langle\cos^2\theta\rangle - \tfrac{1}{3}$ and $S = 0$.

An important distinction between liquid crystals and ordinary anisotropic or isotropic liquids is that, in the isotropic phase, there could exist a so-called short-range order;[1,2] that is, molecules within a short distance of one another are correlated by intermolecular interactions.[4] These molecular interactions may be viewed as remnants of those existing in the nematic phase. Clearly, the closer the isotropic liquid crystal is to the phase transition temperature, the more pronounced the short-range order and its manifestations in many physical parameters will be. Short-range order in the isotropic phase gives rise to interesting critical behavior in the response of the liquid crystals to externally applied fields (electric, magnetic, and optical) (see Section 3.2).

As pointed out at the beginning of this chapter, the physical and optical properties of liquid crystals may be roughly classified into two types: one pertaining to the ordered phase, characterized by long-range order and crystalline like physical properties; the other pertaining to the so-called disordered phase, where a short-range order exists. All these order parameters show critical dependences as the temperature approaches the phase transition temperature T_c from the respective directions.

2.2. MOLECULAR INTERACTIONS AND PHASE TRANSITIONS

In principle, if the electronic structure of a liquid crystal molecule is known, one can deduce the various thermodynamical properties. This is a monumental task in quantum statistical chemistry that has seldom, if ever, been attempted in a quantitative or conclusive way. There are some fairly reliable guidelines, usually obtained empirically, that relate molecular structures with the existence of the liquid crystal mesophases and, less reliably, the corresponding transition temperatures.

One simple observation is that to generate liquid crystals, one should use elongated molecules. This is best illustrated by the nCB homolog[5] ($n = 1, 2, 3,...$). For $n \leq 4$, the material does not exhibit a nematic phase. For $n = 5$–7, the material possesses a nematic range. For $n > 8$, smectic phases begin to appear.

Another reliable observation is that the nematic $\rightarrow$ isotropic phase transition temperature T_c is a good indicator of the thermal stability of the nematic phase;[6] the higher the T_c, the greater the thermal stability of the nematic phase is. In this respect, the types of chemical groups used as substituents in the terminal groups or side chain play a significant role—an increase in the polarizability of the substituent tends to be accompanied by an increase in T_c.

Such molecular-structure-based approaches are clearly extremely complex and often tend to yield contradictory predictions, because of the wide variation in the molecular electronic structures and intermolecular interactions present. In order to explain the phase transition and the behavior of the order parameter in the vicinity of the phase transition temperature, some simpler physical models have been employed.[6] For the nematic phase, a simple but quite successful approach was introduced by Maier and Saupe.[7] The liquid crystal molecules are treated as rigid rods, which are correlated (described by a long-range order parameter) with one another by Coulomb interactions. For the isotropic phase, deGennes introduced a Landau type of phase transition theory,[1–3] which is based on a short-range order parameter.

The theoretical formalism for describing the nematic $\rightarrow$ isotropic phase transition and some of the results and consequences are given in the next section. This is followed by a summary of some of the basic concepts introduced for the isotropic phase.

2.3. MOLECULAR THEORIES AND RESULTS FOR THE LIQUID CRYSTALLINE PHASE

Among the various theories developed to describe the order parameter and phase transitions in the liquid crystalline phase, the most popular and successful one is the theory first advanced by Maier and Saupe and corroborated in studies by others.[8] In this formalism Coulombic intermolecular dipole–dipole interactions are assumed.

The interaction energy of a molecule with its surroundings is then shown to be of the form.[6]

$$W_{int} = -\frac{A}{V^2} S\left(\frac{3}{2}\cos^2\theta - 1\right),$$

(2.11)

where V is the molar volume ($V = M/p$), S is the order parameter, and A is a constant determined by the transition moments of the molecules. Both V and S are functions of temperature. Comparing Equations (2.11) and (2.1) for the definition of S we note that $W_{int} \approx S^2$, so this mean field theory by Maier and Saupe is often referred to as the S^2 interaction theory.[1] This interaction energy is included in the free enthalpy per molecule (chemical potential) and is used in conjunction with an angular distribution function $f(\theta, \phi)$ for statistical mechanics calculations.

2.3.1. Maier–Saupe Theory: Order Parameter Near T_c

Following the formalism of deGennes, the interaction energy may be written as

$$G_1 = -\tfrac{1}{2}U(p,T)S(\tfrac{3}{2}\cos^2\theta - 1).$$

(2.12)

The total free enthalpy per molecule is therefore

$$G(p,T) = G_i(p,T) + K_B T \int f(0,\phi)\log 4\,\pi f(\theta,\phi)d\Omega + G_1(p,T,S),$$

(2.13)

where G_i is the free enthalpy of the isotropic phase. Minimizing $G(p, T)$ with respect to the distribution function f, one gets

$$f(0) = \frac{\exp(m\cos^2\theta)}{4\pi z},$$

(2.14)

where

$$m = \frac{3}{2}\frac{US}{K_B T},$$

(2.15)

and the partition function z is given by

$$z = \int_0^1 e^{mx^2}\,dx.$$

(2.16)

From the definition of $S = -\tfrac{1}{2} + \tfrac{3}{2}\langle\cos^2\theta\rangle$, we have

$$S = -\frac{1}{2} + \frac{3}{2z}\int_0^1 x^2 e^{mx^2}\,dx$$

$$= -\frac{1}{2} + \frac{3}{2}\frac{\partial z}{z\partial m}.$$

(2.17)

The coupled equations (2.15) and (2.17) for m and S may be solved graphically for various values of U/K_BT, the relative magnitude of the intermolecular interaction to the thermal energies. Figure 2.2 depicts the case for T below a temperature T_c defined by

$$\frac{k_BT_c}{U(T_c)} = 4.55. \tag{2.18}$$

Figure 2.2 shows that curves 1 and 2 for S intersect at the origin O and two points N and M. Both points O and N correspond to minima of G, whereas M corresponds to a local maximum of G. For $T < T_c$, the value of G is lower at point N than at point O; that is, S is nonzero and corresponds to the nematic phase. For temperatures above T_c the stable (minimum energy) state corresponds to O; that is, $S = O$ and corresponds to the isotropic phase.

The transition at $T = T_c$ is a first-order one. The order parameter just below T_c is

$$S_c \equiv S(T_c) = 0.44. \tag{2.19}$$

It has also been demonstrated that the temperature dependence of the order parameter of most nematics is well approximated by the expression.[9]

$$S = \left(1 - \frac{0.98TV^2}{T_cV_c^2}\right)^{0.22}, \tag{2.20}$$

where V and V_c are the molar volumes at T and T_c, respectively.

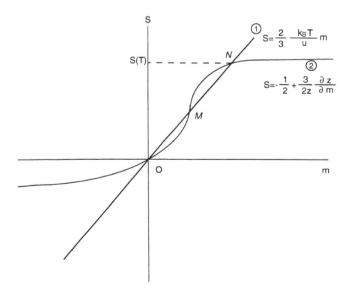

Figure 2.2.   Schematic depiction of the numerical solution of the two transcendental equations for the order parameter for $T < T_c$; there is only one intersection point (at the origin).

In spite of some of these predictions which are in good agreement with the experimental results, the Maier–Saupe theory is not without its shortcomings. For example, the universal temperature dependence of S on T/T_c is really not valid;[10] agreement with experimental results requires an improved theory that accounts for the noncylindrical shape of the molecules.[8] The temperature variation given in Equation (2.20) also cannot account for the critical dependence of the refractive indices. Nevertheless, the Maier–Saupe theory remains an effective, clear, and simple theoretical framework and starting point for understanding nematic liquid crystal complexities.

2.3.2. Nonequilibrium and Dynamical Dependence of the Order Parameter

While the equilibrium statistical mechanics of the nematic liquid crystal order parameter and related physical properties near T_c are now well understood, the dynamical responses of the order parameter remain relatively unexplored. This is probably due to the fact that most studies of the order parameter near the phase transition point, such as the critical exponent, changes in molar volume, and other physical parameters, are directed at understanding the phase transition processes themselves and are usually performed with temperature changes occurring at very slow rates.

There have been several studies in recent years, however, in which the temperature of the nematics are abruptly raised by very short laser pulses.[5,11,12] The pulse duration of the laser is in the nanosecond or picosecond time scale, which, as we shall see, is much shorter than the response time of the order parameter. As a result the nematic film under study exhibits delayed signals.

Figures 2.3a and 2.3b show the observed diffraction from a nematic film in a dynamic grating experiment.[11] In such an experiment, explained in more detail in Chapter 7, the diffracted signal is a measure of the dynamical change in the refractive index, $\Delta n(t)$, following an instantaneous (delta function like) pump pulse. For a nematic liquid crystal the principal change in the refractive index associated with a rise in temperature is through the density and order parameter;[12] that is,

$$\Delta n = \frac{dn}{d\rho}\,d\rho + \frac{dn}{dS}\,dS. \tag{2.21}$$

Unlike the change in order parameter, which is a collective molecular effect, the change in density $d\rho$ arises from the individual responses of the molecules and responds relatively quickly to the temperature change.

These results are reflected in Figures 2.3a and 2.3b. The diffracted signal contains an initial "spike," which rises and decays away in the time scale on the order of the laser pulse. On the other hand, the order parameter contribution to the signal builds up rather slowly. In Figure 2.3a the buildup time is about 30 μs for nanosecond and visible laser pulse excitations, while in Figure 2.3b the buildup time is as long as 175 μs for infrared laser pulse excitation. Figure 2.4 shows the observed "slowing down" in the response of the order parameter as the temperature approaches T_c.

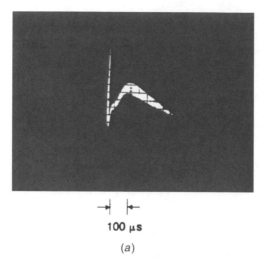

100 μs

(a)

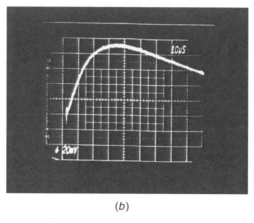

(b)

Figure 2.3. (a) Observed oscilloscope trace of the diffracted signal in a dynamical scattering experiment involving microsecond infrared (CO_2 at 10.6 μm) laser pump pulses. Sample used is a planar aligned nematic (E7) film; (b) observed oscilloscope trace of the diffracted signal from a nematic film under nanosecond visible (Nd: YAG at 0.53 μm) laser pump pulse excitation. Sample used is a planar aligned nematic (E7) film.

One can also see from the relative heights of the density and order parameter components in Figures 2.3a and 2.3b that the overall response of the nematic film is different for the two forms of excitation. The absorption of infrared photons ($\lambda = 10.6$ μm) corresponds to the excitation of the ground (electronic) state's rovibrational manifold, whereas the visible photoabsorption ($\lambda = 0.53$ μm) corresponds to the excitation of the molecules to the electronically excited states (see Fig. 2.5). The electronic molecular structures of these two excited states are different and may therefore account for the different dynamical response behavior of the order parameter, which is dependent on the intermolecular Coulombic dipole–dipole interaction. From this observation one may conclude that the dynamical grating technique would be an interesting technique for probing the different dynamical

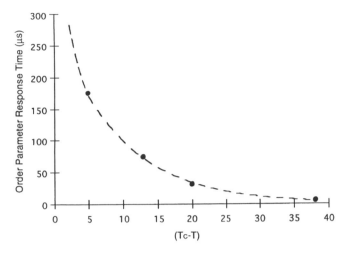

Figure 2.4. Observed buildup times of the diffracted signal associated with order parameter change as a function of the temperature vicinity of T_c; excitation by infrared microsecond laser pulses on E7 nematic film.

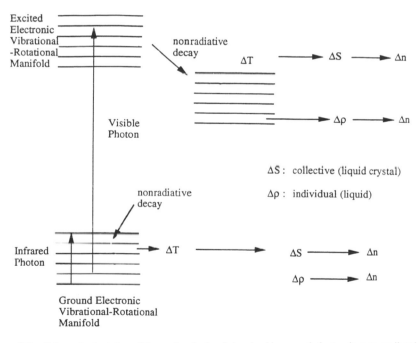

Figure 2.5. Schematic depiction of the molecular levels involved in ground electronic state rovibrational excitations by infrared photoabsorptions and excited electronic state excitation by visible photoabsorptions.

behaviors of the order parameters for molecules in various states of excitation, as well as the critical slowing down of the order parameter near T_c. This information will be important in the operation of practical devices based on the laser-induced order parameter changes in liquid crystals (see Chapter 12).

2.4. ISOTROPIC PHASE OF LIQUID CRYSTALS

Above T_c liquid crystals lose their directional order and behave in many respects like liquids. All bulk physical parameters also assume an isotropic form although the molecules are anisotropic.

The isotropic phase is, nevertheless, a very interesting and important phase for both fundamental and applied studies. It is fundamentally interesting because of the existence of short-range order, which gives rise to the critical temperature dependence of various physical parameters just above the phase transition temperature. These critical behaviors provide a good testing ground for the liquid crystal physics.

On the other hand, recent studies have also shown that isotropic liquid crystals may be superior in many ways for constructing practical nonlinear optical devices (see Section 12.6), in comparison to the other liquid crystalline phases (see Chapter 8). In general, the scattering loss is less and thus allows longer interaction lengths, and relaxation times are on a much faster scale. These properties easily make up for the smaller optical nonlinearity for practical applications.

2.4.1.   Free Energy and Phase Transition

We begin our discussion of the isotropic phase of liquid crystals with the free energy of the system, following deGennes' pioneering theoretical development.[1,2] The starting point is the order parameter, which we denote by Q.

In the absence of an external field, the isotropic phase is characterized by $Q = 0$; the minimum of the free energy also corresponds to $Q = 0$. This means that, in the Landau expansion of the free energy in terms of the order parameter Q, there is no linear term in Q; that is,

$$F = F_0 + \tfrac{1}{2}A(T)\sum_{\rho,\alpha} Q_{\alpha\beta}Q_{\beta\alpha} + \tfrac{1}{3}B(T)\sum_{\alpha,\beta,\gamma} Q_{\alpha\beta}Q_{\alpha\gamma}Q_{\gamma\alpha} + O(Q4), \qquad (2.22)$$

where F_0 is a constant and $A(T)$ and $B(T)$ are temperature-dependent expansion coefficients:

$$A(T) = \alpha(T - T^*), \qquad (2.23)$$

where T^* is very close to, but lower than, T_c. Typically, $T_c - T_c^* = 1\mathrm{K}$.

Note that F contains a nonzero term of order Q^3. This odd function of Q ensures that states with some nonvanishing value of Q (e.g., due to some alignment of molecules) will have different free-energy values depending on the direction of the alignment. For example, the free energy for a state with an order parameter Q of the form

$$Q_1 = \begin{pmatrix} -\xi & 0 & 0 \\ 0 & -\xi & 0 \\ 0 & 0 & 2\xi \end{pmatrix} \qquad (2.24a)$$

(i.e., with some alignment of the molecule in the z direction) is not the same as the state with a negative Q parameter

$$Q_2 = \begin{pmatrix} \xi & 0 & 0 \\ 0 & \xi & 0 \\ 0 & 0 & -2\xi \end{pmatrix} = -Q_1 \tag{2.24b}$$

(which signifies some alignment of the molecules in the x-y plane).

The cubic term in F is also important in that it dictates that the phase transition at $T = T_c$ is of the first order (i.e., the first-order derivative of F, $\partial F/\partial \theta$, is vanishing at $T = T_c$, as shown in Fig. 2.6). The system has two stable minima, corresponding to $Q = 0$ or $Q \neq 0$ (i.e., the coexistence of the isotropic and nematic phases). On the other hand, for $T=T_c^*(< T_c)$, there is only one stable minimum at $Q \neq 0$; this translates into the existence of a single liquid crystalline phase (e.g., nematic or smectic).

2.4.2. Free Energy in the Presence of an Applied Field

In the presence of an externally applied field (e.g., dc or low-frequency electric, magnetic, or optical electric field), a corresponding interaction term should be added to the free energy.

For an applied magnetic field H, the energy associated with it is

$$F_{\text{int}} = -\int_0^H \mathbf{M} \cdot d\mathbf{H}, \tag{2.25}$$

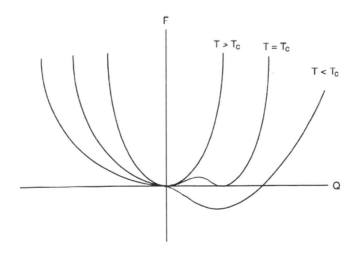

Figure 2.6. Free energies $F(Q)$ for different temperatures T. At $T = T_c$, $\partial F / \partial Q = 0$ at two values of Q, where F has two stable minima. On the other hand, at $T = T_c^*$ ($< T_c$), there is only one stable minimum where $\partial F / \partial Q = 0$.

where $\mathbf{M}$ is the magnetization given by

$$M_\alpha = \sum_\beta \chi_{\alpha\beta} H_\beta. \tag{2.26}$$

Thus

$$F_{\text{int}} = -\frac{1}{2} \sum_\alpha \chi^m_{\alpha\beta} H_\beta H_\alpha. \tag{2.27}$$

Using Equation (2.9), we can rewrite F_{int} as

$$F_{\text{int}} = -\frac{1}{2} \chi^m_\perp \sum_\alpha H_\alpha H_\beta - \frac{1}{2} \sum_{\alpha,\beta} \Delta \chi^m n_\alpha n_\beta H_\alpha H_\beta. \tag{2.28}$$

The first term on the right-hand side of Equation (2.28) is independent of the orientation of the (anisotropic) molecules, and it can therefore be included in the constant F_0.

On the other hand, the second term is dependent on the orientation of the molecules. Using Equation (2.10) for the order parameter $Q_{\alpha\beta}$ we can write it as

$$F^H_{\text{int}} = -\frac{1}{2} \sum_{\alpha,\beta} Q_{\alpha\beta} H_\alpha H_\beta. \tag{2.29}$$

Therefore, the total free energy of a liquid crystal in the isotropic phase, under the action of an externally applied magnetic field, is given by

$$F = F'_0 + \frac{1}{2} A(T) \sum_{\alpha,\beta} Q_{\alpha\beta} Q_{\beta\alpha} + \frac{1}{3} B(T) \sum_{\alpha,\beta,\gamma} Q_{\alpha\beta} Q_{\beta\gamma} Q_{\gamma\alpha} \\ -\frac{1}{2} \sum_{\alpha,\beta} Q_{\alpha\beta} H_\alpha H_\beta. \tag{2.30}$$

Without solving the problem explicitly, we can infer from the magnetic interaction term that a lower energy state corresponds to some alignment of the molecules in the direction of the magnetic field (for $\Delta \chi^m > 0$).

Using a similar approach, we can also deduce that the electric interaction contribution to the free energy is given by (in inks units)

$$F^E_{\text{int}} = -\frac{1}{2} \int_0^E \mathbf{D} \cdot \mathbf{E} = -\frac{1}{2} \varepsilon_\perp E^2 - \frac{1}{2} \Delta \varepsilon (\hat{n} \cdot \mathbf{E})^2. \tag{2.31}$$

The orientation-dependent term is therefore

$$F^E = -\frac{1}{2} \Delta \varepsilon (\hat{n} \cdot \mathbf{E})^2 = -\frac{1}{2} \sum_{\alpha,\beta} Q_{\alpha\beta} E_\alpha E_\beta, \tag{2.32}$$

where $Q_{\alpha\beta}$ is defined in Equation (2.4).

In Chapter 8 we will present a detailed discussion of the isotropic phase molecular orientation effects by an applied optical field from a short intense laser pulse. It is shown that both the response time and the induced order Q depend on the temperature vicinity $(T - T_c)$ in a critical way; they both vary as $(T - T_c)^{-1}$, which becomes very large near T_c. This near-T_c critical slowing down behavior of the order parameter Q of the isotropic phase is similar to the slowing down behavior of the order parameter S of the nematic phase discussed in the previous section. Besides the nematic $\leftrightarrow$ isotropic phase transition, which is the most prominent order $\leftrightarrow$ disorder transition exhibited by liquid crystals, there are other equally interesting phase transition processes among the various mesophases,[13] such as smectic-A $\leftrightarrow$ smectic-C*, which will be discussed in Chapter 4.

REFERENCES

1. deGennes, P. G. 1974. *The Physics of Liquid Crystals*. Oxford: Clarendon Press.

2. deGennes, P. G. 1971. *Mol. Cryst. Liq. Cryst.* 12:193.

3. Landau, L. D. 1965. *Collected Papers*. D. Ter Haar (ed.). New York: Gordon & Breach.

4. Litster, J. D. 1971. *Critical Phenomena*. R. E. Mills (ed.). New York: McGraw-Hill.

5. Khoo, I. C. and S. T. Wu. 1993. *Optics and Nonlinear Optics of Liquid Crystals*. Singapore: World Scientific.

6. See, for example, Blinov, L. M. 1983. *Electro-optical and Magneto-optical Properties of Liquid Crystals*. Chichester: Wiley.

7. Maier W. and A. Saupe. 1959. *Z. Naturforsch.* 14A:882; for a concise account of the theory, see Khoo and Wu.[5]

8. Humphries, R. L., and O. R. Lukhurst. 1972. *Chem. Phys. Lett.* 17:514; Luckhurst, G. R., C. Zannoni, P. L. Nordio, and U. Segré. 1975. *Mol. Phys.* 30:1345; Freiser, M. J. 1971. *Mol. Cryst. Liq. Cryst.* 14:165.

9. Blinov, L. M., V. A. Kizel, V. G. Rumyantsev, and V. V. Titov. 1975. *J. Phys. (Paris)*, 36:Colloq. C1–C69; see also Blinov.[6]

10. DeJeu, W. H. 1980. *Physical Properties of Liquid Crystalline Materials*. New York: Gordon and Breach.

11. Khoo, I. C., R. G. Lindquist, R. R. Michael, R. J. Mansfield, and P. G. LoPresti 1991. *J. Appl. Phys.* 69:3853.

12. Khoo, I. C., and R. Normandin. 1985. *IEEE J. Quantum Electron.* QE21:329.

13. Chandrasekhar, S. 1992. *Liquid Crystals*. 2nd ed. Cambridge: Cambridge University Press; see also deGennes.[1]

3

Nematic Liquid Crystals

3.1. INTRODUCTION

Nematics are the most widely studied liquid crystals. They are also the most widely used. As a matter of fact, nematics best exemplify the dual nature of liquid crystals–fluidity and crystalline structure. To describe their liquidlike properties, one needs to invoke hydrodynamics. On the other hand, their crystalline properties necessitate theoretical formalisms pertaining to solids or crystals. To study their optical properties, it is also necessary that we invoke their electronic structures and properties.

In this chapter we discuss all three aspects of nematogen theory: solid-state continuum theory, hydrodynamics, and electro-optical properties, in that order.

3.2. ELASTIC CONTINUUM THEORY

3.2.1. The Vector Field: Director Axis $\hat{n}(\vec{r})$

In elastic continuum theory, introduced and refined over the last several decades by several workers,[1–3] nematics are basically viewed as crystalline in form. An aligned sample may thus be regarded as a single crystal, in which the molecules are, on the average, aligned along the direction defined by the director axis $\hat{n}(\vec{r})$.

The crystal is uniaxial and is characterized by a tensorial order parameter:

$$S_{\alpha\beta} = S(T)\left(n_\alpha n_\beta - \tfrac{1}{3}\delta_{\alpha\beta}\right). \tag{3.1}$$

As a result of externally applied fields, stresses/constraints from the boundary surfaces, the director will also vary spatially. The characteristic length over which significant variation in the order parameter will occur, in most cases, is much larger than the molecular size. Typically, for distortions of the form shown in Figures 3.1a–3.1c, the characteristic length is on the order of 1 μm, whereas the molecular dimension is on the order of at most a few tens of angstroms. Under this

Liquid Crystals, Second Edition   By Iam-Choon Khoo
Copyright © 2007 John Wiley & Sons, Inc.

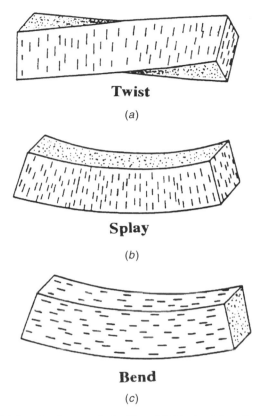

Twist

(a)

Splay

(b)

Bend

(c)

Figure 3.1. (a) Twist deformation in a nematic liquid crystal; (b) splay deformation; (c) bend deformation.

circumstance, as in other similar systems or media (e.g., ferromagnets), the continuum theory is valid.

The first principle of continuum theory therefore neglects the details of the molecular structures. Liquid crystal molecules are viewed as rigid rods; their entire collective behavior may be described in terms of the director axis $\hat{n}(\vec{r})$, a vector field. In this picture the spatial variation of the order parameter is described by

$$S_{\alpha\beta}(\vec{r}) = S(T)\left[n_{\alpha}(\vec{r})n_{\beta}(\vec{r}) - \tfrac{1}{3}\delta_{\alpha\beta}\right]. \tag{3.2}$$

In other words, in a spatially "distorted" nematic crystal, the local optical properties are still those pertaining to a uniaxial crystal and remain unchanged; it is only the orientation (direction) of $\hat{n}$ that varies spatially.

For nematics, the states corresponding to $\hat{n}$ and $-\hat{n}$ are indistinguishable. In other words, even if the individual molecules possess permanent dipoles (actually most liquid crystal molecules do), the molecules are collectively arranged in such a way that the net dipole moment is vanishingly small; that is, there are just as many dipoles up as there are dipoles down in the collection of molecules represented by $\hat{n}$.

3.2.2. Elastic Constants, Free Energies, and Molecular Fields

Upon application of an external perturbation field, a nematic liquid crystal will undergo deformation just as any solid. There is, however, an important difference. A good example is shown in Figure 3.1a, which depicts a "solid" subjected to torsion, with one end fixed. In ordinary solids this would create a very large stress, arising from the fact that the molecules are translationally displaced by the torsional stress. On the other hand, such twist deformations in liquid crystals, owing to the fluidity of the molecules, simply involve a rotation of the molecules in the direction of the torque; there is no translational displacement of the center of gravity of the molecules, and thus, the elastic energy involved is quite small. Similarly, other types of deformations such as splay and bend deformations, as shown in Figures 3.1b and 3.1c, respectively, involving mainly changes in the director axis $\hat{n}(\vec{r})$, will incur much less elastic energy change than the corresponding ones in ordinary solids. It is evident from Figures 3.1a–3.1c that the splay and bend deformations necessarily involve flow of the liquid crystal, whereas the twist deformation does not. We will return to these couplings between flow and director axis deformation in Section 3.5.

Twist, splay, and bend are the three principal distinct director axis deformations in nematic liquid crystals. Since they correspond to spatial changes in $\hat{n}(\vec{r})$, the basic parameters involved in the deformation energies are various spatial derivatives [i.e., curvatures of $\hat{n}(\vec{r})$, such as $\nabla \times \hat{n}(\vec{r})$ and $\nabla \cdot \hat{n}(\vec{r})$, etc.]. Following the theoretical formalism first developed by Frank,[1] the free-energy densities (in units of energy per volume) associated with these deformations are given by

$$\text{splay}: f_1 = \tfrac{1}{2} K_1 (\nabla \cdot \hat{n})^2, \tag{3.3}$$

$$\text{twist}: f_2 = \tfrac{1}{2} K_2 (\hat{n} \cdot \nabla \times \hat{n})^2, \tag{3.4}$$

$$\text{bend}: f_3 = \tfrac{1}{2} K_3 (\hat{n} \times \nabla \times \hat{n})^2, \tag{3.5}$$

where K_1, K_2, and K_3 are the respective Frank elastic constants.

In general, the three elastic constants are different in magnitude. Typically, they are on the order of 10^{-6} dyne in centimeter-gram-second (cgs) units [or 10^{-11} N in meter-kilogram-second (mks) units]. For p-methoxybenzylidene-p'-butylaniline (MBBA), K_1, K_2, and K_3 are, respectively, 5.8×10^{-7}, 3.4×10^{-7}, and 7×10^{-7} dyne. For almost all nematics K_3 is the largest, as a result of the rigid-rod shape of the molecules.

In general, more than one form of deformation will be induced by an applied external field. If all three forms of deformation are created, the total distortion free-energy density is given by

$$F_d = \tfrac{1}{2} K_1 (\nabla \cdot \hat{n})^2 + \tfrac{1}{2} K_2 (\hat{n} \cdot \nabla \times \hat{n})^2 + \tfrac{1}{2} K_3 (\hat{n} \times \nabla \times \hat{n})^2. \tag{3.6}$$

This expression, and the resulting equations of motion and analysis, can be greatly simplified if one makes a frequently used assumption, namely, the one-constant approximation ($K_1 = K_2 = K_3 = K$). In this case Equation (3.6) becomes

$$F_d = \tfrac{1}{2} K \left[(\nabla \cdot \hat{n})^2 + (\nabla \times \hat{n})^2 \right]. \tag{3.7}$$

Equation (3.6) or its simplified version, Equation (3.7), describes the deformation of the director axis vector field $\hat{n}(\vec{r})$ in the bulk of the nematic liquid crystal. A complete description should include the surface interaction energy at the nematic liquid crystal cell boundaries. To accounting for this, the total energy density of the system should be

$$F_d{}' = F_d + F_{surface}, \tag{3.8}$$

where the surface energy term is dependent on the surface treatment. In other words, the equilibrium configuration of the nematic liquid crystal is obtained by a minimization of the total free energy of the system, $F_{total} = \int F_d{}'\, dV$. If external fields (electric, magnetic, or optical) are applied, the corresponding free-energy terms (see the following sections) will be added to the total free-energy expression.

Under the so-called hard-boundary condition, in which the liquid crystal molecules are strongly anchored to the boundary and do not respond to the applied perturbation fields (see Fig. 3.2), the surface energy may thus be regarded as a constant; the surface interactions therefore do not enter into the dynamical equations describing the field-induced effects in nematic liquid crystals.

On the other hand, if the molecules are not strongly anchored to the boundary, that is, the so-called soft-boundary condition (Fig. 3.3), an applied field will perturb the orientation of the molecules at the cell boundaries. In this case a quantitative description of the dynamics of the field-induced effects must account for these surface energy terms. A good account of surface energy interaction may be found in the work of Barbero et al.,[4] which treats the case of optical field-induced effects in a hybrid aligned nematic liquid crystal cell.

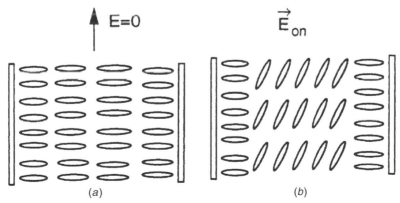

Figure 3.2. A homeotropic nematic liquid crystal with strong surface anchoring: (a) external field off; (b) external field on — only the bulk director axis is deformed.

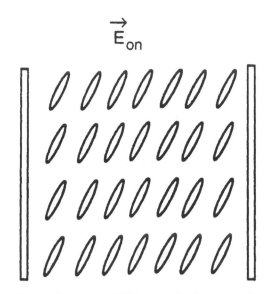

Figure 3.3. Soft boundary condition. Applied field will reorient both the surface and bulk director axis.

From Equation (3.6) for the free energy, one can obtain the corresponding so-called molecular fields $\vec{f}$ using the Lagrange equation.[3] In spatial coordinate component form, we have

$$f_\alpha = \frac{-\partial F}{\partial n_\alpha} + \sum_\beta \frac{\partial}{\partial x_\beta} \frac{\partial F}{\partial g_{\alpha\beta}},$$ (3.9)

where

$$g_{\alpha\beta} = \frac{\partial n_\alpha}{\partial x_\beta}.$$ (3.10)

More explicitly, Equation (3.9) gives, for the total molecular field associated with splay, twist, and bend deformations,

$$\vec{f} = \vec{f}_1 + \vec{f}_2 + \vec{f}_3,$$ (3.11)

and torque $\vec{\Gamma} = \hat{n} \times \vec{f}$, where

$$\text{splay} : \vec{f}_1 = K_1 \nabla (\nabla \cdot \hat{n}),$$ (3.12)

$$\text{twist} : \vec{f}_2 = - K_2 [A \nabla \times \hat{n} + \nabla \times (A \hat{n})],$$ (3.13)

$$\text{bend} : \vec{f}_3 = K_3 [\vec{B} \times (\nabla \times \hat{n}) + \nabla \times (\hat{n} \times \vec{B})],$$ (3.14)

with $A = \hat{n} \cdot (\nabla \times \hat{n})$ and $\vec{B} = \hat{n} \times (\nabla \times \hat{n})$.

3.3. DIELECTRIC CONSTANTS AND REFRACTIVE INDICES

Dielectric constants and refractive indices, as well as electrical conductivities of liquid crystals, are physical parameters that characterize the electronic responses of liquid crystals to externally applied fields (electric, magnetic, or optical). Because of the molecular and energy level structures of nematic molecules, these responses are highly dependent on the direction and the frequencies of the field. Accordingly, we shall classify our studies of dielectric permittivity and other electro-optical parameters into two distinctive frequency regimes: (1) dc and low frequency, and (2) optical frequency. Where the transition from regime (1) to (2) occurs, of course, is governed by the dielectric relaxation processes and the dynamical time constant; typically the Debye relaxation frequencies in nematics is on the order of 10^{10} Hz.

3.3.1. dc and Low-Frequency Dielectric Permittivity, Conductivities, and Magnetic Susceptibility

The dielectric constant ε is defined by the Maxwell equation:[5]

$$\vec{D} = \overset{\leftrightarrow}{\varepsilon} \cdot \vec{E}, \tag{3.15a}$$

where $\vec{D}$ is the displacement current, $\vec{E}$ is the electric field, and $\overset{\leftrightarrow}{\varepsilon}$ is the tensor. For a uniaxial nematic liquid crystal, we have

$$\overset{\leftrightarrow}{\varepsilon} = \begin{bmatrix} \varepsilon_\perp & 0 & 0 \\ 0 & \varepsilon_\perp & 0 \\ 0 & 0 & \varepsilon_\parallel \end{bmatrix}. \tag{3.15b}$$

Equations (3.15a) and (3.15b) yield, for the two principle axes,

$$D_\parallel = \varepsilon_\parallel E_\parallel \tag{3.16}$$

and

$$D_\perp = \varepsilon_\perp E_\perp. \tag{3.17}$$

Typical values of $\varepsilon_\parallel$ and $\varepsilon_\perp$ are on the order of $5\varepsilon_0$, where ε_0 is the permittivity of free space. Similarly, the electric conductivities $\sigma_\parallel$ and $\sigma_\perp$ of nematics are defined by

$$J_\parallel = \sigma_\parallel E_\parallel \tag{3.18}$$

and

$$J_\perp = \sigma_\perp E_\perp, \tag{3.19}$$

where $J_\parallel$ and $J_\perp$ are the currents flowing along and perpendicularly to the director axis, respectively. In conjunction with an applied dc electric field, the conductivity anisotropy could give rise to space charge accumulation and create strong director axis reorientation in a nematic film, giving rise to an orientational photorefractive[6] effect (see Chapter 8).

Most nematics (e.g., E7, 5CB, etc.) are said to possess positive (dielectric) anisotropy ($\varepsilon_\parallel > \varepsilon_\perp$). On the other hand, some nematics, such as MBBA, possess negative anisotropy (i.e., $\varepsilon_\parallel < \varepsilon_\perp$). The controlling factors are the molecular constituents and structures.

In general, $\varepsilon_\parallel$ and $\varepsilon_\perp$ have different dispersion regions, as shown in Figure 3.4 for 4-methoxy-4′-n-butylazoxy-benzene,[7] which possesses negative dielectric anisotropy ($\Delta\varepsilon < 0$). Also plotted in Figure 3.4 is the dispersion of ε_{iso}, the dielectric constant for the isotropic case. Notice that for frequencies of 10^9 Hz or less, $\varepsilon_\perp > \varepsilon_\parallel$. At higher frequencies and in the optical regime, $\varepsilon_\parallel > \varepsilon_\perp$ (i.e., the dielectric anisotropy changes sign).

For some nematic liquid crystals this changeover in the sign of $\Delta\varepsilon = \varepsilon_\parallel - \varepsilon_\perp$ occurs at a much lower frequency (cf. Fig. 3.5 for phenylbenzoates[8]). This changeover frequency f_{co} is lower because of the long three-ringed molecular structure, which is highly resistant to the rotation of molecules around the short axes.

For electro-optical applications, the dielectric relaxation behavior of $\varepsilon_\parallel$ and $\varepsilon_\perp$ for the different classes of nematic liquid crystals, and the relationships between the molecular structures and the dielectric constant, is obviously very important. This topic, however, is beyond the scope of this chapter, and the reader is referred to Blinov[8] and Khoo and Wu[9] and the references quoted therein for more detailed information.

Pure organic liquids are dielectric [i.e., nonconducting ($\sigma = 0$)]. The electric conductivities of liquid crystals are due to some impurities or ions. In general, $\sigma_\parallel$ is larger than $\sigma_\perp$. Electrical conduction plays an important role in electro-optical applications of liquid crystals in terms of stability and instability, chemical degradation,

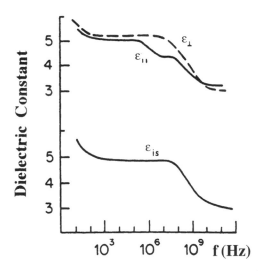

Figure 3.4. Disperson data of the dielectric constant $\varepsilon_\parallel$ and $\varepsilon_\perp$ for the nematic and isotropic phase of the liquid crystal 4-methoxy-4′-n-butylazoxy-benzene.

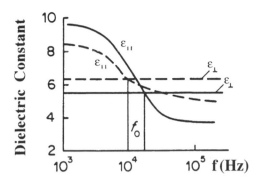

Figure 3.5. Disperson data of the dielectric constant $\varepsilon_\parallel$ and $\varepsilon_\perp$ for the nematic and isotropic phase of the liquid crystal phenylbenzoate.

and lifetime of device.[9] Typically, $\sigma_\parallel$ and $\sigma_\perp$ are on the order of 10^{-10} S^{-1}cm^{-1} for pure nematics. As in almost all materials, these conductivities could be varied by several orders of magnitude with the use of appropriate dopants.

The magnetic susceptibility of a material is defined in terms of the magnetization $\vec{M}$, the magnetic induction $\vec{B}$, and the magnetic strength $\vec{H}$ by

$$\vec{M} = \frac{\vec{B}}{\mu_0} - \vec{H} = \vec{\vec{\chi}} : \vec{H} \tag{3.20}$$

and

$$\vec{B} = \mu_0 (1 + \vec{\vec{\chi}}_m) : \vec{H}. \tag{3.21}$$

The magnetic susceptibility tensor $\vec{\chi}_m$ is anisotropic. For a uniaxial material such as a nematic, the magnetic susceptibility takes the form

$$\vec{\chi}_m = \begin{bmatrix} \chi_\perp^m & 0 & 0 \\ 0 & \chi_\perp^m & 0 \\ 0 & 0 & \chi_\parallel^m \end{bmatrix} \tag{3.22}$$

Note that this is similar to the dielectric constant $\vec{\vec{\varepsilon}}$.

Nematic liquid crystals, in fact liquid crystals in general, are diamagnetic. Therefore,

$$\vec{\vec{\chi}}_m = \begin{bmatrix} \chi_\perp^m & 0 & 0 \\ 0 & \chi_\perp^m & 0 \\ 0 & 0 & \chi_\parallel^m \end{bmatrix} \quad \text{and} \quad \vec{\vec{\chi}}_m = \begin{bmatrix} \chi_\perp^m & 0 & 0 \\ 0 & \chi_\perp^m & 0 \\ 0 & 0 & \chi_\parallel^m \end{bmatrix} \tag{3.22}$$

are small and negative, on the order of 10^{-5} [in International System (SI) of units]. As a result of the smallness of these magnetic susceptibilities, the magnetic interactions among the molecules comprising the liquid crystal are small (in comparison with their interaction with the external applied field). Consequently, the local field acting on the molecules differs very little from the external field, and in general, magnetic measurements are the preferred method to study liquid crystal order parameters and other physical processes.

3.3.2.   Free Energy and Torques by Electric and Magnetic Fields

In this section we consider the interactions of nematic liquid crystals with applied fields (electric or magnetic); we will limit our discussion to only dielectric and diamagnetic interactions.

For a generally applied (dc, low frequency, or optical) electric field $\vec{E}$, the displacement $\vec{D}$ may be written in the from

$$\vec{D} = \varepsilon_\perp \vec{E} + (\varepsilon_\parallel - \varepsilon_\perp)(\hat{n} \cdot \vec{E})\hat{n}. \tag{3.23}$$

The electric interaction energy density is therefore

$$\mu_E = -\int_0^E \vec{D} \cdot d\vec{E} = -\frac{1}{2}\varepsilon_\perp(\vec{E} \cdot \vec{E}) - \frac{\Delta\varepsilon}{2}(\hat{n} \cdot \vec{E})^2. \tag{3.24}$$

Note that the first term on the right-hand side of Equation (3.24) is independent of the orientation of the director axis. It can therefore be neglected in the director axis deformation energy. Accordingly, the free-energy density term associated with the application of an electric field is given by

$$F_E = -\frac{\Delta\varepsilon}{2}(\hat{n} \cdot \vec{E})^2 \tag{3.25}$$

in SI units [in cgs units, $F_E = -(\Delta\varepsilon/8\pi)(\hat{n} \cdot \vec{E})^2$]. The molecular torque produced by the electric field is given by

$$\vec{\Gamma}_E = \vec{D} \times \vec{E}$$

$$= \Delta\varepsilon(\hat{n} \cdot \vec{E})(\hat{n} \times \vec{E}). \tag{3.26}$$

Similar considerations for the magnetic field yield a magnetic energy density term U_m given by

$$U_m = -\int_0^M \vec{B} \cdot d\vec{M}$$

$$= \frac{1}{2\mu_0}\chi_\perp^m B^2 - \frac{1}{2\mu_0}\Delta\chi^m(\hat{n} \cdot \vec{B})^2, \tag{3.27}$$

a magnetic free-energy density (associated with director axis reorientation) F_m given by

$$F_m = \frac{1}{2\mu_0} \Delta \chi^m (\hat{n} \cdot \vec{B})^2, \tag{3.28}$$

and a magnetic torque density

$$\vec{\Gamma}_m = \vec{M} \times \vec{H}$$
$$= \Delta \chi^m (\hat{n} \cdot \vec{H})(\hat{n} \cdot \vec{H}). \tag{3.29}$$

These electric and magnetic torques play a central role in various field-induced effects in liquid crystals.

3.4. OPTICAL DIELECTRIC CONSTANTS AND REFRACTIVE INDICES

3.4.1. Linear Susceptibility and Local Field Effect

In the optical regime, $\varepsilon_\parallel > \varepsilon_\perp$. Typically, $\varepsilon_\parallel$ is on the order of $2.89\varepsilon_0$ and $\varepsilon_\perp$ is $2.25\varepsilon_0$. These correspond to refractive indices $n_\parallel = 1.7$ and $n_\perp = 1.5$. An interesting property of nematic liquid crystals is that such a large birefringence ($\Delta\varepsilon = \varepsilon_\parallel - \varepsilon_\perp \approx 0.2$) is manifested throughout the whole optical spectral regime [from near ultraviolet (≈ 400 nm), to visible (≈ 500 nm) and near infrared ($1-3$ μm), to the infrared regime ($8-12$ μm), i.e., from 400 nm to 12 μm]. Figure 3.6 shows the measured birefringence of three typical nematic liquid crystals from the UV to the far infrared ($\lambda = 16$ μm).

The optical dielectric constants originate from the linear polarization $\vec{P}$ generated by the incident optical field $\vec{E}_{op}$ on the nematic liquid crystal:

$$\vec{P} = \varepsilon_0 \vec{\bar{\chi}} \cdot \vec{E}. \tag{3.30a}$$

From the defining equation

$$\vec{D} = \varepsilon_0 \vec{E} + \vec{P} = \vec{\bar{\varepsilon}} : \vec{E}, \tag{3.30b}$$

we have

$$\vec{\bar{\varepsilon}} = \varepsilon_0 \left[1 + \vec{\bar{\chi}}^{(1)} \right]. \tag{3.30c}$$

Here $\vec{\bar{\chi}}^{(1)}$ is the linear (sometimes termed "first order") susceptibility tensor of the nematics. $\vec{\bar{\chi}}^{(1)}$ is a macroscopic parameter and is related to the microscopic

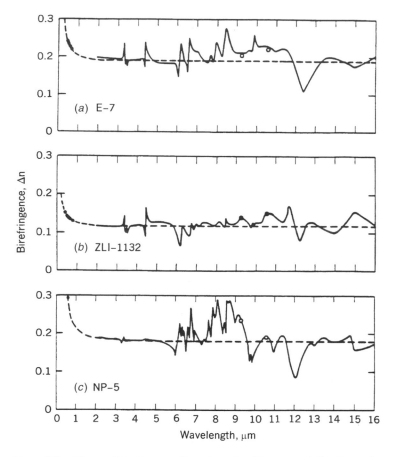

Figure 3.6. Measured birefringence $De = \varepsilon_{\parallel} - \varepsilon_{\perp}$ of three nematic liquid crystals.

(molecular) parameter, the molecular polarizabilities tensor α_{ij}, in the following way:

$$d_i = \alpha_{ij} E_j^{loc},$$

$$\vec{d} = \vec{\vec{\alpha}} : \vec{E}^{loc}, \qquad (3.31a)$$

$$\vec{P} = N\vec{d}, \qquad (3.31b)$$

where d_i is the ith component of the induced dipole $\vec{d}$ and N is the number density. In Chapter 8 a rigorous quantum mechanical derivation of α in terms of the dipole matrix elements or oscillator strengths and the energy levels and level populations will be presented. The connection between the microscopic parameter α_{ij} and the macroscopic parameter χ_{ij} is the local field correction factor (i.e., the difference between the externally applied field and the actual field as experienced by the molecules). Several theoretical formalisms have been developed to evaluate the field correction factor, ranging from simplified to complex and sophisticated ones.

Most of the approaches used to obtain the local field correction factor are based on the Lorentz results,[5] which state that the internal field (i.e., the local field as experienced by a molecule $\vec{E}^{loc}$ in a solid) is related to the applied field $\vec{E}^{app}$ by

$$\vec{E}^{loc} = \frac{n^2 + 2}{3}\vec{E}^{app}. \qquad (3.32a)$$

In particular, Vuks[10] analyzed experimental data and proposed that the local field in an anisotropic crystal may be taken as isotropic and expressed in the form

$$\vec{E}^{loc} = \frac{\langle n^2 \rangle + 2}{3}\langle \vec{E} \rangle, \qquad (3.32b)$$

where $\langle n^2 \rangle = \frac{1}{3}(n_x^2 + n_y^2 + n_z^2)$ and n_x, n_y, and n_z are the principal refractive indices of the crystal. This approach has been employed in the study of liquid crystals.[11] A more generalized expression for anisotropic crystals is given in Dunmar:[12]

$$E_i^{loc} = [1 + \vec{\vec{L}}_{ii}(n_i^2 - 1)]E_i^{app}, \quad i = x, y, z. \qquad (3.33)$$

More generally, one can write Equation (3.33) as

$$\vec{E}^{loc} - \vec{\vec{K}} : \vec{E}^{app}, \qquad (3.34)$$

where $\vec{\vec{K}}$, the local field "factor," is a second-rank tensor, which states that the local field $\vec{E}^{loc}$ is linearly related to the mascroscopic applied field $\vec{E}^{app}$. In general, experimental measurements show that the treatments by Vuks and Dunmar are qualitatively in agreement.

3.4.2. Equilibrium Temperature and Order Parameter Dependences of Refractive Indices

The two principal refractive indices $n_\perp$ and $n_\parallel$ of a uniaxial liquid crystal and the anisotropy $n_\parallel - n_\perp$ have been the subject of intensive studies for their fundamental importance in the understanding of liquid crystal physics and for their vital roles in applied electro-optic devices. Since the dielectric constants ($\varepsilon_\perp$ and $\varepsilon_\parallel$) enter directly and linearly into the constitutive equations [Eqns. (3.30a) – (3.30c)], it is theoretically more convenient to discuss the fundamentals of these temperature dependences in terms of the dielectric constants.

From Equation (3.34) for the local field $\vec{E}^{loc}$ and Equation (3.31) for the induced dipole moments, we can express the polarization $\vec{p} \equiv N\vec{d}$ by

$$\vec{P} = N\vec{\vec{\alpha}} : (\vec{\vec{K}} : \vec{E}), \qquad (3.35)$$

where $\vec{\vec{\alpha}}$ is the polarizability tensor of the molecule, N is the number of molecules per unit volume, and the parentheses denote averaging over the orientations of all molecules.

The dielectric constant $\bar{\bar{\varepsilon}}$ (in units of ε_0) is therefore given by

$$\bar{\bar{\varepsilon}} = 1 + \frac{N}{\varepsilon_0} \bar{\bar{\alpha}} : \bar{\bar{K}} \qquad (3.36)$$

and

$$\Delta\varepsilon = \varepsilon_{\parallel} - \varepsilon_{\perp}$$
$$= \frac{N}{\varepsilon_0} (\langle \bar{\bar{\alpha}} : \bar{\bar{K}} \rangle_{\parallel} - \langle \bar{\bar{\alpha}} : \bar{\bar{K}} \rangle_{\perp}). \qquad (3.37)$$

From these considerations and from observations by deJeu and Bordewijk[13] that

$$\Delta\varepsilon \propto \rho S \qquad (3.38)$$

and

$$\langle \bar{\bar{\alpha}} : \bar{\bar{K}} \rangle_{\parallel} - \langle \bar{\bar{\alpha}} : \bar{\bar{K}} \rangle_{\perp} \propto S, \qquad (3.39)$$

we can write $\varepsilon_{\parallel}$ and $\varepsilon_{\perp}$ as

$$\varepsilon_{\parallel} = n_{\parallel}^2 = 1 + \left(\frac{N}{3\varepsilon_0} \right) [\alpha_l K_l (2S + 1) + \alpha_t K_t (2 - 2S)] \qquad (3.40)$$

and

$$\varepsilon_{\perp} = n_{\perp}^2 = 1 + \left(\frac{N}{3\varepsilon_0} \right) [\alpha_l K_l (1 - S) + \alpha_t K_t (2 + S)], \qquad (3.41)$$

respectively, where K_l and K_t are the values of $\bar{\bar{K}}$ along the principal axis and S is the order parameter.

One can rewrite Equations (3.40) and (3.41) as

$$\varepsilon_{\parallel} = \varepsilon_l + \tfrac{2}{3} \Delta\varepsilon \qquad (3.42)$$

and

$$\varepsilon_{\perp} = \varepsilon_l - \tfrac{1}{3} \Delta\varepsilon, \qquad (3.43)$$

where

$$\Delta\varepsilon = \left(\frac{N}{\varepsilon_0} \right) [\alpha_l K_l - \alpha_t K_t] S$$
$$= \frac{N_A \rho}{\varepsilon_0 M} (\alpha_l K_l - \alpha_t K_t) S \sim \rho S \qquad (3.44)$$

and

$$\varepsilon_l = 1 + \frac{N_A \rho}{3\varepsilon_0 M}(\alpha_l K_l + 2\alpha_t K_t). \tag{3.45}$$

Notice that we have replaced N by $N_A\rho/M$, where N_A is Avogadro's number, ρ is the density, and M is the mass number.

The final explicit forms of the ε's depend on the determination of the internal field tensor. However, it is important to note that, in terms of the temperature dependence of ε's,[14]

$$\varepsilon_l \sim 1 + \mathrm{const}\,\rho = 1 + C_1\rho \tag{3.46}$$

and

$$\Delta\varepsilon \sim \mathrm{const}\,\rho S = C_2\rho S. \tag{3.47}$$

In other words, the temperature (T) dependence of $\varepsilon_{\parallel}$ and $\varepsilon_{\perp}$ (and the corresponding refractive indices $n_{\parallel}$ and $n_{\perp}$) is through the dependences of ρ and S on T.

One of the most striking features of the temperature dependence of the refractive indices of nematic liquid crystals is that the thermal index gradients ($dn_{\parallel}/dT$ and $dn_{\perp}/dT$) become extraordinarily large near the phase transition temperature (Figs. 3.7a and 3.7b). From Equations (3.46), (3.47), (3.40), and (3.41), we can obtain $dn_{\parallel}/dT$ and $dn_{\perp}/dT$ as

$$\frac{dn_{\parallel}}{dT} = \frac{1}{n_{\parallel}}\left(C_1\frac{d\rho}{dT} + \frac{2}{3}C_2 S\frac{d\rho}{dT} + \frac{2}{3}C_2\rho\frac{dS}{dT}\right), \tag{3.48}$$

$$\frac{dn_{\perp}}{dT} = \frac{1}{n_{\perp}}\left(C_1\frac{d\rho}{dT} - \frac{1}{3}C_2 S\frac{d\rho}{dT} - \frac{1}{3}C_2\rho\frac{dS}{dT}\right). \tag{3.49}$$

Figure 3.8 shows the plot of $dn_{\parallel}/dT$ and $dn_{\perp}/dT$ for 5CB as a function of temperature, with experimental data deduced from a more detailed measurement by Horn.[15]

Studies of the optical refractive indices of liquid crystals, as presented previously, are traditionally confined to what one may term as the classical and steady-state regime. In this regime the molecules are assumed to be in the ground state, and the optical field intensity is stationary. Results or conclusions obtained from such an approach, which have been outlined previously and in the next section, have to be considered in the proper context when these fundamental assumptions about the state of the molecules and the applied field are no longer true.

Detailed theories dealing with these quantum mechanical, nonlinear, or transient optical effects are given in Chapters 8 and 10. As an example consider the expression for the (linear) molecular polarizability given in Equation (10.28). Note that the refractive indices of an excited molecule are completely different from those associated with a molecule in the ground state; these differences are due to the fact that a totally different set of dipole matrix elements d_{ij} and frequency denominators ($\omega_i - \omega_j$) are involved.

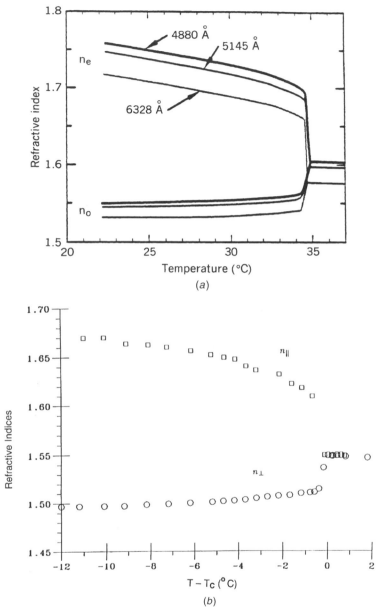

Figure 3.7. (a) Temperature dependence of the refractive indices of 5CB in the visible spectrum. (b) Temperature dependence of the refractive indices of 5CB in the infrared (10.6 μm) region.

Second, if the intensities of the impinging optical fields are fast and oscillatory (e.g., picosecond laser pulses) and their time durations are comparable to the internal relaxation dynamics of the molecules, those optical fields will "see" the transient responses of the molecules. These transient responses in the internal motions

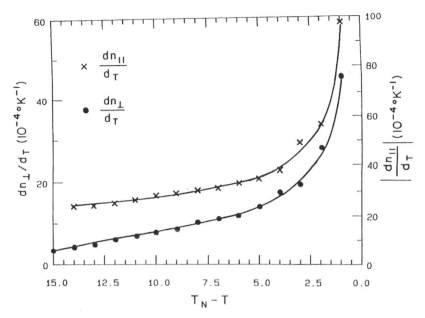

Figure 3.8. Plot of $dn_\parallel/dT$ and $dn_\perp/dT$ for the liquid crystal for temperature near T_c [after Khoo and Normandin (14) and Horn (15)]. Solid curve is for visual aid only.

(sometimes termed "internal temperature") of the molecules are usually manifested in the form of spectral shift; that is, the emission and the absorption spectra of the molecules are momentarily shifted, usually in the picosecond time scale. Accordingly, the "effective" molecular polarizabilities, which translate into the refractive indices, will also experience a time-dependent change. These transient changes in the refractive index associated with the molecular excitations under ultra short laser pulses should be clearly distinguished from the usual temperature effects associated with the stationary or equilibrium state. In the stationary case the fast molecular transients have relaxed and the absorbed energy has been converted to an overall rise in the bulk temperature. As stated earlier in Chapter 1, even in the so-called stationary situation where the internal molecular excitations have relaxed, the order parameter S may still not attain the equilibrium state. This usually happens in the nanosecond or microsecond time scale. Details of these considerations are given in Chapter 9, where we discuss pulsed laser-induced heating effects.

3.5. FLOWS AND HYDRODYNAMICS

One of the most striking properties of liquid crystals is their ability to flow freely while exhibiting various anisotropic and crystalline properties. It is this dual nature of liquid crystals that makes them very interesting materials to study; it also makes the theoretical formalism very complex.

The main feature that distinguishes liquid crystals in their ordered mesophases (e.g., the nematic phase) from ordinary fluids is that their physical properties are dependent on the orientation of the director axis $\vec{n}(\vec{r})$; these orientation flow processes are necessarily coupled, except in very unusual cases (e.g., pure twisted deformation). Therefore, studies of the hydrodynamics of liquid crystals will involve a great deal more (anisotropic) parameters, than studies of the hydrodynamics of ordinary liquids.

We begin our discussion by reviewing first the hydrodynamics of an ordinary fluid. This is followed by a discussion of the general hydrodynamics of liquid crystals. Specific cases involving a variety of flow-orientational couplings are then treated.

3.5.1.   Hydrodynamics of Ordinary Isotropic Fluids

Consider an elementary volume $dV = dx\,dy\,dz$ of a fluid moving in space as shown in Figure 3.9. The following parameters are needed to describe its dynamics:

position vector: $\vec{r}$,

velocity: $\vec{v}(\vec{r}, t)$,

density: $\rho(\vec{r}, t)$,

pressure: $p(\vec{r}, t)$, and

forces in general: $\vec{f}(\vec{r}, t)$.

In later chapters where we study laser-induced acoustic (sound, density) waves in liquid crystals, or generally when one deals with acoustic waves, it is necessary to assume that the density $\rho(\vec{r}, t)$ is a spatially and temporally varying function. In this chapter, however, we "decouple" such density wave excitation from all the processes under consideration and basically limit our attention to the flow and orientational effects of an *incompressible* fluid. In that case we have

$$\rho(\vec{r},t) = \text{const.} \tag{3.50}$$

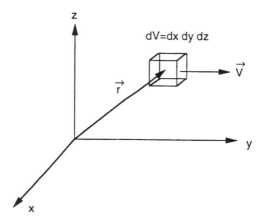

Figure 3.9.   An elementary volume of fluid moving at velocity $v\,(r, t)$ in space.

For all liquids, in fact for all gas particles or charges in motion, the equation of continuity also holds

$$\nabla \cdot (\rho \bar{v}) = -\frac{\partial \rho}{\partial t}. \tag{3.51}$$

This equation states that the total variation of $\rho \bar{v}$ over the surface of an enclosing volume is equal to the rate of decrease of the density. Since $\partial \rho / \partial t = 0$, we thus have, from Equation (3.51),

$$\nabla \cdot \bar{v} = 0. \tag{3.52}$$

The equation of motion describing the acceleration $d\bar{v}/dt$ of the fluid elements is simply Newton's law:

$$\rho \frac{d\bar{v}}{dt} = \bar{f}. \tag{3.53a}$$

Studies of the hydrodynamics of liquids may be said to be centered around this equation of motion, as we identify all the various origins and mechanisms of forces acting on the fluid elements and attempt to solve for their motion in time and space.

We shall start with the left-hand side of Equation (3.53a). Since $\bar{v} = \bar{v}(\bar{r}, t)$,

$$\frac{d\bar{v}}{dt} = \frac{\partial \bar{v}}{\partial t} + (\nabla \cdot \bar{v})\bar{v}. \tag{3.53b}$$

The force on the right-hand side of Equation (3.53a) comes from a variety of sources, including the pressure gradient $-\Delta \rho$, viscous force $\bar{f}_{vis}$, and external fields $\bar{f}_{ext}$ (electric, magnetic, optical, gravitational, etc.). Equation (3.53a) thus becomes

$$\rho \left[\frac{\partial \bar{v}}{\partial t} + (\nabla \cdot \bar{v})\bar{v} \right] = -\nabla \rho + \bar{f}_{vis} + \bar{f}_{ext}. \tag{3.54}$$

Let us ignore the external field for the moment. The formulation of the equation of motion for a fluid element is complete once we identify the viscous forces. Note that, in analogy to the pressure gradient term, the viscous force $\bar{f}_{vis}$ is the space derivation of a quantity which has the unit of pressure (i.e., force per unit area). Such a quantity is termed the stress tensor σ (i.e., the force is caused by the gradient in the stress; see Fig. 3.10). For example, the α component of $\bar{f}$ may be expressed as

$$f_\alpha = \frac{\partial}{\partial x_\beta} \sigma_{\alpha\beta}. \tag{3.55}$$

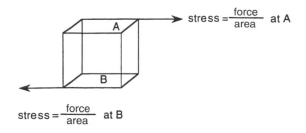

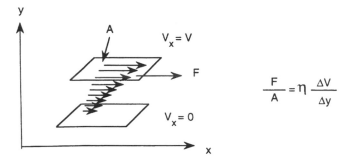

Figure 3.10.   Stresses acting on opposite planes of an elementary volume of fluid.

Accordingly, we may rewrite Equation (3.54) as

$$\rho \left[\frac{\partial v_\alpha}{\partial t} + v_\beta \frac{\partial v_\alpha}{\partial x_\beta} \right] = -\frac{\partial p}{\partial x_\alpha} + \frac{\partial \sigma_{\alpha\beta}}{\partial x_\beta}, \tag{3.56}$$

where summation over repeated indices is implicit.

By consideration of the fact that there is no force acting when the fluid velocity is a constant, the stress tensor is taken to be linear in the gradients of the velocity (see Fig. 3.10), that is,

$$\sigma_{\alpha\beta} = \eta \left(\frac{\partial v_\beta}{\partial x_\alpha} + \frac{\partial v_\alpha}{\partial x_\beta} \right). \tag{3.57}$$

The proportionality constant η in Equation (3.57) is the viscosity coefficient (in units of g cm^{-1} S^{-1}). Note that for a fluid under uniform rotation $\vec{w}_o$ (i.e., $\vec{v} = \vec{w}_o \times \vec{r}$), we have $\partial v_\beta / \partial x_\alpha = -\partial v_\alpha / \partial x_\beta$, which means $\sigma_{\alpha\beta} = 0$.

Equation (3.56), together with Equation (3.57), forms the basis for studying the hydrodynamics of an isotropic fluid. Note that since the viscosity for $\vec{f}_{\text{vis}}$ is a spatial derivative of the stress tensor, which in turn is a spatial derivative of the velocity, $\vec{f}_{vis}$ is of the form $\eta \nabla^2 \vec{v}$. Equation (3.54) therefore may be written as

$$\frac{\partial \vec{v}}{\partial t} + (\nabla \cdot \vec{v})\vec{v} = -\frac{\nabla p}{\rho} + \frac{\eta \nabla^2 \vec{v}}{\rho} + \frac{\vec{f}_{ext}}{\rho}, \tag{3.58}$$

which is usually referred to as the Navier–Stokes equation for an incompressible fluid.

3.5.2. General Stress Tensor for Nematic Liquid Crystals

The general theoretical framework for describing the hydrodynamics of liquid crystals has been developed principally by Leslie[16] and Ericksen.[17] Their approaches account for the fact that the stress tensor depends not only on the velocity gradients, but also on the orientation and rotation of the director. Accordingly, the stress tensor is given by

$$\sigma_{\alpha\beta} = \alpha_1 n_\gamma n_\delta A_{\gamma\delta} n_\alpha n_\beta + \alpha_2 n_\alpha n_\beta + \alpha_3 n_\beta n_\alpha + \alpha_4 A_{\alpha\beta} + \alpha_5 n_\gamma A_{\gamma\beta} + \alpha_6 n_\beta n_\gamma A_{\gamma\alpha}, \tag{3.59}$$

where the $A_{\alpha\beta}$'s are defined by

$$A_{\alpha\beta} = \frac{1}{2} \left[\frac{\partial v_\beta}{\partial x_\alpha} + \frac{\partial v_\alpha}{\partial x_\beta} \right]. \tag{3.60}$$

Note that all the other terms on the right-hand side of Equation (3.59) involve the director orientation, except the fourth term, $\alpha_4 A_{\alpha\beta}$. This is the same term as that for an isotropic fluid [cf. Eq. (3.57)], that is, $\alpha_4 = 2\eta$.

Therefore, in this formalism, there are six so-called Leslie coefficients, $\alpha_1, \alpha_2, \ldots, \alpha_6$, which have the dimension of viscosity coefficients. It was shown by Parodi[18] that

$$\alpha_2 + \alpha_3 = \alpha_6 - \alpha_5 \tag{3.61}$$

and so there are really five independent coefficients.

In the next few sections we will study particular cases of director axis orientation and deformation and we will show how these Leslie coefficients are related to other commonly used viscosity coefficients.

3.5.3. Flows with Fixed Director Axis Orientation

Consider here the simplest case of flows in which the director axis orientation is held fixed. This may be achieved by a strong externally applied magnetic field (see Fig. 3.11), where the magnetic field is along the direction $\hat{n}$. Consider the case of shear flow, where the velocity is in the z direction and the velocity gradient is along the x direction. This process could occur, for example, in liquid crystals confined by two parallel plates in the y-z plane.

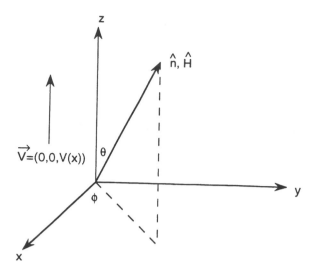

Figure 3.11.   Sheer flow I the presence of an applied magnetic field.

In terms of the orientation of the director axis, there are three distinct possibilities involving three corresponding viscosity coefficients:

1. η_1: $\hat{n}$ is parallel to the velocity gradient, that is, along the x axis ($\theta=90°$, $\phi=0°$).
2. η_2: $\hat{n}$ is parallel to the flow velocity, that is, along the z axis and lies in the shear plane x-z ($\theta=0°$, $\phi=0°$).
3. η_3: $\hat{n}$ is perpendicular to the shear plane, that is, along the y axis ($\theta=0°$, $\phi=90°$).

These three configurations have been investigated by Miesowicz,[19] and the η's are known as Miesowicz coefficients. In the original paper, as well as in the treatment by deGennes,[3] the definitions of η_1 and η_3 are interchanged. In deGennes notation, in terms of η_a, η_b, and η_c, we have $\eta_a = \eta_1$, $\eta_b = \eta_2$, and $\eta_c = \eta_3$. The notation used here is attributed to Helfrich,[6] which is now the conventional one.

To obtain the relations between $\eta_{1,2,3}$ and the Leslie coefficients $\alpha_{1,2,...,6}$, one could evaluate the stress tensor $\sigma_{\alpha\beta}$ and the shear rate $A_{\alpha\beta}$ for various director orientations and flow and velocity gradient directions. From these considerations, the following relationships are obtained:[3]

$$\begin{aligned}
\eta_1 &= \tfrac{1}{2}(\alpha_4 + \alpha_5 - \alpha_2), \\
\eta_2 &= \tfrac{1}{2}(\alpha_3 + \alpha_4 + \alpha_6), \\
\eta_3 &= \tfrac{1}{2}\alpha_4.
\end{aligned} \tag{3.62}$$

In the shear plane x-z, the general effective viscosity coefficient is actually more correctly expressed in the form[20]

$$\eta_{\text{eff}} = \eta_1 + \eta_2 \cos^2 \theta + \eta_2 \tag{3.63}$$

in order to account for angular velocity gradients. The coefficient $\eta_{1,2}$ is related to the Leslie coefficient α_1 by

$$\eta_{1,2} = \alpha_1. \tag{3.64}$$

3.5.4. Flows with Director Axis Reorientation

The preceding section deals with the case where the director axis is fixed during fluid flow. In more general situations director axis reorientation often accompanies fluid flows and vice versa. Taking into account the moment of inertia I and the torque $\vec{\Gamma} = \hat{n} \times \vec{f}$, where $\vec{f}$ is the molecular internal elastic field defined in Equation (3.11), $\vec{\Gamma}_{ext}$ is the torque associated with an externally applied field, and $\vec{\Gamma}_{vis}$ is the viscous torque associated with the viscous forces, the equation of motion describing the angular acceleration $d\Omega/dt$ as the director axis may be written as

$$I \frac{d\Omega}{dt} = (\hat{n} \times \vec{f} + \vec{\Gamma}_{ext}) - \vec{\Gamma}_{vis}. \tag{3.65}$$

The viscous torque $\vec{\Gamma}_{vis}$ consists of two components:[3] one arising from pure rotational effect (i.e., no coupling to the fluid flow) given by $\gamma_1 \hat{n} \times \vec{N}$ and another arising from coupling to the fluid motion given by $\gamma_2 \hat{n} \times \hat{A}\hat{n}$. Therefore, we have

$$\vec{\Gamma}_{vis} = \hat{n} \times \left[\gamma_1 \vec{N} + \gamma_2 \hat{A}\hat{n} \right]. \tag{3.66}$$

Here $\vec{N}$ is the rate of change of the director with respect to the immobile background fluid, given by

$$\vec{N} = \frac{d\hat{n}}{dt} - \hat{\omega} \times \hat{n}, \tag{3.67}$$

where $\hat{\omega}$ is the angular velocity of the liquid. In Equation (3.66) $\hat{A}$ is the velocity gradient tensor defined by Equation (3.60).

The viscosity coefficients γ_1 and γ_2 are related to the Leslie coefficient α's by[3]

$$\gamma_1 = \alpha_3 - \alpha_2, \tag{3.68a}$$
$$\gamma_2 = \alpha_2 + \alpha_3 = \alpha_6 - \alpha_5. \tag{3.68b}$$

Consider the flow configuration depicted in Figure 3.11. Without the magnetic field and setting $\phi = 0$, we have

$$\vec{v} = [0, 0, v(x)], \tag{3.69a}$$

$$\hat{n} = [\sin\theta, 0, \cos\theta], \tag{3.69b}$$

$$A_{xz} = \frac{1}{2}\frac{dv}{dx}, \tag{3.69c}$$

$$N_z = \omega_y n_x = -A_{xz} n_x, \tag{3.69d}$$

$$N_x = -\omega_y n_z = A_{xz} n_z. \tag{3.69e}$$

From Equation (3.66) the viscous torque along the y direction is given by

$$\begin{aligned}
\Gamma_{\text{vis}} &= -\gamma_1(n_z N_x - n_x N_z) - \gamma_2(n_z n_\mu A_{\mu x} - n_x n_\mu A_{\mu z}) \\
&= -\frac{1}{2}\frac{dv}{dx}[\gamma_1 + \gamma_2(\cos^2\theta - \sin^2\theta)] \\
&= -\frac{dv}{dx}[\alpha_3 \cos^2\theta - \alpha_2 \sin^2\theta].
\end{aligned} \tag{3.70}$$

In the steady state, from which the shear torque vanishes, a stable director axis orientation is induced by the flow with an angle θ_{flow} given by

$$\cos 2\theta_{\text{flow}} = -\frac{\gamma_1}{\gamma_2}. \tag{3.71}$$

For more complicated flow geometries, the director axis orientation will assume correspondingly complex profiles.

3.6.   FIELD-INDUCED DIRECTOR AXIS REORIENTATION EFFECTS

We now consider the process of director axis reorientation by an external static or low-frequency field. Optical field effects are discussed in Chapter 6. The following examples will illustrate some of the important relationships among the various torques and dynamical effects discussed in the preceding sections. We will consider the magnetic field as it does not involve complicated local field effects and other electric phenomena (e.g., conduction). The electric field counterparts of the results obtained here for the magnetic field can be simply obtained by the replacement of $\Delta\chi^m H^2$ by $\Delta\varepsilon E^2$ [cf. Eq. (3.26) and (3.29)].

3.6.1.   Field-Induced Reorientation without Flow Coupling: Freedericksz Transition

The following example demonstrates how the viscosity coefficient γ_1 comes into play in field-induced reorientational effects. Consider pure twist deformation caused by an externally applied field $\vec{H}$ on a planar sample as depicted in Figure 3.12. Let θ denote the angle of deformation. The director axis $\hat{n}$ is thus given by $\hat{n} = (\cos\theta, \sin\theta, 0)$.

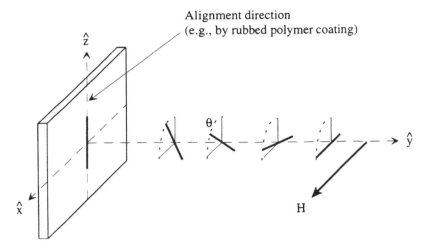

Figure 3.12. Pure twist deformation induced by an external magnetic field H on a planar sample; there is no fluid motion.

From this and the preceding equations, the free energy [Eq. (3.4)] and elastic torque [Eq. (3.13)] are, respectively,

$$F_2 = \frac{K_2}{2}\left(\frac{\partial\theta}{\partial z}\right)^2 \tag{3.72a}$$

and

$$\mathbf{l} = K_2 \frac{\partial^2\theta}{\partial z^2}\hat{z}. \tag{3.72b}$$

The viscous torque is given by

$$\Gamma_{vis} = -\gamma_1 \frac{d\theta}{dt}. \tag{3.73}$$

The torque exerted by the external field $\bar{H}$ (applied perpendicular to the initial director axis), from Equation (3.29), becomes

$$\Gamma_{ext} = \Delta\chi^m H^2 \sin\theta \cos\theta. \tag{3.74}$$

Hence, the torque balance τ equation gives

$$\gamma_1 \frac{d\theta}{dt} = K_2 \frac{\partial^2\theta}{\partial z^2}\hat{z} + \Delta\chi^m H^2 \sin\theta \cos\theta. \tag{3.75a}$$

In the equilibrium situation, $\gamma_1 d\theta/dt = 0$, and Equation (3.75a) becomes

$$K_2 \frac{\partial^2 \theta}{\partial z^2} \hat{z} + \Delta \chi^m H^2 \sin\theta \cos\theta = 0. \tag{3.75b}$$

An interesting result from this equation is the so-called Freedericksz transition.[3] For an applied field strength less than a critical field H_F, $\theta = 0$. For $H > H_F$, reorientation occurs. The expression for H_F is given by

$$H_F = \left(\frac{\pi}{d}\right)\left(\frac{K}{\Delta \chi^m}\right)^{1/2} \tag{3.76}$$

assuming that the reorientation obeys the hard-boundary (strong anchoring) condition (i.e., $\theta = 0$ at $z = 0$ and at $z = d$). For H just above H_F, θ is given approximately by

$$\theta = \theta_0 \sin\left(\frac{\pi z}{d}\right), \tag{3.77a}$$

where

$$\theta_0 \sim 2\frac{(H - H_F)^{1/2}}{H_F}. \tag{3.77b}$$

For the case where H is abruptly reduced from its value above H_F, to 0, Equation (3.75a) becomes

$$A\gamma_1 \frac{d\theta}{dt} = K_2 \frac{\partial^2 \theta}{\partial z^2}. \tag{3.78}$$

Writing $\theta(z, t) = \theta_0 \sin(\pi z/d)$ gives

$$\dot{\theta} = \frac{-\pi^2 K_2}{d^2 \gamma_1} \theta, \tag{3.79}$$

that is,

$$\theta_0(t) = \theta_0 e^{-t/\tau}, \tag{3.80}$$

where the relaxation time constant τ is given by

$$\tau = \frac{\gamma_1 d^2}{\pi^2 K_2}. \tag{3.81}$$

Most practical liquid crystal devices employ ac electric field. Accordingly, the Freedericksz transition field E_F is given by simply replacing $\Delta\chi^m$ with $\Delta\varepsilon$; i.e., we have

$$E_F = \left(\frac{\pi}{d}\right)\left(\frac{K}{\Delta\varepsilon}\right)^{1/2},\qquad(3.82a)$$

$$V_F = \pi\left(\frac{K}{\Delta\varepsilon}\right)^{1/2}.\qquad(3.82b)$$

For 5CB,[20, 21] $k\sim10^{-11}$ N, $\Delta\varepsilon\sim11$ ($\varepsilon_{\parallel}\sim16$, $\varepsilon_{\perp}\sim5$), $\varepsilon_0 = 8.85\times10^{-12}$ F/m, $\Delta\sigma/\sigma_{\perp}\sim0.5$, and $V_F\sim1$ V.

In Chapters 6 and 7, we discuss these field-induced nematic director axis reorientations in detail in the context of electro-optical switching and display applications.

3.6.2. Reorientation with Flow Coupling

Field-induced director axis reorientation, accompanied by fluid flow, is quite complicated as it involves much more physical parameters.

Consider the interaction geometry shown in Figure 3.13. A homeotropically aligned nematic liquid crystal film is acted on by an electric or a magnetic field in the x direction. Let ϕ denote the director axis reorientation angle from the original alignment direction z. Assume hard-boundary conditions at the two cell walls at $z = 0$ and at $z = d$. The flow is in the x direction, with a z dependence.

The following are the pertinent parameters involved:

director axis: $\hat{n} = (\sin\phi, 0, \cos\phi)$, $\qquad(3.83a)$

velocity field: $\vec{v} = [v(z), 0, 0]$, $\qquad(3.83b)$

free energies: $F = \dfrac{1}{2}K_1(\nabla\cdot\hat{n})^2 + \dfrac{1}{2}K_3[\hat{n}\times(\nabla\times\hat{n})]^2$

$$= \frac{1}{2}K_1\sin^2\phi\left(\frac{d\phi}{dz}\right)^2 + \frac{1}{2}K_3\cos^2\phi\left(\frac{d\phi}{dz}\right)^2,\quad(3.83c)$$

elastic torques $= [K_1\sin^2\phi + K_3\cos^2\phi]\dfrac{d^2\phi}{dz^2}$

$$+ [(K_1 - K_3)\sin\phi\cos\phi]\left(\frac{d\phi}{dz}\right)^2,\qquad(3.83d)$$

field-induced torques $= \varepsilon_0\,\Delta\varepsilon E^2\sin\phi\cos\phi$ $\qquad(3.83e)$

rotation viscous torques $= \gamma_1\dfrac{d\phi}{dt}$, $\qquad(3.83f)$

flow-orientational viscous torques $= \dfrac{dv}{dz}[\alpha_2\sin^2\phi - \alpha_3\cos^2\phi]$. $\qquad(3.83g)$

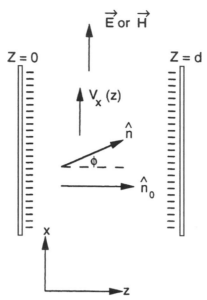

Figure 3.13. Director axis reorientation causing flows.

Using Equation (3.65), the equation of motion taking into account these torques, as well as the moment of inertia I of molecules involved, is given by

$$I\frac{d^2\phi}{dt^2} + \gamma_1 \frac{d\phi}{dt} = [K_1 \sin^2 \phi + K_3 \cos^2 \phi]\frac{d^2\phi}{dz^2} + [(K_1 - K_3)\sin\phi\cos\phi]\left(\frac{d\phi}{dz}\right)^2$$

$$+[\alpha_2 \sin^2 \phi - \alpha_3 \cos^2 \phi]\frac{dv}{dz} + \varepsilon_0 \Delta \varepsilon E^2 \sin\phi\cos\phi. \qquad (3.84)$$

This equation may be solved for various experimental conditions. Optically induced director axis reorientation and flow effects have been studied by two groups[21,22] using picosecond laser pulses. A solution of the previous equation is also presented in the work of Eichler and Macdonald.[22]

REFERENCES

1. Frank, F. C. 1958. *Discuss. Faraday Soc.* 25:19.

2. Ericksen, J. L. 1969. *Liquid Crystals.* G. Brown (ed.). New York: Gordon and Breach.

3. deGennes, P. G. 1974. *Physics of Liquid Crystals.* Oxford: Clarendon Press.

4. See, for example, Barbero, G., and F. Simoni. 1992. *Appl. Phys. Lett.* 41:504; Barbero, G., F. Simoni, and P. Aiello. 1984. *J. Appl. Phys.* 55:304; see also Faetti, S. 1991. *Physics of Liquid Crystalline Materials.* I. C. Khoo and F. Simoni (eds.). Philadelphia: Gordon and Breach.

5. Jackson, J. D. 1975. *Classical Electrodynamics.* New York: Wiley.

6. Helfrich, W. 1969. *J. Chem. Phys.* 51:4092; see also Ref. 8, Chapter 5.

7. Parneix, J. P., A. Chapoton, and E. Constant. 1975. *J. Phys. (Paris).* 36:1143.

8. Blinov, L. M. 1983. *Electro optical and Magneto-optical Properties of Liquid Crystals.* Chichester: Wiley (Interscience).

9. Khoo, I. C., and S. T. Wu. 1993. *Optics and Nonlinear Optics of Liquid Crystals.* Singapore: World Scientific.

10. Vuks, M. F. 1966. *Opt. Spektrosk.* 60:644.

11. Chandrasekhar, S., and N. V. Madhusudana. 1969. *J. Phys. (Paris) Colloq.* 30:C4.

12. Dunmar, D. A. 1971. *Chem. Phys. Lett.* 10:49.

13. deJeu, W. H., and P. Bordewijk. 1978. *J. Chem. Phys.* 68:109.

14. Khoo, I. C., and R. Normandin. 1985. *IEEE J. Quantum Electron.* QE-21:329.

15. Horn, R. G. 1978. *J. Phys. (Paris).* 39:105.

16. Leslie, F. M. 1966. *Quantum J. Mech. Appl. Math.* 19:357.

17. Ericksen, J. L. 1966. *Phys. Fluids.* 9:1205.

18. Parodi, O. 1970. *J. Phys. (Paris).* 31:581.

19. Miesowicz, M. 1935. *Nature (London).* 17:261; 1946. 158:27.

20. See, for example, deJeu, W. H. 1980. *Physical Properties of Liquid Crystalline Materials.* New York: Gordon and Breech.

21. Khoo, I. C., R. G. Lindquist, R. R. Michael, R. J. Mansfield, and P. G. LoPresti. *J. Appl. Phys.* 69:3853; Khoo, I. C., and R. Normandin. 1984. *ibid.* 55:1416.

22. Eichler, H. J., and R. Macdonald. 1991. *Phys. Rev. Lett.* 67:2666.

4

Cholesteric, Smectic, and Ferroelectric Liquid Crystals

4.1. CHOLESTERIC LIQUID CRYSTALS

The physical properties of cholesteric liquid crystals are in almost all aspects similar to nematics, except that the director axis assumes a helical form (Fig. 4.1a) with a finite pitch $p = 2\pi/q_0$. Figures 4.1b and 4.1c show two commonly occurring director axis alignments: planar twisted and fingerprint, respectively. From this point of view, we may regard nematics as a special case of cholesterics with $p \to \infty$. Since the optical property of the nematic, a uniaxial material, is integrally related to the director axis, the helical arrangement of the latter in a cholesteric certainly introduces new optical properties, particularly in the propagation and reflection of light from cholesteric liquid crystal cells. In this section we summarize the main physical properties associated with the helical structure.

4.1.1. Free Energies

Since the equilibrium configuration of a cholesteric liquid crystal is a helical structure with a pitch wave vector q_0 ($q_0 = 2\pi/p_0$), its elastic free energy will necessarily reflect the presence of q_0. The evolution of a cholesteric to a nematic liquid crystal may be viewed as the "untwisting" of the helical structure (i.e., the twist deformation energy is involved). This is indeed rigorously demonstrated in the Frank elastic theory,[1] where consideration of the absence of mirror symmetry in cholesterics results in the addition of another factor to the twist deformation energy:

$$K_2(\hat{n} \cdot \nabla \times \hat{n})^2 \to K_2(\hat{n} \cdot \nabla \times \hat{n} + q_0)^2.$$
$$\text{nematic} \qquad\qquad \text{cholesteric} \tag{4.1}$$

In Figure 4.1, the director axis is described by $\hat{n} = (n_x, n_y, n_z)$, where $n_x = \cos \theta(z)$, $n_y = \sin \theta(z)$, and $n_z = 0$. Note that this configuration corresponds to a state

Liquid Crystals, Second Edition By Iam-Choon Khoo
Copyright © 2007 John Wiley & Sons, Inc.

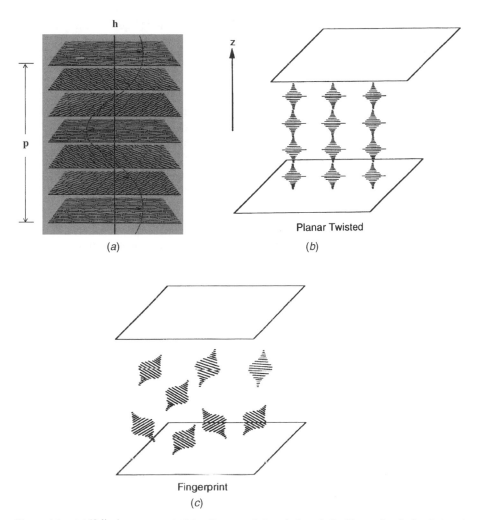

Figure 4.1. (a) Helical arrangement of the director axis in a cholesteric liquid crystal; **p** is the pitch and **h** is the helix direction. Two typical cholesteric liquid crystal cell shown in (b) planar twisted and (c) fingerprint.

of minimum free energy

$$F_2 = \frac{1}{2} K_2 (\hat{n} \cdot \nabla \times \hat{n})^2 = \frac{1}{2} K_2 \left(-\frac{\partial \theta}{\partial z} + q_0 \right)^2 = 0 \tag{4.2}$$

if $q_0 = -\partial\theta/\partial z$, that is,

$$\theta = q_0 z. \tag{4.3}$$

For a general distortion therefore the free energy of a cholesteric liquid crystal is given by

$$F_d = \tfrac{1}{2}K_1(\nabla \cdot \hat{n})^2 + \tfrac{1}{2}K_2(\hat{n} \cdot \nabla \times \hat{n})^2 + \tfrac{1}{2}K_3(\hat{n} \times \nabla \times \hat{n})^2. \tag{4.4}$$

If the pitch of the cholesteric is not changed by an external probing field (electric, magnetic, or optical), the physical properties of cholesterics are those of locally uniaxial crystals. In other words, the anisotropies in the dielectric constant ($\Delta\varepsilon = \varepsilon_\parallel - \varepsilon_\perp$), electric conductivity ($\Delta\sigma = \sigma_\parallel - \sigma_\perp$), magnetic susceptibility ($\Delta\chi^m = \chi_\parallel^m - \chi_\perp^m$), and so on, are defined with respect to the local director axis direction. On the other hand, if one refers to the helical axis (cf. Fig. 4.1, z direction), an applied probing field along z will "see" the $\perp$ components (i.e., $\varepsilon_\perp$, $\sigma_\perp$, $\chi_\perp$, etc.). If the probing field is along a direction perpendicular to z, it will effectively see the average of the $\perp$ and $\parallel$ components [i.e., $\tfrac{1}{2}(\varepsilon_\perp + \varepsilon_\parallel)$, $\tfrac{1}{2}(\sigma_\perp + \sigma_\parallel)$, $\tfrac{1}{2}(\chi_\perp + \chi_\parallel)$, etc.].

Just as in the nematic case, the application of an applied magnetic or electric field gives rise to additional terms in the free energy given by

$$F_{\text{mag}} = -\tfrac{1}{2}\Delta\chi^m(\hat{n} \cdot \vec{H})^2 \tag{4.5a}$$

and

$$F_{\text{el}} = -\tfrac{1}{2}\Delta\varepsilon(\hat{n} \cdot \vec{E})^2, \tag{4.5b}$$

respectively, as well as some terms which are independent of the orientation of the director axis. $\Delta\chi^m = \chi_\parallel^m - \chi_\perp^m$ for cholesterics are usually quite small (about 10 in cgs units) in magnitude and negative in sign. In other words, the directors tend to align normal to the magnetic field.

4.1.2.   Field-Induced Effects and Dynamics

In the purely dielectric interaction picture (i.e., no current flow), the realignment or alignment of a cholesteric liquid crystal in an applied electric or magnetic field, results from the system's tendency to minimize its total free energy.

Clearly, the equilibrium configuration of the director axis depends on its initial orientation and the direction of the applied fields, as well as the signs of $\Delta\chi^m$ and $\Delta\varepsilon$. We will not delve into the various possible cases as they all involve the same basic mechanism; that is, the director axis tends to align parallel to the field for positive dielectric anisotropies, and normal to the field for negative dielectric anisotropies.

In the case of positive dielectric anisotropies, the field-induced reorientation process is analogous to that discussed for nematics (cf. Sect. 3.6). In cholesterics, however, the realignment of the director axis in the direction of the applied field will naturally affect the helical structure.

Magnetic Field.   Figure 4.2a shows the unperturbed director axis configuration in the bulk of an ideal cholesteric liquid crystal. Upon the application of a magnetic field, some molecules situated in the bulk regions A, A', and so on are preferentially

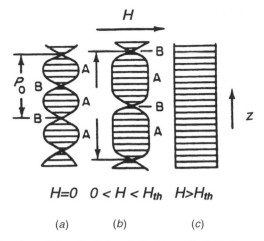

Figure 4.2. Field induced untwisting of a cholesteric liquid crystal: (a) ideal sinusoidal helix; (b) increase of pitch and deviation from ideal helix; (c) complete alignment along applied field and infinite pitch.

aligned along the direction of the field; others, situated in regions B, B', and so forth are not and would tend to reorient themselves along the field direction. As a result, the pitch of the helical structure will be increased; the helix is no longer of the ideal sinusoidal form. Finally, when the field is sufficiently high, this untwisting effect is complete as the pitch approaches infinity; that is, the cholesteric liquid crystal is said to have undergone a transition to the nematic phase.[2]

This process can be described by the free-energy minimization process. The total free energy of the system is given by

$$F_{total} = \frac{1}{2} \int dz \left[K_2 \left(\frac{\partial \phi}{\partial z} - q_0 \right)^2 - \Delta \chi^m H^2 \sin^2 \phi \right].$$ (4.6)

This, upon minimization, yields the Euler equation:

$$K_2 \frac{d^2\phi}{dz^2} + \Delta \chi^m H^2 \sin\phi \cos\phi = 0,$$ (4.7)

which is analogous to Equation (3.75b). One can define a coherence length ξ_H by

$$\xi_H = \frac{K_2}{\Delta \chi^m H^2}.$$ (4.8)

Equation (4.7) thus becomes

$$\xi_H \frac{d^2\phi}{dz^2} = \sin\phi \cos\phi.$$ (4.9)

Upon integration, it yields

$$\xi_H \left(\frac{d\phi}{dz} \right)^2 = \sin^2 \phi. \tag{4.10}$$

The solution for $\phi(z)$ is given by an elliptic function:

$$\sin \phi(z) = S_n(u, k), \tag{4.11}$$

where $u = z/\xi_H k$ and k are the argument and modulus of the elliptic function, respectively.

The condition for minimum free-energy is given by

$$q_0 \xi_H = \frac{2E(k)}{\pi k} \tag{4.12}$$

and the field-dependent pitch $p(H)$ is given by

$$p(H) = p_0 \left(\frac{2}{\pi} \right)^2 F(k)E(k) = 4\xi k F(k), \tag{4.13}$$

where $F(k)$ and $E(k)$ are complete elliptic integrals of the first and second kind. When $k = 1$, $E(k) = 1$, and $F(k)$ diverges logarithmically [i.e., $p(H) \to \infty$]. When $k \to 1$, Equation (4.12) becomes

$$q_0 \xi_H = \frac{2}{\pi} \tag{4.14}$$

or

$$q_0 \frac{K_2}{\Delta\chi^m H^2} = \frac{2}{\pi}. \tag{4.15}$$

This shows that above a critical field H_c defined by

$$H_c^2 = \frac{\pi}{2} \left(\frac{K_2}{\Delta\chi^m} \right) q_0 = \pi^2 \left(\frac{K_2}{\Delta\chi^m} \right) \frac{1}{p_0} \tag{4.16}$$

(where p_0 is the unperturbed pitch), the helix will be completely untwisted ($q \to 0$, $p \to \infty$); the system is essentially nematic. For a typical value of $K_2 = 10^{-6}$ dyne,

$\Delta\chi^m = 10^{-6}$ in cgs units, $P_0 = 20$ μm, and $H_c \sim 15{,}000$ G. For $H < H_c$, the variation of p with H is well approximated by the expressions

$$p = p_0 \left[1 + \frac{(\Delta\chi^m)^2 p_0^4 H^4}{32(2\pi)^4 K_2^2} + \cdots \right]. \tag{4.17}$$

It should be noted here that the preceding treatment of field-induced pitch change assumes that the cholesteric liquid crystal cell is thick and in an initially ideally twisted arrangement, and there is negligible influence from the cell walls. For thin cells or other initial director axis arrangements (e.g., fingerprint or focal-conic texture, etc.), the process will be more complicated. Nevertheless, this example serves well to illustrate the field-induced director axis reorientation and pitch change effect in cholesteric liquid crystals. Experimental measurements[3] in such systems have shown very good agreement with the theories.

Electric Field. The preceding and the following discussions of magnetic field induced effect in cholesteric liquid crystals can be applied to the case of electric field if we replace $\Delta\chi^m$ by $\Delta\varepsilon$, and H by E, as pointed out in the previous chapter.

4.1.3. Twist and Conic Mode Relaxation Times

For situations where an applied field is abruptly turned off, the relaxation constants depend on the kind of deformation involved. There are two distinct forms of deformation: the pure twist one discussed previously and the so-called umbrella or conic mode.[1] The first form of deformation involves fluid motion, whereas the latter does not.

The dynamics of the field-induced twist deformation in cholesterics is described by an equation analogous to the equation for nematics:

$$\gamma_1 \frac{\partial\phi}{\partial t} - K_2 \frac{d^2\phi}{dz^2} - \Delta\chi^m H^2 \sin\phi\cos\phi = 0. \tag{4.18}$$

The dynamical equation for the conic distortion is much more complicated[4,5] and involves the other two elastic constants K_3 and K_1.

The corresponding relaxation times are as follows:

$$\tau_{twist} = \frac{\gamma_1}{K_2 q_0^2} \tag{4.19}$$

for the twist mode, and

$$\tau_{conic} = \frac{\gamma_1}{K_3 q_0^2 + K_1 q^2} \tag{4.20}$$

for the conic mode, where $\vec{q}$ is the wave vector characterizing the wave vector distortion. These relaxation times are modified if an externally applied bias field is present and they depend on the field and director axis configurations.[5]

4.2. LIGHT SCATTERING IN CHOLESTERICS

In terms of their optical properties, a prominent feature of cholesterics is the helical structure of their director axes. Such helicity gives rise to selective reflection and transmission of circularly polarized light. These processes may be quantitatively analyzed, using the electromagnetic approach given in the next section and in Chapter 7. We begin here with some general observations.

Consider a right-handed helix, whose pitch is on the order of the optical wavelength, as depicted in Figure 4.3. A normally incident right circularly polarized light will be reflected as a right circularly polarized light, as the optical field follows the director axis rotation; that is, it follows the helix (Fig. 4.3a). (Note that the right-handed or left-handed circular polarization is defined by an observer looking at the incoming light.)

Under the Bragg condition:

$$p_0 = \frac{2\pi}{q_0} = \lambda \tag{4.21}$$

the reflection is total, and there is no transmission.

On the other hand, an incident left circularly polarized light will be totally transmitted (Fig. 4.3b). In the case of oblique incidence a similar analysis shows that higher-order diffractions are possible. The Bragg diffraction becomes

$$p_0 \cos\gamma = m\lambda, \qquad m = 1, 2, 3, ..., \tag{4.22}$$

where γ is the angle of the refracted light in the cholesteric liquid crystal. In this case the polarization states are elliptical.

We shall now examine light propagation in CLC in detail.

4.2.1. General Optical Propagation and Reflection: Normal Incidence

Consider a light wave propagating along the direction of the helix ($\hat{z}$ direction). For the locally uniaxial system, the electric displacement $\vec{D}$ and electric field $\vec{E}$ are related by the following constitutive equation:

$$D = \varepsilon E = \varepsilon_\perp E + \varepsilon_a n(n \cdot E), \tag{4.23}$$

where $\varepsilon_a = \varepsilon_\parallel - \varepsilon_\perp$. Let us now consider specifically the propagation, along z, of an electromagnetic wave of frequency ω. If the optical electric fields are represented in

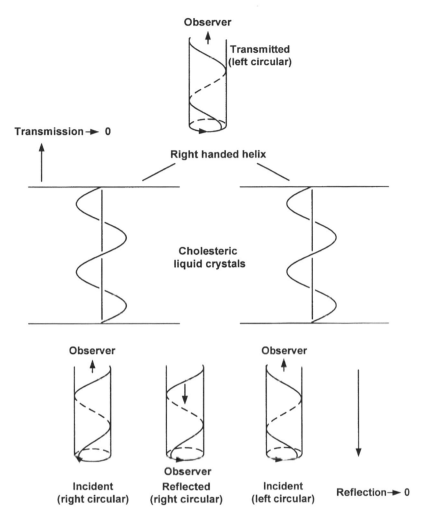

Figure 4.3. Reflection and transmission of circularly polarized light in a cholesteric liquid crystal with positive pitch. (a) Incident light is right circularly polarized; reflected light is right circularly polarized. (b) Left circularly polarized light is totally transmitted.

linearly polarized states:

$$E_x(z,t) = \text{Re}\left[E_x(z)e^{-i\omega t}\right],$$
$$E_y(z,t) = \text{Re}\left[E_y(z)e^{-i\omega t}\right],$$
(4.24)

the Maxwell equation becomes

$$-\frac{d^2}{dz^2}\begin{pmatrix} E_x \\ E_y \end{pmatrix} = \left(\frac{\omega}{c}\right)^2 \hat{\varepsilon}(z)\begin{pmatrix} E_x \\ E_y \end{pmatrix},$$
(4.25)

in which the dielectric tensor is

$$\hat{\varepsilon}(z) = \varepsilon \begin{pmatrix} 1 & 0 \\ 0 & 1 \end{pmatrix} + \overline{\Delta\varepsilon} \begin{pmatrix} \cos(2q_0 z) & \sin(2q_0 z) \\ \sin(2q_0 z) & -\cos(2q_0 z) \end{pmatrix} \tag{4.26}$$

with $\varepsilon = (\varepsilon_{\parallel} + \varepsilon_{\perp})/2, \overline{\Delta\varepsilon} = (\varepsilon_{\parallel} - \varepsilon_{\perp})/2$ (assume positive anisotropy), and $q_0 = 2\pi/P$ (P is the pitch of cholesteric liquid crystal). Substituting Equation (4.26) in Equation (4.25), we obtain the eigenlike equation

$$-\frac{d^2}{dz^2}\begin{pmatrix} E_x \\ E_y \end{pmatrix} = \left(\frac{\omega}{c}\right)^2 \begin{pmatrix} \varepsilon + \overline{\Delta\varepsilon}\cos(2q_0 z) & \overline{\Delta\varepsilon}\sin(2q_0 z) \\ \overline{\Delta\varepsilon}\sin(2q_0 z) & \varepsilon - \overline{\Delta\varepsilon}\cos(2q_0 z) \end{pmatrix}\begin{pmatrix} E_x \\ E_y \end{pmatrix}. \tag{4.27}$$

Since we are dealing with a helical structure, it is more appropriate to express the fields in terms of right-handed and left-handed circular waves:

$$E_R = E_x + iE_y$$
$$E_L = E_x - iE_y, \tag{4.28}$$

respectively.

From Equations (4.27) and (4.28), we derive

$$-\frac{d^2}{dz^2}\begin{pmatrix} E_R \\ E_L \end{pmatrix} = \omega_r^2 \begin{pmatrix} \varepsilon & \overline{\Delta\varepsilon}\,e^{i2q_0 z} \\ \overline{\Delta\varepsilon}\,e^{-i2q_0 z} & \varepsilon \end{pmatrix}\begin{pmatrix} E_R \\ E_L \end{pmatrix}, \tag{4.29a}$$

where ω_r is the reduced frequency, $\omega_r = \omega/c$. Equation (4.29a) can be more explicitly written as

$$-\frac{d^2}{dz^2}E_R = \omega_r^2\left(\varepsilon E_R + \overline{\Delta\varepsilon}\,e^{i2q_0 z}E_L\right),$$

$$-\frac{d^2}{dz^2}E_L = \omega_r^2\left(\varepsilon E_L + \overline{\Delta\varepsilon}\,e^{-i2q_0 z}E_R\right). \tag{4.29b}$$

To solve the problem, we assume that the solutions of E_R and E_L are of the forms

$$E_R = a\exp\left[i(l + q_0)z\right],$$
$$E_L = b\exp\left[i(l - q_0)z\right]. \tag{4.30}$$

(By doing that, we actually transformed the field variable into the coordinate frame rotating spatially with the cholesteric liquid crystal dielectric helix.) Substituting

Equation (4.30) in (4.29b), we have

$$
\left[(l + q_0)^2 - \omega_r^2 \varepsilon \right] a - \left[\omega_r^2 \overline{\Delta\varepsilon} \right] b = 0,
$$
$$
\left[-\omega_r^2 \overline{\Delta\varepsilon} \right] a + \left[(l - q_0)^2 - \omega_r^2 \varepsilon \right] b = 0.
\tag{4.31}
$$

A nontrivial solution requires that

$$
\det \begin{vmatrix} (l + q_0)^2 - \omega_r^2 \varepsilon & \omega_r^2 \overline{\Delta\varepsilon} \\ -\omega_r^2 \overline{\Delta\varepsilon} & (l - q_0)^2 - \omega_r^2 \varepsilon \end{vmatrix} = 0,
\tag{4.32}
$$

which yields

$$
\left[\varepsilon^2 - (\overline{\Delta\varepsilon})^2 \right] \omega_r^4 - \left[2\varepsilon(l^2 + q_0^2) \right] \omega_r^2 + \left[(l^2 - q_0^2)^2 \right] = 0
\tag{4.33}
$$

or

$$
\omega = \sqrt{ \frac{\varepsilon(l^2 + q_0^2) + \sqrt{4\varepsilon^2 l^2 q_0^2 + (\overline{\Delta\varepsilon})^2 (l^2 - q_0^2)^2}}{\varepsilon^2 - (\overline{\Delta\varepsilon})^2} }.
\tag{4.34}
$$

Quantitative analysis of various propagation modes and their polarization states and other optical parameters can be quite involved,[1,6] as one can see from the next section. Here we summarize some of the pertinent results.

Consider Figure 4.4a which plots ω as a function of l in the case where $\varepsilon = 3$ and $\overline{\Delta\varepsilon} = 0.4$. There are two distinct branches, which are called upper [for + square root ω_+ in Eq. (4.34)] and lower [for − square root ω in Eq. (4.34)] branches in typical liquid crystal literature.

For $\omega_- < \omega < \omega_+$ (i.e., $cq_0/n_\parallel < \omega < cq_0/n_\perp$), only one wave with circular polarization may propagate. This is the Bragg reflection regime discussed in the previous section. The spectral width $\Delta\lambda$ of this reflection band is proportional to the optical dielectric anisotropy:

$$
\Delta\lambda = P_0 \, \Delta n.
\tag{4.35}
$$

Outside this selective reflection band, in general, there are two roots l_1 and l_2 with positive group velocity ($V_g = \partial\omega/\partial t > 0$); that is, there are two forward propagating waves and two backward propagating waves ($-l_1$ and $-l_2$). These mode structures are very sensitive to the parameter:[1]

$$
x = \frac{2q_0 l}{k_1^2}.
\tag{4.36}
$$

where $k_1^2 = (\omega/c)^2 \, \overline{\Delta\varepsilon}$.

For $x \ll 1$ (i.e., $\lambda \ll \Delta n P_0$), the optical wave is "guided" by the system; the electric vector of a linearly polarized wave (ordinary or extraordinary) follows the rotation of the director, and the angle of its rotation corresponds to the number of turns of the helix. This is also known as the Mauguin regime.

For the case where n_e and n_o are close (i.e., k_1^2 is small compared to $q_0 l$; x is large), the eigenmodes are nearly circular. In this case the two eigenmodes l_1 and l_2 are given by:[1]

$$l_1 = k_0 + q_0 + \frac{k_1^4}{8k_0 q_0 (k_0 + q_0)} + O(k_1^3) \tag{4.37}$$

$$l_2 = k_0 - q_0 + \frac{k_1^4}{8k_0 q_0 (-k_0 + q_0)} + O(k_1^3), \tag{4.38}$$

where $k_0^2 = (\omega/c)^2 \varepsilon$.

Writing $l_1 - q_0 = (\omega/c)n_1$ and $l_2 + q_0 = (\omega/c)n_2$, the optical rotation per unit length $[\psi/d = (\omega/c)(n_1 - n_2]$ is given by the following two equivalent expressions:

$$\frac{\psi}{d} = \frac{k_1^4}{8q_0 (k_0^4 - q_0^2)} \tag{4.39}$$

or

$$\frac{\psi}{d} = \frac{q_0}{32}\left(\frac{n_e^2 - n_o^2}{n_e^2 + n_o^2}\right)\frac{1}{\bar{\lambda}^{-2}}(1 - \bar{\lambda}^{-2}), \tag{4.40}$$

where $\bar{\lambda} = \lambda/P$. Note that in the vicinity of $\bar{\lambda} \approx 1$ (for $\lambda \approx P$, in the vicinity of the selective reflection band), the rotation per unit length can be very large. Also, the rotation changes sign at $\bar{\lambda} = 1$.

4.2.2. Cholesteric Liquid Crystal as a One-Dimensional Photonic Crystal

In this section, we shall analyze the dispersion relationship from the point of view of band diagram frequently employed in photonic crystal study.[7] From Equations (4.28) and (4.30), one can see that the electric field associated with the eigenmode is in general elliptically polarized depending on the values of a and b. We can define a parameter[1] ρ as

$$\rho = \frac{b - a}{b + a}. \tag{4.41}$$

The axial ratio of the ellipse is $|\rho|$, and the sign of ρ gives the sign of rotation (+ sign corresponds to the left–handed rotation, and − sign for the right–handed rotation). From Equations (4.31), (4.34), and (4.41), we have

$$\rho = \frac{-2lq_0}{\pm\sqrt{\overline{\Delta\varepsilon}^2 \omega_r^4 + 4q_0^2 l^2} - \overline{\Delta\varepsilon}\omega_r^2} \tag{4.42}$$

with $\pm$ corresponding to the upper and lower branches. Figure 4.4b shows the ρ values in terms of l for upper (solid line) and lower (dashed line) branches.

Using the results from Figure 4.4b, we can replot Figure 4.4a by distinguishing the polarization handedness associated with various modes. The result is shown in Figure 4.5, with the dashed line corresponding to left-handed polarization ($\rho \leq 0$) and the solid line for right-handed polarization ($\rho \geq 0$).

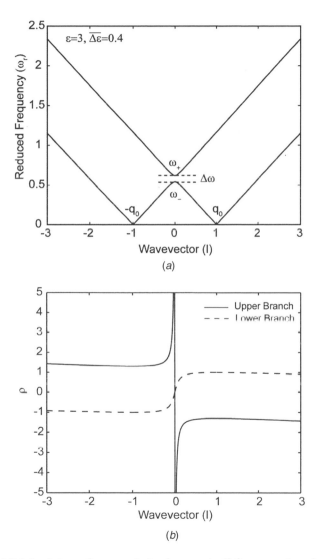

Figure 4.4. (a) Relation between frequency (ω_r) and wavevector (l) for propagation of electromagnetic modes in a cholesteric spiral with $\varepsilon = 3$ and $\overline{\Delta\varepsilon} = 0.4$. (b) Axial ratio and sign of rotation of the ellipses associated with various modes in terms of l.

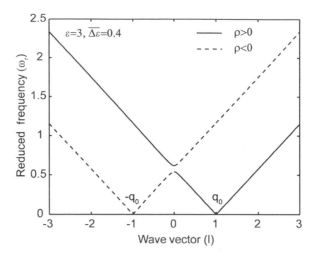

Figure 4.5. Replot of Figure 1(a) by distinguishing the different polarization handness associated with various mode. The dashed line corresponding to right handed polarized ($\rho \leq 0$) and solid line for left handed polarized ($\rho \geq 0$).

In typical photonic crystal analysis, a plane wave is described by $\exp^{[i(kz - \omega t)]}$. Returning to Equation (4.30), we can see that the k wave vector is related to l by

$$
\begin{aligned}
k &= l + q_0 \quad \text{if } E = E_R, \\
k &= l - q_0 \quad \text{if } E = E_L.
\end{aligned}
\tag{4.43}
$$

Using these relationships for k in terms of l, we can "translate" Figure 4.5 into a plot of ω versus k to yield the familiar dispersion relationship $\omega(k)$ used in photonic crystal literature[7] as plotted in Figure 4.6.

Now we would like to reexamine the polarization handedness of various modes in the "language" of photonic crystal, that is, using the $\omega_r - k$ band diagram. From Equation (4.28) and considering the above $k - l$ relation, the general electromagnetic (EM) wave solutions can be written as

$$
E = \begin{pmatrix} E_x \\ E_y \end{pmatrix} = \begin{pmatrix} \left(\dfrac{a+b}{2}\right)\exp[ikz] \\ \left(\dfrac{a-b}{2i}\right)\exp[ikz] \end{pmatrix} = \begin{pmatrix} \left(\dfrac{b+a}{2}\right)\exp[ikz] \\ \left(\dfrac{b-a}{2}\right)\exp[ikz]\exp\left[i\dfrac{\pi}{2}\right] \end{pmatrix}.
\tag{4.44}
$$

Note that if $(b + a)$ and $(b - a)$ take the same sign, E_y is $\pi/2$ in advance of E_x resulting to left-handed polarization, whereas if $(b + a)$ and $(b - a)$ take opposite signs, right-handed polarization occurs. Now let us calculate $(b - a)/(b + a)$ for different branches indicated in Figure 4.6.

For branch A,

$$
\frac{b-a}{b+a} = \frac{-2(k+q_0)q_0}{-\sqrt{\overline{\Delta \varepsilon}^2 \omega_r^4 + 4q_0^2(k+q_0)^2} - \overline{\Delta \varepsilon}\,\omega_r^2},
\tag{4.45}
$$

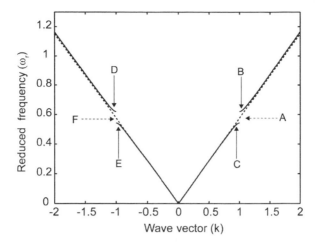

Figure 4.6. Different branches of $\omega_r - k$ relation for a cholesteric spiral with $\varepsilon=3$ and $\Delta\varepsilon=0.4$.

$$\omega_r = \sqrt{\frac{\varepsilon\big((k+q_0)^2 + q_0^2\big) - \sqrt{4\varepsilon^2(k+q_0)^2 q_0^2 + (\overline{\Delta\varepsilon})^2\big((k+q_0)^2 - q_0^2\big)^2}}{\varepsilon^2 - (\overline{\Delta\varepsilon})^2}}. \tag{4.46}$$

For branch B,

$$\frac{b-a}{b+a} = \frac{-2(k-q_0)q_0}{\sqrt{\overline{\Delta\varepsilon}^2\,\omega_r^4 + 4q_0^2(k-q_0)^2} - \overline{\Delta\varepsilon}\,\omega_r^2}, \tag{4.47}$$

$$\omega_r = \sqrt{\frac{\varepsilon\big((k-q_0)^2 + q_0^2\big) + \sqrt{4\varepsilon^2(k-q_0)^2 q_0^2 + (\overline{\Delta\varepsilon})^2\big((k-q_0)^2 - q_0^2\big)^2}}{\varepsilon^2 - (\overline{\Delta\varepsilon})^2}}. \tag{4.48}$$

For branch C,

$$\frac{b-a}{b+a} = \frac{-2(k-q_0)q_0}{-\sqrt{\overline{\Delta\varepsilon}^2\,\omega_r^4 + 4q_0^2(k-q_0)^2} - \overline{\Delta\varepsilon}\,\omega_r^2}, \tag{4.49}$$

$$\omega_r = \sqrt{\frac{\varepsilon\big((k-q_0)^2 + q_0^2\big) - \sqrt{4\varepsilon^2(k-q_0)^2 q_0^2 + (\overline{\Delta\varepsilon})^2\big((k-q_0)^2 - q_0^2\big)^2}}{\varepsilon^2 - (\overline{\Delta\varepsilon})^2}}. \tag{4.50}$$

Similar equations can be found for branches D–F. The results of ratio $(b-a)/(b+a)$ for branches A–F are shown in Figure 4.7.

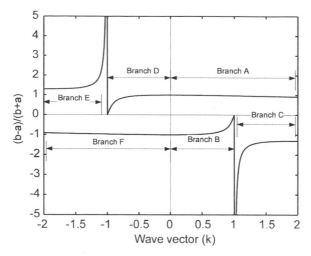

Figure 4.7.   The value of $(b-a)/(b+a)$ for different branches.

Note that for $k \geq 0$ (propagation along $+z$ direction), branches B and C are associated with $(b-a)/(b+a) \leq 0$, such that right-handed polarization (RP) and the electromagnetic modes experience a frequency gap (solid curve in Fig. 4.6). On the other hand, modes in branch A $[(b-a)/(b+a) > 0]$ have left-handed polarization (LP) and experience no frequency gap (dashed line in Fig. 4.6). Similar arguments can be made for the $k \leq 0$ case. The two polarizations (LP and RP) have distinctive behaviors: one experiences frequency gap at $k = \pm q_0$, while the other has no such gap. These confirm the statements made earlier regarding reflection and transmission of right and left circularly polarized light by cholesteric liquid crystals.

4.2.3.   Cholesteric Liquid Crystals with Magneto-Optic Activity: Negative Refraction Effect

In this section, we investigate the optical properties of cholesteric liquid crystals that exhibit magneto-optic activity[8–10] in the presence of an external static magnetic field:

$$\varepsilon' = \begin{pmatrix} \varepsilon + \overline{\Delta\varepsilon} & 0 \\ 0 & \varepsilon - \overline{\Delta\varepsilon} \end{pmatrix} + \begin{pmatrix} 0 & -i\gamma \\ i\gamma & 0 \end{pmatrix} \tag{4.51}$$

with, $\varepsilon = (\varepsilon_\| + \varepsilon_\perp)/2, \overline{\Delta\varepsilon} = (\varepsilon_\|-\varepsilon_\perp)/2$ (assuming positive anisotropy), and $\varepsilon'_{12} = -i\gamma = -\varepsilon'_{12}$ comes from magneto-optic activity.[10] Following the procedure in the previous section, the eigenvalue problem becomes

$$\begin{bmatrix} \omega_r^2(\varepsilon + \overline{\Delta\varepsilon}) - k_z^2 - q_0^2 & -i\omega_r^2\gamma - 2iq_0k_z \\ i\omega_r^2\gamma + 2iq_0k_z & \omega_r^2(\varepsilon - \overline{\Delta\varepsilon}) - k_z^2 - q_0^2 \end{bmatrix} \begin{bmatrix} E_1' \\ E_2' \end{bmatrix} = 0. \tag{4.52}$$

To get nontrivial solutions, we require

$$\begin{vmatrix} \omega_r^2(\varepsilon + \overline{\Delta\varepsilon}) - k_z^2 - q_0^2 & -i\omega_r^2\gamma - 2iq_0k_z \\ i\omega_r^2\gamma + 2iq_0k_z & \omega_r^2(\varepsilon - \overline{\Delta\varepsilon}) - k_z^2 - q_0^2 \end{vmatrix} = 0, \tag{4.53}$$

which yields the dispersion relation for the cholesteric liquid crystal (CLC) with magneto-optic activity,

$$\omega_r^4\left[\varepsilon^2 - (\overline{\Delta\varepsilon})^2 - \gamma^2\right] - \omega_r^2\left[2\varepsilon(k_z^2 + q_0^2) - 4\gamma k_z q_0\right] + (k_z^2 - q_0^2)^2 = 0. \tag{4.54}$$

Note that if $\gamma = 0$ (no magneto-optic activity), Equation (4.54) returns to the typical cholesteric liquid crystal dispersion relation; cf. Equation (4.33). As we will see, the term $4\gamma k_z q_0$ in Equation (4.54) gives rise to nonreciprocal optical properties.

From Equation (4.54), we get

$$\omega_r^2 =$$

$$\frac{\varepsilon(k_z^2 + q_0^2) - 2\gamma k_z q_0 \pm \sqrt{4\varepsilon^2 k_z^2 q_0^2 + (\overline{\Delta\varepsilon})^2(k_z^2 - q_0^2)^2 + \gamma^2(k_z^2 + q_0^2)^2 - 4\gamma k_z q_0 \varepsilon(k_z^2 + q_0^2)}}{\varepsilon^2 - (\overline{\Delta\varepsilon})^2 - \gamma^2} \tag{4.55}$$

Figure 4.8a depicts the numerically calculated photonic band structure for axial propagation $[\omega(k_z, k_x = k_y = 0)]$ in cholesteric liquid crystal ($\varepsilon = 1.7$, $\overline{\Delta\varepsilon} = 0.3$). The solid curve represents the band structure for CLC with magneto-optic activity ($\gamma = 0.01$). For comparison, the case for $\gamma = 0$ is also shown (dotted line). It is clear from Figure 4.8a that the presence of magneto-optic activity leads to a non-reciprocal dispersion relation, as the band structure is not symmetric with regard to the center of the reduced Brillouin zone ($k_z = 0$) This asymmetric property can be more clearly observed in Figure 4.8b, which is an enlarged view of the band struc-ture near $k_z = 0$ and Figure 4.8c near the band edge ($k_z = q_0$).

Besides the nonreciprocal alternation of CLC band structure, the presence of magneto-optic activity can also cause the group velocity (v_g) of the eigenmodes to either increase or decrease or even change direction, depending on their polarization handedness and direction of propagation. Figure 4.9 shows the behavior of the right circularly polarized (RCP) eigenmode at band edge for various γ values (magneto-optic strength). With the increase of γ value, the band slope (or v_g) at band gap edge changes from $v_g = 0$ (standing circularly polarized) to $v_g \neq 0$. The change of group velocity at band edge can generate interesting results. Figure 4.10 shows the calcu-lated RCP photonic band structure for $\gamma = 0$ and $\gamma = 0.2$. As shown in the figure, for $\omega = \omega_a$, there is only one mode for the $\gamma = 0$ case ($k = k_a$). The mode has group velocity in the $+z$ direction. On the other hand, for $\gamma = 0.2$, two modes are associated with $\omega = \omega_a$. One of the modes at $k = k_{a,2}$ with phase velocity along the $+z$ direction has an opposite group velocity, pointing along the $-z$ direction. These results show that some RCP modes in cholesteric liquid crystal with sufficient magneto-optic activ-ity could acquire antiparallel group and phase velocity at a certain frequency range (the so-called negative index phenomena). CLC is but an example of chiral materials which, in general, could exhibit a negative index of refraction.[10]

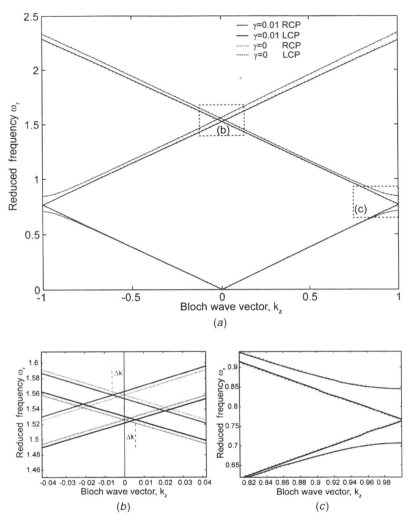

Figure 4.8 (a) Numerical calculated photonic band structure for axial propagation [$\omega(k_z, k_x = k_y = 0)$] in cholesteric liquid crystal [$\varepsilon = 1.7, \overline{\Delta\varepsilon} = 0.3$] with (solid curve $\gamma = 0.01$) and without (dotted curve $\gamma = 0$) magneto-optic activity, (b) and (c) show the magnification of the photonic band structure near $k_z = 0$ and $k_z = q_0$.

4.3. SMECTIC AND FERROELECTRIC LIQUID CRYSTALS: A QUICK SURVEY

Smectic liquid crystals possess a higher degree of order than nematics; they exhibit both positional and directional orderings in their molecular arrangements. Long-range positional ordering in smectics is manifested in the form of layered structures, in which the director axis is aligned in various directions depending on the smectic phase. To date, at least nine distinct smectic phases, bearing the designation smectic-A, smectic-B,

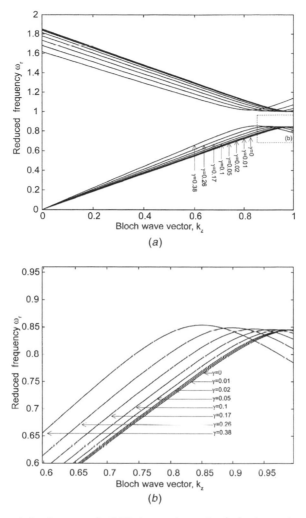

Figure 4.9. Photonic band structure for RCP eigenmode near band edge $k_z = q_0$ for various magneto-optic activity strength (γ).

smectic-C through smectic-I, have been identified, in chronological order.[11] Some smectic liquid crystals, for example, smectic-C* materials, are ferroelectric; their molecules possess permanent dipole moments.[12,13] A good example of a room-temperature smectic-A is 4,4'-n-octylcyanobiphenyl (OCB), whose molecular structure is shown in Figure 4.11a. This material has also been studied in the context of nonlinear optical pulse propagation and optical wave mixing phenomena.[14,15] The liquid crystal nCB ($n = 8 - 12$) also exhibits the smectic-A phase.[13] The well-studied liquid crystal 4-n-octyloxy-4'-cyanobiphenyl (OOCBP), whose molecular structure is shown in Figure 4.11b, exhibits the smectic-C, smectic-A, and nematic phases as functions of temperature. Smectic liquid crystals that are ferroelectric include HOBACPC,[13] DOBAMBC,[16]

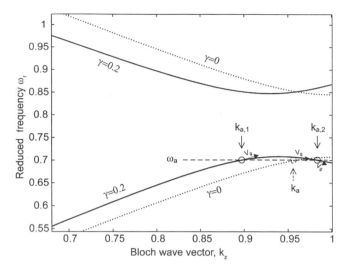

Figure 4.10. Dispersion relationship near the band-edge for cholesteric liquid crystal without (dotted lines) and with (full lines) magneto-optics coupling showing the possibilities of negative group velocity in the latter case.

XL13654,[17,18] and SCE9.[19] A well-studied ferroelectric liquid crystal is DOBAMBC; its molecular structure is shown in Figure 4.11c. Because of these differences in the degree of order and molecular arrangement and the presence of a permanent dipole moment, the physical properties of smectic liquid crystals are quite different from those of the nematic phase. In this and the following sections we examine the pertinent physical theories and the optical properties of three exemplary types of smectics: smectic-A, smectic-C, and (ferroelectric) smectic-C*.

4.4. SMECTIC-A LIQUID CRYSTALS

4.4.1. Free Energy

The molecular arrangement of a smectic-A (SmA) liquid crystal is shown in Figure 4.lla. The physical properties of SmA are analogous to nematics in many ways. However, because of the existence of the layered structures, there are important differences in the dynamics and types of elastic deformation that could be induced by applied fields.

In an ideal single-domain SmA sample, in which the layers are parallel and equidistantly separated, the director axis components n_x and n_y are related to the layer displacement $u(x,y,z)$, in the limit of small distortion, by the following relationships:

$$n_x = -\frac{\partial u}{\partial x},$$
$$(4.56a)$$

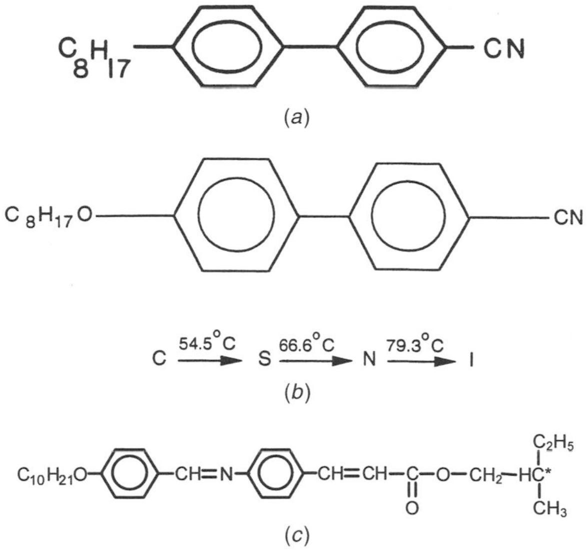

Figure 4.11. (a) A smectic-A liquid crystal. (b) A liquid crystal that exhibits smectic-C, smectic-A, and nematic phases. (c) Molecular structure of a ferroelectric liquid crystal.

$$n_y = -\frac{\partial u}{\partial y}. \tag{4.56b}$$

In the equilibrium case only $u(x,y,z)$ and its spatial derivatives are needed to describe elastic distortion in SmA. For example, a small director axis reorientation may be represented by

$$\theta(\bar{r}) \sim \frac{\partial u}{\partial z}. \tag{4.57}$$

The energy associated with this distortion, which corresponds to a compression of the layer, is given by

$$F_{\text{comp}} = \frac{1}{2}\bar{B}\left(\frac{\partial u}{\partial z}\right)^2. \tag{4.58}$$

This process is analogous to the compressibility of an isotropic liquid crystal. Typically, the compressibility $\bar{B}$ is on the order of 10^7–10^8 erg cm^{-3}. Assuming further that (i) there is no long-range transitional order in the plane, (ii) z and $-z$ are equivalent (no ferroelectricity), and (iii) the deformation is small so that the molecules at any point remain perpendicular to the plane of the layer, the total free energy of the system can be derived:[1]

$$F_{\text{total}} = F_0 + \frac{1}{2}\bar{B}\left(\frac{\partial u}{\partial z}\right)^2 + \frac{1}{2}K_1\left(\frac{\partial^2 u}{\partial x^2} + \frac{\partial^2 u}{\partial y^2}\right) + \frac{1}{2}\Delta\chi^m H^2\left[\left(\frac{\partial u}{\partial x}\right)^2 + \left(\frac{\partial u}{\partial y}\right)^2\right].$$

(4.59)

The first term on the right-hand side is the unperturbed free energy. The second term is the energy associated with layer compression. The third term is the splay distortion energy $\frac{1}{2}K_1(\nabla\cdot\hat{n})^2$, which is identical in form to that in nematics. Similarly, the fourth term is the field-induced distortion energy $\frac{1}{2}\Delta\chi^m(\nabla\cdot\bar{H})^2$, as in nematics.

Note that Equation (4.59) contains only the splay distortion, which preserves the layer spacing (see Fig. 4.12). The other two distortions, bend and twist, allowed in nematics are prohibited in the SmA phase as they involve extremely high distortion energy. This is also manifested in the form of a divergence in the value of the corresponding elastic constants K_3 and K_2 in the nematic phase as the temperature approaches the nematic $\rightarrow$ SmA transition,[20] as shown in Figure 4.13.

4.4.2.   Light Scattering in SmA Liquid Crystals

Light-scattering processes in SmA liquid crystals are governed by fluctuations in the layer displacement; they are analogous to scattering in nematics, which will be discussed in detail in the next chapter. In terms of the Fourier U_q components of the layer displacement, the free energy in the absence of an external field is given by[1]

$$F = \sum_q \frac{1}{2}\left[\bar{B}q_z^2 + kq_\perp^4\right]\left|u_q\right|^2,$$

(4.60)

where q_z and $q_\perp$ are the scattering wave vector components parallel and perpendicular to the z axis, respectively.

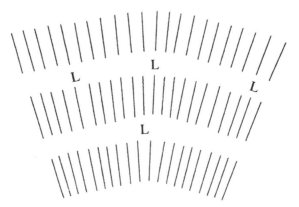

Figure 4.12.   Splay distortion in a SmA liquid crystal.

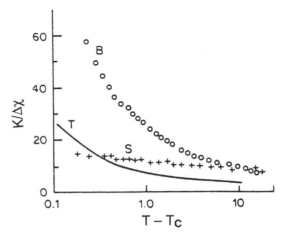

Figure 4.13. Measured values of bend and splay elastic constant in the nematic phase as a function of the vicinity of nematic to SmA transition in CBOOA (after Cheung et al. (20)).

Following the derivation given in Chapter 5, the intensity of the scattered wave is given by

$$I(q) = \left\langle \left| \hat{i} \cdot \delta\varepsilon(q) \cdot \hat{f} \right|^2 \right\rangle,$$

(4.61)

where $\hat{i}$ and $\hat{f}$ are the polarizations of the incoming and outgoing waves and $\delta\varepsilon(q)$ is the fluctuation in the dielectric constant tensor:

$$\delta\varepsilon = \Delta\varepsilon[\delta n : \hat{n} + \hat{n} : \delta n],$$

(4.62)

where $\Delta\varepsilon = \varepsilon_{\parallel} - \varepsilon_{\perp}$.

Using Equations (4.56a), and (4.56b), this gives

$$\delta\varepsilon_{xz} - - \Delta\varepsilon \frac{\partial u}{\partial x},$$

(4.63a)

$$\delta\varepsilon_{yz} = - \Delta\varepsilon \frac{\partial u}{\partial y}.$$

(4.63b)

Accordingly, the scattered intensity can be derived:[1]

$$I = \frac{K_B T}{B} \Delta\varepsilon^2 \frac{q_{\perp}^2}{q_z^2 + \lambda^2 q_{\perp}^4},$$

(4.64)

where $\lambda \equiv (K_1/B)^{1/2}$ is on the order of the layer thickness. There are two distinct regimes: (1) $q_z \sim q_{\perp} \neq 0$ and (2) $q_z = 0$.

In case (1), note that $q\lambda \ll 1$ and we thus have

$$I = \Delta\varepsilon^2 \frac{K_B T}{B} \frac{q_{\perp}^2}{q_z^2},$$

(4.65)

which is on the order of $K_B T/E_{comp}$, similar to isotropic liquid crystals (see Chapter 5). This is the usual case involving the transmission of light through a smectic film; the scattering is much smaller than in the nematic phase.

Case (2), however, corresponds to large scattering. If $q_z = 0$, Equation (4.64) becomes

$$I = \Delta \varepsilon^2 \frac{K_B T}{B \lambda^2 q_\perp^4} = \frac{\Delta \varepsilon^2 K_B T}{K_1 q_\perp^2} \tag{4.66}$$

which is analogous to scattering in the nematic phase. The mode of excitation causing large scattering corresponds to pure undulation for which the interlayer spacing is fixed.

4.5. SMECTIC-C LIQUID CRYSTALS

4.5.1. Free Energy

The finite tilt of the director axis from the layer normal (taken as the $\hat{z}$ axis) introduces a new degree of freedom, namely, a rotation around the z axis, compared to the SmA phase. This rotation preserves the layer spacing and therefore does not require too much energy. Since $\partial u/\partial y$ and $\partial u/\partial x$ are equivalent to rotations around the x and y axes, respectively, we may express the free energy in Smectic-C (SmC) liquid crystals in terms of the rotation components:

$$\Omega_x = \frac{\partial u}{\partial y} \quad \Omega_y = -\frac{\partial u}{\partial x} \quad \Omega_z. \tag{4.67}$$

Taking into account all the energy terms associated with director axis rotation, interlayer distortion, and possible coupling between them, the total free energy of the system is given by[1]

$$F = F_c + F_d + F_{cd}, \tag{4.68}$$

where

$$F_c = \frac{1}{2} B_1 \left(\frac{\partial \Omega_z}{\partial x} \right)^2 + \frac{1}{2} B_2 \left(\frac{\partial \Omega_z}{\partial y} \right)^2 + \frac{1}{2} B_3 \left(\frac{\partial \Omega_z}{\partial z} \right)^2 + B_{13} \frac{\partial \Omega_z}{\partial x} \frac{\partial \Omega_z}{\partial z}, \tag{4.69}$$

$$F_d = \frac{1}{2} A \left(\frac{\partial \Omega_x}{\partial x} \right)^2 + \frac{1}{2} A_{12} \left(\frac{\partial \Omega_y}{\partial x} \right)^2 + \frac{1}{2} A_{21} \left(\frac{\partial \Omega_x}{\partial y} \right)^2 + \frac{1}{2} \bar{B} \left(\frac{\partial u}{\partial z} \right)^2, \tag{4.70}$$

and

$$F_{cd} = C_1 \frac{\partial \Omega_x}{\partial x} \frac{\partial \Omega_z}{\partial x} + C_2 \frac{\partial \Omega_x}{\partial y} \frac{\partial \Omega_z}{\partial y}. \tag{4.71}$$

Here F_c is the free energy associated with director axis rotation without change of layer spacing, F_d is due to layer distortions, and F_{cd} is the cross term describing the coupling of these layer distortions and the free-rotation process.

4.5.2. Field-Induced Director Axis Rotation in SmC Liquid Crystals

In practical implementations or switching devices, the logical thing to do is to involve only one or a small number of these distortions. If an external field is applied, the field-dependent terms [cf. Eq. (4.5a) and (4.5b)] should be added to the total free-energy expression. The process of field-induced director axis distortion in SmC is analogous to the nematic case. For example, the first three terms on the right-hand side of Equation (4.70) correspond to the splay term in nematics:

$$\frac{1}{2} K_1 \left(\frac{\partial^2 u}{\partial x^2} + \frac{\partial^2 u}{\partial y^2} \right)^2 = \frac{1}{2} K_1 \left(\frac{\partial \Omega_x}{\partial y} - \frac{\partial \Omega_y}{\partial x} \right)^2. \tag{4.72}$$

Accordingly, if only such distortions (i.e., no layer displacement or coupling effects) are induced in a SmC sample by an applied field, Freedericksz transitions (discussed in the previous chapter for nematics) will occur.

Consider, for example, the effect caused by a magnetic field as depicted in Figure 4.14. The applied field has three components, H_1, H_2, and H_3, and the respective diamagnetic susceptibility components are χ_1^m, χ_2^m, and χ_3^m; χ_3^m corresponds to the director axis, usually denoted as the C axis, χ_2^m is along y, and χ_1^m is in a direction orthogonal to both C and y axis.

If the applied field is along the C axis (i.e., H_3), a Freedericksz transition is possible for $\chi_2^m > \chi_3^m$. The director should rotate around the z axis so that, in the strong field limit, its projection onto the smectic layer coincides with the y axis. There is no change in the tilt angle θ. The threshold field for this process is

$$H_{3c} = \frac{\pi \sin \theta}{d} \left(\frac{B}{\chi_2^m - \chi_3^m} \right)^{1/2}, \tag{4.73}$$

where B is the appropriate elastic constant defined by Equation (4.69).

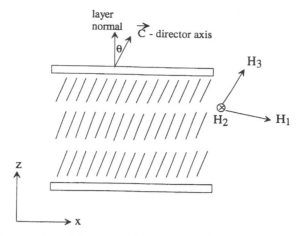

Figure 4.14.   Directions of the applied field components of H relative to the smectic C axis.

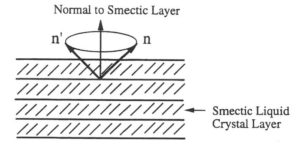

Figure 4.15.   Rotation of the C axis around the layer normal; no change in layer spacing.

If the applied field is along the y axis (i.e., H_2), a rotation of the C axis around z as depicted in Figure 4.15 will occur with a threshold field of

$$H_{2c} = \frac{\pi \sin \theta}{d} \left(\frac{B}{\chi_1^m \cos^2 \theta + \chi_3^m \sin^2 \theta - \chi_2^m} \right)^{1/2}. \tag{4.74}$$

For an applied field along H_1 and $\chi_2^m > \chi_1^m$, the threshold field is given by

$$H_{1c} = \frac{\pi \sin \theta}{d \cos \theta} \left(\frac{B}{\chi_2^m - \chi_1^m} \right)^{1/2}. \tag{4.75}$$

Usually $\chi_3^m > \chi_2^m \approx \chi_1^m$ and therefore case (2) involves the least field strength.

4.6.   SMECTIC-C* AND FERROELECTRIC LIQUID CRYSTALS

Smectic-C*, or ferroelectric liquid crystals, which possess nonzero spontaneous polarization $\vec{P}$, may be classified into two categories. In the case of unwound SmC*

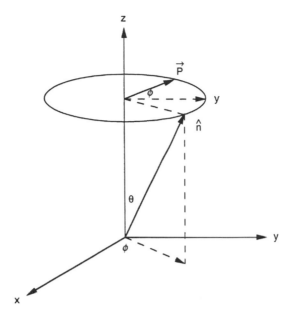

Figure 4.16. A helically modulated smectic-C* liquid crystal; the directions of both **n** and **p** vary spatially in a helical manner.

liquid crystals, the director axis $\hat{n}$ is tilted at a fixed angle θ with respect to the layer normal. It follows that the direction of $\vec{P}$ is also fixed, as shown in Figure 1.12d. It is optically homogeneous. On the other hand, the director axis of helically modulated SmC* liquid crystals varies in a helical manner from layer to layer, as shown in Figure 4.16. The director axis precesses around the normal to the layer with a pitch that is much larger than the layer thickness. The magnitude of the pitch is on the order of the optical wavelength, and thus helically modulated SmC* liquid crystals are optically inhomogeneous.

In helically modulated SmC* liquid crystals, the bulk polarization is vanishingly small. The helicity can be unwound by an external field applied parallel to the smectic layers. It can also be unwound by surface effects if the samples are sufficiently thin (thickness $\ll$ pitch), leading to the so-called surface-stabilized ferroelectric liquid crystal (SSFLC) with a nonvanishing macroscopic polarization.

Both the tilt angle θ and the spontaneous polarization $\vec{P}$ decrease in magnitude as the temperature of the system is increased. Above a critical temperature T_{C*A}, a phase transition to the untilted SmA phase takes place.[21]

Various optical effects arise as a result of the presence of the spontaneous polarization. Electro-optical effects will be discussed in Chapter 6. In this section our attention will be focused on their basic physical and optical properties.

4.6.1. Free Energy of Ferroelectric Liquid Crystals

Recalling the free-energy expression for the nonferroelectric SmC liquid crystals discussed in the preceding section, one should not be too surprised to find that the free energy of ferroelectric liquid crystals is even more complicated. Besides the elastic energy F_E, several others play an equally important role in determining the response of the ferroelectric-liquid crystal to an external field. These include the surface energy density F_S, the spontaneous polarization density F_P, and the dielectric interaction energy density F_{diel} with the applied field. These interactions have been studied by various workers; here we summarized the main results.

The elastic part of the free energy is analogous to the chiral nematic phase. It has been derived by Nakagawa et al.[22] and can be expressed as follows:

$$F_K = \tfrac{1}{2}\{K_1(\nabla \cdot \hat{n})^2 + K_2(\hat{n} \cdot (\nabla \times \hat{n}) + Q_T)^2 + K_3[\hat{n} \times (\nabla \times \hat{n}) + Q_B]^2\}, \quad (4.76)$$

where Q_T and Q_B are the inherent twist and bend wave numbers and $b = \hat{n} \times \hat{k}$, where $\hat{k}$ is the layer normal unit vector.

Using the geometry shown in the Figure 4.16, the director axis $\hat{n}$ becomes

$$\hat{n} = (\sin\theta\cos\phi, \sin\theta\sin\phi, \cos\theta). \quad (4.77)$$

Equation (4.59) becomes

$$F_E = F_0 + \tfrac{1}{2}A(1 + v\sin^2\phi)\left(\frac{\partial\phi}{\partial y}\right)^2 + A(1 + v)Q_0\sin\phi\left(\frac{\partial\phi}{\partial y}\right), \quad (4.78)$$

where

$$F_0 = \tfrac{1}{2}[K_2 Q_T^2 + K_3 Q_B^2], \quad (4.79a)$$

$$A = K_1 \sin^2\theta \quad (4.79b)$$

$$B = (K_2 \cos^2\theta + K_3 \sin^2\theta)\sin^2\theta, \quad (4.79c)$$

$$v = \frac{B - A}{A}, \quad (4.79d)$$

$$Q_0 = \frac{K_2 Q_T \sin\theta\cos\theta + K_3 Q_B \sin^2\theta}{A(1 + v)}. \quad (4.79e)$$

The dielectric interaction energy density is simply given by $F_{\text{diel}} = -\tfrac{1}{2}\vec{D}\cdot\vec{E}$. For an applied electric field along the y axis, for example, in the plane of the smectic layer, the interaction energy is given by

$$F_{\text{diel}} = -\tfrac{1}{2}\varepsilon_{yy}E_y^2, \quad (4.80)$$

where

$$\varepsilon_{yy} = \varepsilon_\perp \cos^2 \phi + \varepsilon_\parallel' \sin^2 \phi, \tag{4.81a}$$

$$\varepsilon_\parallel' = \varepsilon_\perp \cos^2 \theta + \varepsilon_\parallel \sin^2 \theta. \tag{4.81b}$$

More generally, F_{diel} may be written as

$$F_{diel} = -\tfrac{1}{2}\varepsilon_\perp (1 + \Delta\varepsilon' \sin^2 \phi)E^2, \tag{4.82}$$

where $\Delta\varepsilon' = (\varepsilon_\parallel' - \varepsilon_\perp)/\varepsilon$.

The interaction of an electric field $\vec{E}$ with the permanent polarization (sometimes also termed spontaneous polarization) $\vec{P}_s$ is simply $-\vec{P}_s \cdot \vec{E}$. For the configuration given in Figure 4.16, we have

$$F_P = -P_s E_y \cos\phi. \tag{4.83}$$

As in the other phases of liquid crystals, the free energy associated with the surface interaction is the most complicated one. It takes on further significance in the case of surface-stabilized ferroelectric liquid crystals. These surface interactions have been studied by various workers.[23,24] In their treatments the surface energy is expressed as

$$F_s = -g_1 \cos^2(\phi - \phi_s) - g_2 \cos(\phi - \phi_s), \tag{4.84}$$

where ϕ_s is the pretilt angle of the molecules. g_1 and g_2 are the respective coefficients for the nonpolar and polar surface interaction terms. Nakagawa et al.[22] later improved upon this expression:

$$F_s(\phi) = \sum_{1,2} h^{1,2}(\phi), \tag{4.85}$$

where

$$h^{1,2}(\phi) = -g^{1,2}\left\{ C^{1,2} \exp\left[-\frac{\alpha'}{2}\sin^2(\phi - \phi^{1,2}) \right] \right. $$
$$\left. + (1 - C^{1,2})\exp\left[-\frac{\alpha'}{2}\cos^2(\phi + \phi^{1,2}) \right] \right\}. \tag{4.86}$$

Here the superscripts 1 and 2 refer to the two cell boundary plates, α is a parameter characterizing the anchoring potential, and $C^{1,2}$ relates the relative stability between the $\phi^{1,2}$ and the $\pi - \phi^{1,2}$ states. These expressions satisfy the symmetry requirement $h^{1,2}(\pi - \phi) = h^{1,2}(\phi)$ for $C^1 = C^2 = 1/2$.

The total free energy of the ferroelectric liquid crystal system depicted in Figure 4.16 is therefore given by

$$G = \int_{\text{vol}} F \, dv, \tag{4.87}$$

where $F = F_E + F_{\text{diel}} + F_p + F_s$.

The equilibrium configuration of the system is obtained by minimizing the total energy with respect to ϕ:

$$\frac{\partial G}{\partial \phi} = 0. \tag{4.88}$$

This yields

$$\frac{\partial}{\partial y}(1 + \Delta \varepsilon' \sin^2 \phi) \frac{\partial \phi}{\partial y} = \frac{-P_s}{\varepsilon_\perp} \sin \phi \frac{\partial \phi}{\partial y}. \tag{4.89}$$

The dynamics of the molecular reorientation process is described by the torque balance equation:

$$\frac{\partial G}{\partial \phi} + \gamma_1 \frac{\partial \phi}{\partial t} = 0, \tag{4.90}$$

where γ_1 is the rotational viscosity coefficient.

In the bulk of the FLC, we thus have

$$\gamma_1 \frac{\partial \phi}{\partial t} = A(1 + \nu \sin^2 \phi)\left(\frac{\partial^2 \phi}{\partial y^2}\right) + \frac{\nu}{2} A \sin 2\phi \left(\frac{\partial \phi}{\partial y}\right)^2$$

$$+ \frac{\Delta \varepsilon'}{2} \varepsilon_\perp E^2 \sin 2\phi + P_s E \sin \phi. \tag{4.91}$$

This equation is greatly simplified under the one-constant approximation ($K_1 = K_2 = K_3 = K$). This gives

$$K \sin^2 \theta \frac{\partial^2 \phi}{\partial y^2} + \frac{\Delta \varepsilon'}{2} \varepsilon_\perp E^2 \sin 2\phi + P_s E \sin \phi = \gamma_1 \frac{\partial \phi}{\partial t}. \tag{4.92}$$

From this, one can see that the dynamics is controlled by the elastic torque (first term on the left-hand side), the optical dielectric torque (second term on the left-hand side), and the polarization torque (third term on the right-hand side).

The polarization part of the dynamics is governed by a time constant given by

$$\tau = \frac{\gamma_1}{P_s E}.$$ (4.93)

Using typical values of P_s and γ_1 for ferroelectric liquid crystals ($P_s \sim 10^{-5}$ cm^{-2}, $\gamma_1 \approx 10^{-2}$ mks units, and $E = 10^{-7}$ Vm^{-1}), τ is on the order of 100 μs. Ferroelectric liquid crystals of much faster response time have by now been developed by several research groups.[25] In Chapter 6, we will discuss further details on electro-optical switching in ferroelectric liquid crystals.

4.6.2. Smectic-C*–Smectic-A Phase Transition

The principal parameter that distinguishes smectic-C* from smectic-A is the tilt angle θ_0. Because of the chiral character of the molecule, the tilt processes around the normal to the smectic layers, together with the transverse electric polarization $\vec{P}$ (cf. Fig. 4.16). In the theories developed to describe the phase transition phenomena from the smectic-C* phase to the smectic-A phase, the tilt angle is treated as a primary order parameter of the system, very much as the director axis $\hat{n}$ in the nematic or cholesteric phase, while $\vec{P}$ is regarded as a secondary one.

Writing the two components of $\vec{P}$ and the tilt angle in the x-y plane as $\vec{P} = (P_x, P_y)$ and $\theta = (\theta_1, \theta_2)$ the free-energy density $f_0(z)$ of the system can be expressed as a Landau type of expansion in terms of θ_1, θ_2, P_x, and P_y, in the following form:[26,27]

$$f_0(z) = \frac{1}{2} A(\theta_1^2 + \theta_2^2) + \frac{1}{4} B(\theta_1^2 + \theta_2^2)^2$$
$$- \Lambda \left(\theta_1 \frac{d\theta_2}{dz} - \theta_2 \frac{d\theta_1}{dz} \right) + \frac{1}{2} K_3 \left[\left(\frac{d\theta_1}{dz} \right)^2 + \left(\frac{d\theta_2}{dz} \right)^2 \right]$$
$$+ \frac{1}{2\varepsilon} (P_x^2 + P_y^2) - \mu \left(P_x \frac{d\theta_1}{dz} + P_y \frac{d\theta_2}{dz} \right) + C(P_x \theta_2 - P_y \theta_1).$$ (4.94)

In this expression only the coefficient of the term quadratic in the primary parameter is temperature dependent, whereas the coefficient of the P^2 term is constant; this is so because it is not the interaction between the electric polarization that leads to a phase transition. The coefficient A is of the form $A = A_0(T - T_{CA})$, where T_{CA} is the smectic-C–smectic-A transition temperature, K_3 is the elastic constant, and Λ is the coefficient of the so-called Lifshitz term responsible for the helicoidal structure. μ and C are the coefficients of the flexoelectric and piezoelectric bilinear couplings between the tilt and the polarization. The coefficients Λ and C are dependent on the chiral character of the molecules. For nonchiral molecules, Λ and C are zero; minimization of the free energy given in Equation (4.94) yields a system where the director axis is homogeneously tilted below the transition temperature T_c. There is no linear coupling between the tilt and the polarization, and thus, $\vec{P}$. For temperatures below T_{CA}, the smectic-C–smectic-A transition temperature, the magnitude of the tilt angle θ is given by a square-root dependence:

$$\theta_0 = \sqrt{\frac{A_0}{B}(T - T_{CA})}.$$ (4.95)

On the other hand, for chiral molecules, Λ and C are nonzero. In this case the free energy is minimized if θ and $\hat{P}$ are described by the helical functions

$$\theta_1 = \theta_0 \cos q_0 z, \qquad \theta_2 = \theta_0 \sin q_0 z,$$ (4.96)

$$P_x = -P_0 \sin q_0 z, \qquad P_y = P_0 \cos q_0 z.$$ (4.97)

Note that $\hat{P}$ is locally perpendicular to the tilt. Both process around z with a pitch wave vector q_0 given by

$$q_0 = \frac{\Lambda + \varepsilon \mu C}{K_3 - \varepsilon \mu^2}.$$ (4.98)

The magnitude of the spontaneous polarization P_0 is proportional to the tilt angle θ_0:

$$P_0 = \varepsilon(C + \mu q_0)\theta_0,$$ (4.99)

where θ_0 is similar to that given in Equation (4.95):

$$\theta_0 = \sqrt{\frac{A_0}{B}(T_{C^*A} - T)}.$$ (4.100)

The smectic-C* smectic–A transition temperature T_{C^*A} is given by

$$T_{C^*A} = T_{CA} + \frac{1}{A_0}(\varepsilon C^2 + K_3 q_0^2).$$ (4.101)

Above T_{C^*A}, in the smectic-A phase, the two order parameters P_0 and θ_0 vanish. If an external field is present, as pointed out earlier, the helical structure can be unwound. From this point of view, the effect of the electric field is equivalent to "canceling" the elastic torque term $K_3 q_0^2$ in Equation (4.101), resulting in a homogeneous (i.e., non-helical) C* system. The corresponding phase transition temperature for the unwound system is thus given by

$$T_{C^*A}^{\text{unwound}} = T_{CA} + \frac{1}{A_0}\varepsilon C^2.$$ (4.102)

Experimentally, T_{CA}, T_{C*A}, and $T_{C*A}^{unwound}$ are found to be very close to one another. This is expected as the chiral terms in the free-energy expression are basically small-perturbation terms. Their optical and electro-optical properties, however, are considerably modified by the presence of the chirality and spontaneous polarization. As we remarked earlier, ferroelectric liquid crystals provide a faster electro-optical switching mechanism.[13] In the context of nonlinear optics, the noncentrosymmetry caused by the presence of $\vec{P}$ allows the generation of even harmonic light.

REFERENCES

1. deGennes P. G. 1974. *Physics of Liquid Crystals*. Oxford: Clarendon Press.
2. Meyer, R. B. 1968. *Appl. Phys. Lett.* 14:208; deGennes, P. G. 1968. *Solid State Commun.* 6:163.
3. Sackmann, E., S. Meiboom, and L. C. Snyder. 1967. *J. Am. Chem. Soc.* 89:5982; Wysocki, J., J. Adams, and W. Haas. 1968. *Phys. Rev. Lett.* 20:1025; Durand, G., L. Leger, F. Rondele, and M. Veyssie. 1969. *Ibid.* 22: 227; Meyer, R. B. 1969. *Appl. Phys. Lett.* 14:208.
4. Fan, C., L. Kramer, and M. J. Stephen. 1970. *Phys. Rev. A.* 2: 2482.
5. Parson, J. D., and C. F. Hayes. 1974. *Phys. Rev. A.* 9:2652; see also deGennes.[1]
6. Yeh, P., and C. Gu. 1999. *Optics of Liquid Crystal Displays*. New York: Wiley Interscience.
7. Joannopoulos, J. D., R. Meade, and J. Winn. 1995. *Photonic Crystals*. Princeton, NJ: Princeton University Press.
8. Zvedin, K., and V. A. Kotov. 1997. *Modern Magnetooptics and Magnetooptical Materials*. Philadelphia, PA: Institute of Physics.
9. Eritsyan, S. 2000. *J. Exp. Theor. Phys.* 90:102–108.
10. Bita, I., and E. L. Thomas. 2005. Structurally chiral photonic crystals with magneto-optic activity: Indirect photonic bandgaps, negative refraction, and superprism effect. *J. Opt. Soc. Am. B.* 22:1199–1210.
11. Gray, G. W., and J. Goodby. 1984. *Smectic Liquid Crystals: Textures and Structures*. London: Leonard Hill.
12. Meyer, B., L. Liebert, L. Strzelecki, and P. J. Keller. 1975. *J. Phys. (France) Lett.* 36:69.
13. Khoo, I. C., and S. T. Wu. 1993. *Optics and Nonlinear Optics of Liquid Crystals*. Singapore: World Scientific.; see also Anderson, G., I. Dahl, L. Komitov, S. T. Lagerwall, K. Sharp, and B. Stebler. 1989. *J. Appl. Phys.* 66:4983 for other chiral smectics.
14. Khoo, I. C., and R. Normandin. 1984. *J. Appl. Phys.* 55:1416.
15. Khoo, I.C., R. R. Michael, and P. Y. Yan. 1987. *IEEE J. Quantum Electron.* QE-23:1344.
16. Shtykov, N. M., M. I. Barnik, L. M. Blinov, and L. A. Beresnev. 1985. *Mol. Cryst. Liq. Cryst.* 124:379.
17. Taguchi, A., Y. Oucji, H. Takezoe, and A. Fukuda. 1989. *Jpn. J. Appl. Phys., Part 2.* 28: L997.
18. Macdonald, R., J. Schwartz, and H. J. Eichler. 1992. *J. Nonlinear Opt. Phys.* 1:103.
19. Liu, J., M. G. Robinson, K. M. Johnson, and D. Doroski. 1990. *Opt. Lett.* 15:267.
20. Cheung, L., R. B. Meyer, and H. Gruler. 1973. *Phys. Rev. Lett.* 31:349.

21. Pikin, S.A., and V. L. Indenbom. 1978. *Ferroelectrics*. 20:151.

22. Nakagawa, M., M. Ishikawa, and I. Akahane. 1988. *Jpn. J. Appl. Phys., Part 1*. 27:456.

23. Handschy, M. A., and N. A. Clark. 1984. *Ferroelectrics*. 59:69.

24. Yamada, Y., T. Tsuge, N. Yamamoto, M. Yamawaki, H. Orihara, and Y. Ishibashi. 1987. *Jpn. J. Appl. Phys, Part 1*. 26:1811.

25. See, for example, commercial information leaflets by the E. Merck (Germany) or BDH (UK) companies; see also Khoo and Wu[13].

26. Pikin, S. A., and V. L. Indenbom. 1978. *Usp. Fiz. Nauk*. 125:251.

27. Blinc, R., and B. Zeks. 1978. *Phys. Rev. A*. 18:740.

5

Light Scattering

5.1. INTRODUCTION

In earlier chapters we discussed some specific light-scattering processes in the mesophases of liquid crystals. In particular, we found that the ability of the molecules to scatter light is very much dependent on the orientations and fluctuations of the director axis and their reconfiguration under applied fields. There are, however, light-scattering processes that occur on the molecular level that involve the electronic responses of the molecules. In this chapter we discuss the general approaches and techniques used to analyze light-scattering processes in liquid crystals that are applicable in many respects to other media as well.

Approaches to the problems of light scattering in liquid crystals may be classified into two categories. In one category, such as Brillouin and Raman scatterings, knowledge of the actual molecular physical properties, such as resonances and energy level structures, is needed. On the other hand, in the electromagnetic formalism for light-scattering phenomena, one needs to invoke only the optical dielectric constants and their fluctuations. This latter approach is generally used to analyze orientational fluctuations in liquid crystals.

The process of light scattering can also be divided into linear and nonlinear regimes. In linear optics the properties of liquid crystals are not affected by the incident light, which may be regarded as a probe or signal field. The resulting scattered or transmitted light, in terms of its spatial or temporal frequency spectrum and intensity, reflects the physical properties of the material. On the other hand, in the nonlinear optical regime the incident light interacts strongly with and modifies the properties of the liquid crystals. The resulting scattered or transmitted light will reflect these strong interactions.

In this and the preceding chapters, our attention is focused on linear optical scattering processes, which are nevertheless quite important in nonlinear optical phenomena. Nonlinear optics and the nonlinear optical properties of liquid crystals will be presented in Chapters 8–12. We begin with a review of the electromagnetic theory of light-scattering terms associated with fluctuations of optical dielectric constants associated with temperature effects. Raman and Brillouin scatterings, which involve molecular energy levels and rovibrational excitations, will be given later in this chapter.

Liquid Crystals, Second Edition   By Iam-Choon Khoo
Copyright © 2007 John Wiley & Sons, Inc.

5.2. GENERAL ELECTROMAGNETIC FORMALISM OF LIGHT SCATTERING IN LIQUID CRYSTALS

Scattering of light in a medium is caused by fluctuations of the optical dielectric constants $\delta\varepsilon(\vec{r}, t)$. In isotropic liquids $\delta\varepsilon(\vec{r}, t)$ are mainly due to density fluctuations caused by fluctuations in the temperature. For liquid crystals in their ordered phases, an additional and important contribution to $\delta\varepsilon(\vec{r}, t)$ arises from director axis fluctuations.

As a result of $\delta\varepsilon(\vec{r}, t)$, an incident light will be scattered. The direction, polarization, and spectrum of the scattered light depends on the optical-geometrical configuration.

In general, for a uniaxial birefringent medium such as a liquid crystal, the dielectric constant tensor may be written as

$$\varepsilon_{\alpha\beta} = \varepsilon_{\perp}\delta_{\alpha\beta} + \left(\varepsilon_{\|} - \varepsilon_{\perp}\right)n_{\alpha}n_{\beta}, \tag{5.1}$$

where n_{α} and n_{β} are the components of a unit vector $\hat{n}$ along the optical axis. (In liquid crystals $\hat{n}$ is the director axis.) Fluctuations in $\varepsilon_{\alpha\beta}$ come from changes in $\varepsilon_{\perp}$ and $\varepsilon_{\|}$ due to density and temperature fluctuations and from fluctuations in the directions of $\hat{n}$. Light scattering in liquid crystals was first quantitatively analyzed by deGennes[1] using classical electromagnetic theory and assuming that $\Delta\varepsilon$ is small.

Consider a small volume dV located at a position $\mathbf{r}$ from the origin as shown in Figure 5.1. The induced polarization $\delta\vec{P}$ (dipole moment per unit volume) at this location by an incident field $\vec{E}_{inc} = \hat{i}\, E\, \exp[i(\vec{k}_i \cdot \vec{r} - \omega t)]$ is given by

$$\delta\mathbf{P} = \varepsilon_0\, \bar{\bar{\chi}} : \mathbf{E}_{inc}$$

$$= \delta\bar{\bar{\varepsilon}} : \mathbf{E}_{inc}, \tag{5.2}$$

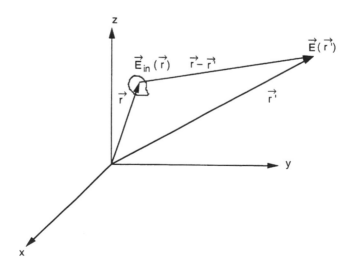

Figure 5.1.   Scattering of light from an elementary volume located at **r**.

where $\delta\bar{\varepsilon}$ is the change in the dielectric constant tensor. This contributes an outgoing field at $\mathbf{r}'$ given by[2]

$$\mathbf{E}(\mathbf{r}') = \frac{\omega^2}{c^2 R}[\exp i(\mathbf{k}_f \cdot \mathbf{R} - \omega t)]\mathbf{P}_\omega(\mathbf{r}), \tag{5.3}$$

where $\bar{P}_\omega(\bar{r})$ is the component of $\delta\bar{P}$ normal to the direction $\mathbf{R} = \mathbf{r}' - \mathbf{r}$. The total out-going field $\mathbf{E}_{\text{out}}$ at a position $\mathbf{r}'$ in the so-called far-field zone ($|\mathbf{r}' - \mathbf{r}| \gg \lambda$) is given by the integral of all the radiated contributions from the volume.

Let $\hat{f}$ denote the polarization vector of the outgoing field and let $\vec{k}_f$ denote its wave vector; that is, writing

$$\mathbf{E}_{\text{out}} = \hat{f} E_{\text{out}} \exp i(\mathbf{k}_f \cdot \mathbf{r}' - \omega t) \tag{5.4}$$

we have

$$\hat{f} \cdot \mathbf{E}_{\text{out}} = \frac{\omega^2 E}{c^2 R} \exp(i\mathbf{k}_f \cdot \mathbf{r}') \int_{\text{vol}} [\hat{f} \cdot \delta\bar{\bar{\varepsilon}} : \hat{i}] \exp[-i(\mathbf{q} \cdot \mathbf{r})] dV, \tag{5.5}$$

where $\mathbf{q} = \mathbf{k}_f - \mathbf{k}_i$. The scattering amplitude a_{fi} is defined by

$$\hat{f} \cdot \mathbf{E}_{\text{out}} = \alpha_{fi} \left[\frac{E}{R} \exp(i\mathbf{k}_f \cdot \mathbf{r}') \right], \tag{5.6a}$$

that is,

$$\alpha_{fi} = \frac{\omega^2}{c^2} \int_{\text{vol}} \delta\varepsilon_{fi}(\mathbf{r}) \exp(-i\mathbf{q} \cdot \mathbf{r}) dV \tag{5.6b}$$

where $\delta\varepsilon_{fi}(\bar{r}) = \hat{f} : \delta\varepsilon : \hat{i}$. Writing

$$\delta\varepsilon_{fi}(\mathbf{q}) = \frac{1}{V} \int_{\text{vol}} \delta\varepsilon_{fi}(\mathbf{r}) \exp(-i\mathbf{q} \cdot \mathbf{r}) dV \tag{5.7}$$

we have

$$\alpha_{fi} = \frac{\omega^2}{c^2} V \delta\varepsilon_{fi}(\mathbf{q}). \tag{5.8}$$

The differential scattering cross section $d\sigma/d\Omega = \langle|\alpha_{fi}|^2\rangle$ (per solid angle) is given by the *thermal* average of $|\alpha_{fi}|^2$; that is,

$$\frac{d\sigma}{d\Omega} = \langle|\alpha_{fi}|^2\rangle. \tag{5.9}$$

The thermal average is used because the fluctuations are attributed to the temperature.

5.3. SCATTERING FROM DIRECTOR AXIS FLUCTUATIONS IN NEMATIC LIQUID CRYSTALS

Scattering of light in nematic liquid crystals is a complicated problem, since so many vector fields are involved. The crucial parameters are the wave vectors $\mathbf{k}_i$ and $\mathbf{k}_f$, the scattering wave vector $\mathbf{q}$, the director axis orientation $\hat{n}$, and its fluctuations $\delta\mathbf{n}$ from its equilibrium direction $\hat{n}_0$. As developed by deGennes,[1] the problem of analyzing light scattering in nematic liquid crystals can be greatly simplified if the coordinate system is properly defined in terms of the initial orientation of the director axis $\hat{n}_0$ with respect to the scattering wave vector $\mathbf{q}$.

As shown in Figure 5.2, the director axis fluctuation $\delta\mathbf{n}$ (which is normal to $\hat{n}_0$ since $|\hat{n}| = 1$) is decomposed into two orthogonal components $\delta\mathbf{n}_1$ and $\delta\mathbf{n}_2$, along the unit vectors $\hat{e}_1$ and $\hat{e}_2$, respectively. Note that one of them, $\delta\mathbf{n}_1$, is in the plane defined by $\mathbf{q}$ and $\hat{n}_0$ (taken as $\hat{z}$), and the other, $\delta\mathbf{n}_2$, is perpendicular to the q-z plane.

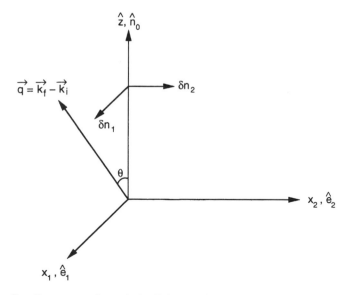

Figure 5.2.   Coordinate system for analyzing light scattering in nematic liquid crystals in terms of two normal modes.

In this case one can express $\delta\mathbf{n}_1(\mathbf{r})$ and $\delta\mathbf{n}_2(\mathbf{r})$ in terms of their Fourier components as

$$\delta n_{1,2}(\mathbf{r}) = \sum_q n_{1,2}(\mathbf{q})\exp(i\mathbf{q}\cdot\mathbf{r}). \tag{5.10}$$

The inverse Fourier transform is given by

$$n_{1,2}(\mathbf{q}) = \frac{1}{V}\int n_{1,2}(\mathbf{r})e^{-i\mathbf{q}\cdot\mathbf{r}}\,dV. \tag{5.11}$$

The total free energy associated with this director axis deformation is, from Equation (3.6), given by

$$F_{\text{total}} = \frac{1}{2}\int K_1\left(\frac{\partial n_1}{\partial x_1} + \frac{\partial n_2}{\partial x_2}\right)^2 + K_2\left(\frac{\partial n_1}{\partial x_2} - \frac{\partial n_2}{\partial x_1}\right)^2$$
$$+ K_3\left[\left(\frac{\partial n_1}{\partial x_3}\right)^2 + \left(\frac{\partial n_2}{\partial x_3}\right)^2\right]dV. \tag{5.12}$$

Substituting Equation (5.10) in Equation (5.12), we thus have

$$F_{\text{total}} = \frac{V}{2}\sum_q\sum_{1,2}|n_{1,2}(q)|^2\left(K_{1,2}q_1^2 + K_3q_3^2\right). \tag{5.13}$$

An important feature of the total free energy is that it is the sum of the energies associated with the two normal modes $\delta\mathbf{n}_1$ and $\delta\mathbf{n}_2$; these modes are not coupled to each other. From classical mechanics, at thermal equilibrium, the thermally averaged energy of each normal mode is $\frac{1}{2}k_BT$, where k_B is the Boltzmann constant. In other words, we have

$$\left\langle\frac{V}{2}|n_{1,2}(q)|^2\left[K_{1,2}q_1^2 + K_3q_3^2\right]\right\rangle = \frac{1}{2}k_BT, \tag{5.14}$$

that is,

$$\left\langle|n_{1,2}(q)|^2\right\rangle = \frac{K_BT/V}{K_3q_3^2 + K_{1,2}q_1^2}. \tag{5.15}$$

From Equation (5.1) the change in $\varepsilon_{\alpha\beta}$ associated with the director axis fluctuation comes from the second term:

$$\delta\varepsilon_{\alpha\beta} = \Delta\varepsilon(n_\alpha\delta n_\beta + \delta n_\alpha n_\beta). \tag{5.16}$$

This means

$$\delta\varepsilon_{fi} = \Delta\varepsilon[(\hat{n}_0\cdot\hat{i})(\delta\mathbf{n}\cdot f) + (\hat{n}_0\cdot\hat{f})(\delta\hat{n}\cdot i)], \tag{5.17}$$

where $\hat{i}$ and $\hat{f}$ are the incident and outgoing optical field polarization directions, assumed to be orthogonal to one another (cf. Fig. 5.3). If we express $\delta\mathbf{n}(\mathbf{r})$ in terms of its Fourier components, that is,

$$\delta n(\mathbf{q}) = \hat{e}_i \, n_i(\mathbf{q}) + \hat{e}_2 \, n_2(\mathbf{q}), \tag{5.18}$$

Equation (5.17) becomes

$$\delta\varepsilon_{fi} = \Delta\varepsilon \sum_{\alpha=1,2} n_\alpha(q) [(\hat{i}\cdot\hat{n}_0)(\hat{f}\cdot\hat{e}_\alpha) + (\hat{f}\cdot\hat{n}_0)(\hat{i}\cdot\hat{e}_\alpha)]. \tag{5.19}$$

Using Equations (5.8) and (5.9) and $\delta\varepsilon_{fi}$ given previously and noting that when we square α_{fi} cross terms are decoupled, we finally have

$$\frac{d\sigma}{d\Omega} = \left(\frac{\Delta\varepsilon\omega^2}{c^2}\right)^2 V \sum_{\alpha=1,2} \langle| n_\alpha(q)|^2\rangle[(\hat{i}\cdot\hat{n}_0)(\hat{f}\cdot\hat{e}_\alpha) + (\hat{f}\cdot\hat{n}_0)(\hat{i}\cdot\hat{e}_\alpha)]^2. \tag{5.20}$$

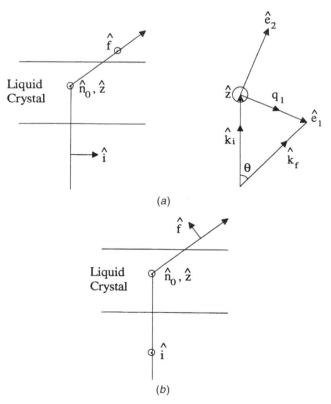

(a)

(b)

Figure 5.3. Geometries for intense scattering of the incident light. Note that the incident and scattered light are cross polarized.

Using Equation (5.15) for $\langle |n_{12}(q)|^2 \rangle$, the differential cross section thus becomes

$$\frac{d\sigma}{d\Omega} = \left(\frac{\Delta\varepsilon\omega^2}{c^2}\right)^2 V \sum_{\alpha=1,2} \frac{k_B T}{K_3 q_3^2 + K_\alpha q_1^2} [f_\alpha(\hat{i}\cdot\hat{n}_0) + i_\alpha(\hat{f}\cdot\hat{n}_0)]^2. \qquad (5.21)$$

Consider the scattering geometry shown in Figure 5.3a. Note that $q_1 \approx 2k \sin(\theta/2)$ (since $k_i \approx k_f \approx k$), $q_3 = 0$, $i_3 = f_1 = f_2 = 0$, $f_3 = 1$, $i_1 = \cos(\theta/2)$, and $i_2 = \sin(\theta/2)$. Inserting these parameters into Equation (5.21) yields

$$\frac{d\sigma}{d\Omega} = \frac{\Delta\varepsilon^2\omega^4}{c^4} V k_B T \left[\frac{\cos^2(\theta/2)}{K_1 q_1^2} + \frac{\sin^2(\theta/2)}{K_2 q_1^2}\right]. \qquad (5.22)$$

Using $q_l \sim 2k \sin(\theta/2)$ gives

$$\frac{d\sigma}{d\Omega} \sim \left[\cot^2\left(\frac{\theta}{2}\right) + \frac{K_1}{K_2}\right]. \qquad (5.23)$$

One can similarly deduce the differential scattering cross sections for other geometries. Equations (5.22) and (5.23) give good agreement with experimental observations:[3, 4]

1. Scattering is intense for crossed polarizations (i.e., when the incident and outgoing optical electric fields are orthogonal to each other).
2. Scattering is particularly strong at small scattering wave vector q, small θ.

If an external field is present (e.g., a magnetic field applied in the z direction), the director axis fluctuations may be reduced. Quantitatively, this may be estimated by including in the free-energy equation [Eq. (5.12)] the magnetic interaction term

$$F_{\text{mag}} = \tfrac{1}{2}\int \Delta\chi^m H^2 (n_1^2 + n_2^2)\,dV. \qquad (5.24)$$

The reader can easily show that this will modify the result for $\langle |n_{1,2}(q)|^2 \rangle$ [cf. Eq. (5.15)] to yield

$$\langle |n_{1,2}(q)|^2 \rangle = \frac{k_B T/V}{K_3 q_3^2 + K_{1,2}q_1^2 + \Delta\chi H^2}, \qquad (5.25)$$

which shows that the fluctuation-induced scattering will be quenched at high field.[5]

5.4. LIGHT SCATTERING IN THE ISOTROPIC PHASE OF LIQUID CRYSTALS

In the isotropic phase director axis orientations are random. The optical dielectric constant, a thermal average, is therefore a scalar parameter. The fluctuations in this case are due mainly to fluctuations in the density of the liquid caused by temperature fluctuations.

Denoting the average dielectric constant in the isotropic phase by ε and denoting the local change in the volume by $u(r)$, the dielectric constant may be expressed as

$$
\begin{aligned}
\varepsilon &= \bar{\varepsilon} + \frac{d\varepsilon}{dV} u(\mathbf{r}) \\
&= \bar{\varepsilon} + \varepsilon' u(\mathbf{r}).
\end{aligned}
\tag{5.26}
$$

The compressional energy associated with the volume change is

$$
F_u = \tfrac{1}{2} \int W \, |u(\mathbf{r})|^2 \, dV
\tag{5.27}
$$

$$
= \frac{VW}{2} \sum_q |u(\mathbf{q})|^2,
\tag{5.28}
$$

where $u(\mathbf{q})$ is the Fourier transform of $u(\mathbf{r})$, in analogy to our previous analysis of $n(\mathbf{r})$ and $n(\mathbf{q})$, and W is the isothermal compressibility. Applying the equipartition theorem, we get

$$
\langle |u(\mathbf{q})|^2 \rangle = \frac{k_B T}{WV}.
\tag{5.29}
$$

From Equation (5.8) and noting that $\delta\varepsilon_{fi}(\mathbf{q}) \equiv \hat{f} \cdot \hat{i} \, \varepsilon' u\,(\mathbf{q})$, the scattering amplitude a_{fi} is given by

$$
\alpha_{fi}^2 = \left(\frac{\omega^2}{c^2} V \hat{f} \cdot \hat{i} \, \varepsilon' \right)^2 |u(\mathbf{q})|^2.
\tag{5.30}
$$

This finally gives the differential cross section

$$
\left(\frac{d\sigma}{d\Omega} \right)_{\text{iso}} = V \left(\frac{\varepsilon' \omega^2}{c^2} \hat{f} \cdot \hat{i} \right)^2 \frac{k_B T}{W}.
\tag{5.31}
$$

Light scattering in the isotropic phase is considerably less than in the nematic phase, as one may see from the photographs of a transmitted laser beam through a nematic liquid crystal in Figures 5.4a and 5.4b for temperatures below and above, respectively, the phase transition temperature T_c. Because of such a drastic change in the scattering loss, T_c is sometimes called the "clearing temperature."

More quantitatively, one can compare the corresponding scattering cross sections [cf. Eq. (5.21). and (5.31)]. The ratio σ_R of $d\sigma/d\Omega$ (nematic) over (divided by) $d\sigma/d\Omega$ (isotropic) is on the order of

$$\sigma_R \sim \frac{\Delta\varepsilon}{\varepsilon'}\frac{W}{Kq^2}. \tag{5.32}$$

Letting a be the characteristic length of the binding energy of the molecule u, we have $K \sim u/a$, $W \sim u/a^3$, $1/q \sim \lambda$ (optical wavelength), and $\Delta\varepsilon/\varepsilon' \sim 1$. Therefore,

$$\sigma_R \sim \left(\frac{\lambda}{a}\right)^3. \tag{5.33}$$

Typically, $\lambda \sim 5000$ Å and $a \approx 20$ Å, so $\sigma_R \sim 10^6$; that is, scattering in the nematic phase is about six orders of magnitude larger than scattering in the isotropic phase (cf. Fig. 5.4c). This is attributed to the fact that dilation leading to density change in the isotropic phase is characterized by W; this involves much more energy than rotation of the director axis, characterized by K, in the nematic phase.

As the temperature of the isotropic liquid crystal is lowered toward T_c, the molecular correlation becomes appreciable, and the scattering contributed by the orientational fluctuations will begin to dominate. The scattering cross section is proportional to $\langle|\delta\varepsilon_{fl}(\mathbf{q})|^2\rangle$ [cf. Eq. (5.9)]. Ignoring the tensorial nature of ε, the quantity $|\delta\varepsilon(\mathbf{q})|^2$ in the $\mathbf{q}$ variable may be expressed in terms of the correlation of $\varepsilon(\mathbf{r})$; that is, the scattering cross section is proportional to

$$\sigma_{iso} \sim \int \langle\varepsilon(\mathbf{r}_1)\varepsilon(\mathbf{r}_2)\rangle \exp\left[i\mathbf{q}\cdot(\mathbf{r}_1 - \mathbf{r}_2)\right]dV_{1,2}. \tag{5.34}$$

Using the Landau theory,[1,6]

$$\langle\varepsilon(0)\varepsilon(R)\rangle \approx \frac{e^{-R/\xi}}{R}, \tag{5.35}$$

where ξ is the so-called correlation length:

$$\xi = \xi_0\left(\frac{T^*}{T-T^*}\right)^{1/2}. \tag{5.36}$$

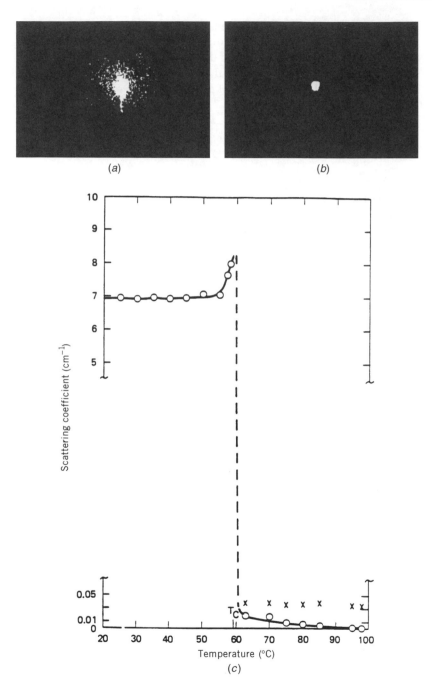

Figure 5.4. Photograph of the transmitted laser beam through a nematic liquid crystal film (100 nm thick): (a) below and (b) above the nematic–isotropic phase transition temperature; (c) experimentally measured scattering coefficient below and above T_c of an unaligned E7 sample (after Khoo and Wu[8]).

Substituting Equation (5.35) in Equation (5.34) gives

$$\sigma_{iso} \approx 4\pi\xi^2 \tag{5.37}$$

$$\approx (T - T^*)^{-1}. \tag{5.38}$$

This is in agreement with the experimental observation of Lister and Stinson.[7]

5.5. TEMPERATURE, WAVELENGTH, AND CELL GEOMETRY EFFECTS ON SCATTERING

While the ability to scatter light easily is a special advantage of nematics for various optical and electro-optical applications, it also imposes a severe limitation on the optical path length through the liquid crystal. The actual amount of scattering experienced by a light beam depends on a variety of factors, including the cell thickness, geometry, boundary surface conditions, external fields, and optical wavelength. Here we summarize some of the general observations.

As experimentally observed by Chatelain,[3] light scattering in the nematic phase is remarkably independent of the temperature vicinity to T_c (i.e., independent of $T_c - T$), in spite of the dependence of the scattering cross section on several highly temperature-dependent parameters such as the dielectric anisotropy $\Delta\varepsilon$ and the elastic constants K_1, K_2, and K_3. This is actually due to the fortuitous cancellation of the temperature dependences of K and $\Delta\varepsilon$. In the Landau−deGennes theory, K is proportional to the square of the order parameter, S^2, whereas $\Delta\varepsilon$ is proportional to S. From Equation (5.13) note that $d\sigma/d\Omega \sim \Delta\varepsilon^2/K$ and is therefore independent of S. In other words, the scattering cross section $d\sigma/d\Omega$ is relatively constant with respect to variation in $T_c - T$.

This temperature independence is also noted in nonlinear scattering[4] involving the mixing of two coherent laser beams in nematic liquid crystals. The coherently scattered wave intensity is also shown to be proportional to $\Delta\varepsilon^2/K$ and is thus independent of $T_c - T$.

The differential scattering cross sections given in Equations (5.21) and (5.31) allow us to deduce the optical wavelength dependence. Note that, basically, for nematics

$$\frac{d\sigma}{d\Omega} \sim \left(\frac{\omega^2}{c^2}\right)^2 \frac{1}{q^2} \sim \frac{1}{\lambda^2} \tag{5.39}$$

since $\omega/c = K = 2\pi/\lambda$ and $q \sim 1/\lambda$ for a fixed scattering angle. For isotropic liquid crystals

$$\left(\frac{d\sigma}{d\Omega}\right)_{iso} \sim \left(\frac{\omega^2}{c^2}\right)^2 \frac{1}{\lambda^4}, \tag{5.40}$$

which is a well-known dependence for an isotropic medium.

The wavelength dependence of the scattering cross section of nematic liquid crystals, in terms of losses experienced by a laser beam in traversing an aligned cell, has been measured in our laboratory by Liu.[8] Figure 5.5 shows a plot of the loss constant α in an aligned E46 sample (defined by $I_{\text{transmitted}} = I_{\text{incident}}e^{-\alpha d}$, where d is the cell thickness) for several argon and He-Ne laser lines. In general, a dependence of the form λ^{-n} ($n = 2.39$) is observed.

Scattering of light in nematic cells is highly dependent on the cell geometry, especially the cell thickness. This may be seen by including the boundary elastic restoring energy term, on the order of K/d^2 (d is the cell thickness) into the calculation for the director axis fluctuations. For $d > q^{-1}$, this boundary energy term is insignificant. On the other hand, for $d \ll q^{-1}$, one would expect that the scattering would be reduced by these boundary effects.

An important manifestation of the dependence of light scattering on the liquid crystal cell geometry is seen in studies on nematic optical fibers.[9] These are made by filling a microcapillary tube with nematic liquid crystals. It is found that if the core diameter of the nematic fiber is on the order of 10 μm or less, the scattering loss is dramatically reduced to 1 or 2 dB/cm, compared to the usual value of about 20 dB/cm for a flat cell.

A rigorous calculation for the director axis fluctuations in a cell of finite thickness (as opposed to the theory developed by deGennes previously for bulk film) has been performed by Zeldovich and Tabiryan.[10] Furthermore, their theory also treats the problem of phase coherence of the light just after passing the cell (i.e., in the so-called near zone). It is found that, in general, very strong transverse phase fluctuations, caused by thermal fluctuations of the director axis, are experienced only by an obliquely incident extraordinary wave; on the other hand, an ordinary wave undergoes no phase fluctuation for any

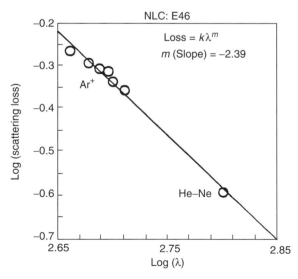

Figure 5.5. Experimentally measured scattering loss in a nematic liquid crystal (E46) that is homeotropically aligned. In a 90-μm-thick sample, the typical scattering loss was measured to be about 18 cm^{-1}.

alignment of the nematic liquid crystal cell (homeotropic, planar, or hybrid) at normal or oblique incidence. These phase fluctuations across the beam's cross section are manifested in strong speckled scatterings of the transmitted beam.

5.6. SPECTRUM OF LIGHT AND ORIENTATION FLUCTUATION DYNAMICS

Just as the scattered light is distributed over a spectrum of wave vectors k_f, for a given incident wave vector k_i, the frequency of the scattered light is distributed over a spectrum of frequencies ω_f, for a given incident light frequency ω_i. This spread in frequency $\Delta\omega = \omega_f - \omega_i$ is inherently related to the fact that the director axis fluctuations are characterized by finite relaxation time constants.

In general, if the relaxation dynamics is in an exponential form [e.g., $\Delta n(t) = \Delta n(0)e^{-t/\tau}$], the frequency spectrum associated with this process, which is obtained by the Fourier transform of $\Delta n(t)$, is a Lorentzian. The half-width of the Lorentzian $\Delta\omega$ is related to τ by $\Delta\omega = 2/\tau$.

As shown in Chapter 3, the dynamics of the orientational fluctuations depends on the distortion modes involved and the corresponding viscosity coefficient.

For a pure twist-type distortion, the equation of motion governing the elastic restoring energy and the viscous force is of the form

$$K_2 \frac{\partial^2 \Delta n(\mathbf{q})}{\partial x^2} - \gamma_1 \frac{\partial \Delta n(\mathbf{q})}{\partial t} = 0, \qquad (5.41)$$

where x is the direction of the wave vector $\mathbf{q}$ of the fluctuation. We may write

$$\Delta n(\mathbf{q}) = \Delta n(t)\exp(iqx) \qquad (5.42)$$

Equation (5.41) could be solved to give

$$\Delta n(q) \sim \Delta n(0)\exp\left(-\frac{q^2 K_2}{\gamma_1}t\right)\exp(iqx). \qquad (5.43)$$

The time constant for the twist deformation is thus

$$\tau_{\text{twist}} = \frac{\gamma_1}{K_2 q^2}, \qquad (5.44)$$

and the scattered light will possess a frequency spectrum, centered at ω_i, with a half-width

$$\Delta\omega_{\text{twist}} = \frac{2K_2 q^2}{\gamma_1}. \qquad (5.45)$$

The dynamics of the other two types of orientation distortions, bend and splay deformations, are more complicated because such distortions are necessarily accompanied by flow (i.e., physical translational motion of the liquid crystal); this phenomenon is sometimes called the backflow effect, which may be regarded as the reverse effect of the flow-induced reorientation effect discussed in Chapter 3. A quantitative analysis of these processes[11] shows that the half-widths of the spectra associated with pure splay and pure bend deformations are given, respectively, by

$$\Delta\omega_{splay} = \frac{2K_1 q^2}{\gamma_1 - (\alpha_3^2)/\eta_2} \tag{5.46a}$$

and

$$\Delta\omega_{bend} = \frac{2K_3 q^2}{\gamma_1 - (\alpha_2^2)/\eta_1}, \tag{5.46b}$$

where α_2 and α_3 are Leslie coefficients, and η_1 and η_2 are viscosity coefficients defined in Chapter 3.

Using typical values (in cgs units) of $K_1 \sim 6 \times 10^{-7}$, $K_2 \sim 3 \times 10^{-7}$, $K_3 \sim 8 \times 10^{-7}$, $\gamma_1 \sim 76 \times 10^{-2}$, $\eta_2 \sim 103 \times 10^{-2}$, $\eta_1 \sim 41 \times 10^{-2}$ for MBBA, $\gamma_1 - \alpha_3^2/\eta_2 = 19 \times 10^{-2}$, and $\gamma_1 - \alpha_3^2/\eta_1 = 126 \times 10^{-2}$, these linewidths are estimated to be on the order of

$$\begin{aligned} \Delta\omega_{twist} &\sim 10^{-6} q^2, \\ \Delta\omega_{splay} &\sim 10^{-5} q^2, \\ \Delta\omega_{bend} &\sim 10^{-6} q^2. \end{aligned} \tag{5.47}$$

Depending on the value of q, the scattering wave vector, these frequency bandwidths are on the order of a few to $\sim 10^2$ Hz. For example, if the wavelength of light $\lambda = 0.5$ μm $= 0.5 \times 10^{-4}$ cm and the scattering angle is 10^{-2}, we have $q^2 \approx 1.6 \times 10^6$ and so $\Delta\omega_{splay} \sim 16$, whereas $\Delta\omega_{twist} \sim \Delta\omega_{bend} \sim 1.6$. These are usually referred to as the "slow" mode spectrum.

On the other hand, the relaxation process associated with the backflow is a comparatively faster one. This flow process is characterized by a time constant:[12]

$$\tau_f \approx \frac{p}{\eta_2 q^2}, \tag{5.48}$$

and thus a frequency bandwidth $\Delta\omega_f$ of

$$\Delta\omega_f \sim \frac{2\eta_2 q^2}{\rho}. \qquad (5.49)$$

Using $\eta_2 \sim 1 \times 10^{-2}$ and $\rho \sim 1$, we have $\Delta\omega_f \sim 2 \times 10^{-2} q^{-2}$, which is about four or five orders of magnitude larger than the slow-mode spectrum.

5.7. RAMAN SCATTERINGS

5.7.1. Introduction

Raman scatterings provide very useful spectroscopic and molecular structural information.[13] The process involves the interaction of the incident lasers with the vibrational or rotational excitations of the material. These Raman transitions involve two energy levels of the material connected by a two-photon process, as shown in Figure 5.6. Actually, of course, there are much more levels involved, but the following two-level model is sufficient for describing the basic physics of the interaction.

When an incident laser of frequency ω_1 propagates through the material, two distinct Raman processes could happen. In the so-called Stokes scatterings, the electron in level 1 will be excited to level 2, and light emission at a frequency $\omega_s = \omega_1 - \omega_R$ will occur. In anti-Stokes scatterings, the electron in level 2 makes a transition to level 1, owing to its interaction with the incident laser, and light emission at $\omega_\alpha = \omega_1 + \omega_R$ will take place.

The population of level 1 is usually greater than that of level 2, and Stokes scattering will occur more efficiently than anti-Stokes scattering. However, if level 2 is unusually populated (e. g., as a result of a laser-induced transition from level 1), this

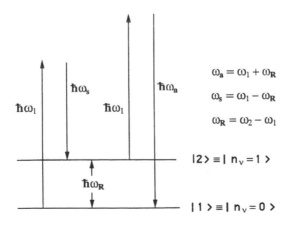

Figure 5.6. Raman scattering involving the generation of Stokes and anti-Stokes light.

notion regarding Stokes and anti-Stokes scattering efficiency is not valid, especially when intense laser fields are involved.

Since the process involves the emission of light by the material, it may occur in either spontaneous or stimulated fashion. In Chapter 9 stimulated Raman (as well as Brillouin) scattering processes, where the generated radiations are phase coherent with respect to the incident wave, will be discussed in the context of nonlinear optics. Here we present a discussion of spontaneous and stimulated Raman scattering processes in terms of quantum mechanics.

5.7.2. Quantum Theory of Raman Scattering: Scattering Cross Section

Raman scatterings are due to the interaction of light with material vibrations. Let q be the normal vibrational coordinate and let $U(q)$ be the potential energy associated with this mode. Because of the optical field displacement q_v on q, the polarizability $\alpha(q)$ of the atom (or molecules comprising the nonlinear material) becomes

$$\alpha(q) = \alpha(q_0) + \left(\frac{\partial \alpha}{\partial q}\right)_{q_0} q_v. \tag{5.50}$$

This gives an induced dipole moment $\mathbf{d}_{\text{ind}}$:

$$\mathbf{d}_{\text{ind}} = \left[\varepsilon_0 \left(\frac{\partial \alpha}{\partial q}\right)_{q_0} q_v\right] \mathbf{E}, \tag{5.51}$$

and an interaction Hamiltonian H_R:

$$H_R = -\mathbf{d}_{\text{ind}} \cdot \mathbf{E} = -\varepsilon_0 \left(\frac{\partial \alpha}{\partial q}\right)_{q_0} q_v \mathbf{E} \cdot \mathbf{E}. \tag{5.52}$$

In a quantized harmonic oscillator representation (cf. Yariv[14]), q_v and E can be expressed in terms of annihilation (a) and creation operators (a^+):

$$q_v \sim (a_v^+ + a_v), \tag{5.53a}$$

$$E \sim \sqrt{\omega_l}(a_l^+ - a_l) + \sqrt{\omega_s}(a_s^+ - a_s). \tag{5.53b}$$

The first term on the right-hand side of Equation (5.53b) comes from the incident laser field E_l, and the second term comes from the Stokes field E_S.

The action of the annihilation a and creation operators a^+ is as follows: Let $|n_l\rangle$, $|n_s\rangle$, and $|n_v\rangle$ and denote the energy states of the laser, the Stokes, and the molecular vibrations, respectively. Then we have $a_l|n_l\rangle \to |n_l-1\rangle$, $a_l^+|n_l\rangle \to |n_l+1\rangle$;

$a_s |n_s\rangle \rightarrow |n_s-1\rangle$; $a_s^+ |n_s\rangle \rightarrow |n_s+1\rangle$; and $a_v |n_v\rangle \rightarrow |n_v-1\rangle$, $a_v^+ |n_v\rangle \rightarrow |n_v+1\rangle$. In other words, the action of the operators a and a^+ is to decrease or to increase, respectively, the number of excitations in the corresponding energy state. In this picture the total wave function of the system is represented by a product of the laser, the Stokes, and the vibrational states, that is,

$$|\psi\rangle = |n_l\rangle |n_s\rangle |n_v\rangle. \tag{5.54}$$

From Equations (5.52) and (5.53) we can see that, in general, the interaction Hamiltonian H_R for Raman scattering consists of a triple product in the a's and a^+'s. In particular, Stokes scattering is associated with the action of a term $a_l a_s^+ a_v^+$ on the initial total wave function of the system $|i\rangle_s = |n_l\rangle |n_s\rangle |0\rangle_v$, which yields a final state where the vibration state (initially a vacuum state) is increased by one unit, while a photon is removed from the incident laser and a Stokes photon is created:

$$a_l a_s^+ a_v^+ \{|n_l\rangle |n_s\rangle |n_v = 1\rangle\} \rightarrow \{|n_l - 1\rangle |n_s + 1\rangle |n_v = 1\rangle\}.$$
$$\{|i_s\rangle\} \qquad\qquad\qquad \{|f_s\rangle\} \tag{5.55}$$

Similarly, the inverse scattering process corresponds to the action of $a_l^+ a_s a_v$ on an initial state $|i\rangle_a = |n_l\rangle |n_s\rangle |1\rangle$, that is,

$$a_l^+ a_s a_v \{|n_l\rangle |n_s\rangle |n_v = 1\rangle\} \rightarrow \{|n_l + 1\rangle |n_s - 1\rangle |n_v = 0\rangle\}.$$
$$\{|i_a\rangle\} \qquad\qquad\qquad \{|f_a\rangle\} \tag{5.56}$$

In accordance with perturbation theory,[14] the probability of these transitions is proportional to the square moduli of the matrix elements for these processes. For Stokes scattering,

$$W_s \alpha |\langle f_s | a_l a_s^+ a_v^+ | i_s \rangle|^2 = Dn_l(n_s + 1). \tag{5.57}$$

For inverse scattering,

$$W_a \alpha |\langle f_a | a_l^+ a_s a_v | i_a \rangle|^2 = D(n_l + 1)n_s, \tag{5.58}$$

where D is the proportional constant, which may be shown by a rigorous quantum mechanical calculation[14] to be the same for these two processes.

Considering the growth in time of the Stokes photons and using Equations (5.57) and (5.58), we have

$$\frac{dn_s}{dt} = DN_1 n_l (n_s + 1) - DN_2 n_s (n_l + 1), \tag{5.59}$$

where N_1 and N_2 are the respective probabilities of occupation of the ground state $|1\rangle(n_v=0)$ and the excited state $|2\rangle(n_v=1)$.

In ordinary (spontaneous) Raman scattering, $n_s \ll 1$ and $N_2 \ll N_1$. We thus have

$$\frac{dn_s}{dt} \sim DN_1 n_l. \tag{5.60}$$

Since the number of photons is conserved, $dn_s/dt = -dn_l/dt$, Equation (5.60) yields

$$\frac{dn_l}{dt} = -DN_1 n_l \tag{5.61}$$

or

$$\frac{dn_l}{dz} = -\frac{D}{c/n(\omega_1)} N_1 n_l, \tag{5.62}$$

where $n(w_l)$ is the refractive index. This gives

$$n_l(z) = n_l(0)e^{-\beta_l z}, \tag{5.63}$$

where

$$\beta_l = \frac{DN_1 n(\omega_l)}{c} \tag{5.64}$$

is the attenuation constant for the incident laser at ω_l. Note that β_l is *independent of the incident laser intensity*. Also, from Equation (5.60), the number of Stokes photons scattered is proportional to the incident laser intensity at any given time. By considering the number of Stokes photons scattered into a differential solid angle $d\Omega$, the proportional constant D in Equation (5.64) is shown by Yariv[14] to be related to the differential cross section $d\sigma/d\Omega$ by

$$\left.\frac{d\sigma}{d\Omega}\right|_{90°,\phi} = \frac{3v_s^3 n^3(\omega_s)n(\omega_l)DV \Delta v N_1}{v_l N c^4}, \tag{5.65}$$

where N is the density of the molecules, V is the volume, $n(\omega_s)$ and $n(\omega_l)$ are the respective refractive indices at Stokes and incident laser frequencies, and Δv is the natural linewidth of the transition.

From Equation (5.60) we can see that the rate of production of Stokes photons is proportional to the incident laser intensity (i.e., proportional to n_l). Hence, if the incident laser is intense enough, the Stokes photon number n_s can be substantial ($\gg 1$).

In this case the scattering process will take on a totally different form, and the Stokes wave will grow exponentially. To see this, consider Equation (5.59) for the case n_l, $n_s \gg 1$. We have

$$\frac{dn_s}{dt} = D(N_1 - N_2)n_l n_s. \tag{5.66}$$

This gives

$$\frac{dn_s}{dz} = \frac{Dn(\omega_s)}{c}(N_1 - N_2)n_l n_s \tag{5.67}$$

or, for the Stokes intensity,

$$I_s(z) = I_s(0)e^{g_s z}, \tag{5.68}$$

where

$$g_s = \frac{Dn(\omega_s)}{c}(N_1 - N_2)n_l. \tag{5.69}$$

Since n_l is related to I_l by

$$n_l = \frac{n(\omega_l)\,I_l}{h\nu_l c}$$

we have

$$g_s = \frac{Dn(\omega_s)n(\omega_l)}{h\nu_l c^2}(N_1 - N_2)I_l. \tag{5.70}$$

In other words, the Stokes wave experiences an exponential gain constant (if $N_1 > N_2$) that is proportional to the incident laser intensity. If this gain is larger than the loss α owing to absorption, random scatterings, and so on experienced by the Stokes wave, it is possible to have laser oscillations (i.e., the generation of Stokes lasers). Such a process is called stimulated Raman scattering.

For a more detailed discussion of stimulated scatterings as nonlinear optical wave mixing processes, the reader is referred to Chapter 11.

5.8. BRILLOUIN AND RAYLEIGH SCATTERINGS

In general, for pure nonabsorbing media, the spectrum of spontaneously scattered light from an incident laser centered at ω_l is of the form given in Figure 5.7. The

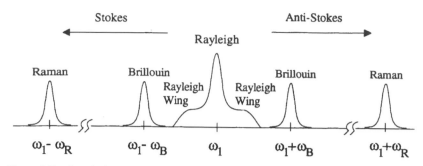

Figure 5.7. A typical spectrum of light spontaneously scattered from a liquid or liquid crystal.

Raman-Stokes radiations (at $\omega = \omega_l - \omega_R$) and the anti-Stokes radiations (at $\omega = \omega_l - \omega_R$) are far to the side of the central frequency.

Around the central frequency ω_l, in general, the spectrum consists of a Rayleigh scattering peak centered at ω_l, two Brillouin scattering peaks at ($\omega_l \pm \omega_\beta$), and a broad background that is generally referred to as the Rayleigh wing.

Rayleigh scattering is due to the entropy or temperature fluctuations, which cause nonpropagating density (and therefore dielectric constant) fluctuations. On the other hand, Brillouin scattering is due to the propagating pressure (i.e., sound) waves, and is often referred to as the electrostrictive effect. An incident laser can generate copropagating or counterpropagating sound waves at a frequency ω_B. Hence, the spectrum of the scattered light consists of a doublet centered at $\omega_l \pm \omega_B$.

Rayleigh wing scattering is due to the orientational fluctuations of the anisotropic molecules. For typical liquids these orientational fluctuations are characteristic of the *individual* molecules' movements and occur in a very short time scale ($\leq 10^{-12}$ s). Consequently, the spectrum is quite broad. In liquid crystals studies of individual molecular orientation dynamics have shown that the relaxation time scale is on the order of picoseconds[15] and thus the Rayleigh wing spectrum for liquid crystals is also quite broad.

From the discussion given in earlier chapters, we note that in liquid crystals, the main scattering is due to collective or correlated orientational fluctuations. These, of course, are much slower processes than individual molecular motions. The spectrum from these correlated or collective orientational fluctuations is therefore very sharp and is embedded in the central Rayleigh peak region.

The dynamics of orientational scattering has been discussed on various occasions in the preceding chapter. We shall discuss here the mechanism of Brillouin and Rayleigh scatterings.

The origin of these scattering processes is the dielectric constant change $\Delta\varepsilon$ associated with a change $\Delta\rho$ in the density, that is,

$$\Delta\varepsilon = \frac{\partial\varepsilon}{\partial\rho}\Delta\rho. \tag{5.71}$$

The fundamental independent variables for the density are the pressure P and the entropy S. Therefore, we can express $\Delta\rho$ as

$$\Delta\rho = \left(\frac{\partial\rho}{\partial P}\right)_s \Delta P + \left(\frac{\partial\rho}{\partial S}\right)_P \Delta S, \tag{5.72}$$

where the terms on the right-hand side of Equation (5.72) correspond to the adiabatic and isobaric density fluctuations, respectively.

5.8.1. Brillouin Scattering

The pressure wave ΔP obeys an equation of motion of the form:[16]

$$\frac{\partial^2\Delta P}{\partial t^2} - \Gamma\nabla^2\frac{\partial\Delta P}{\partial t} - v_s^2\nabla^2\Delta P = 0, \tag{5.73}$$

where v_s is the velocity of sound and Γ is the damping constant.

The velocity of sound can be expressed in terms of the compressibility C_s or the bulk modulus β:

$$C_s = \frac{1}{\beta} = -\frac{1}{v}\frac{\partial V}{\partial P} = \frac{1}{\rho}\left(\frac{\partial\rho}{\partial P}\right)_s. \tag{5.74}$$

Then

$$v_s^2 = \left(\frac{\partial P}{\partial\rho}\right)_s = \frac{1}{C_s\rho}. \tag{5.75}$$

The damping constant can be further expressed as

$$\Gamma = \frac{1}{\rho}\left(\frac{4}{3}\eta_s + \eta_b\right), \tag{5.76}$$

where η_s is the shear viscosity coefficient and η_b is the bulk viscosity coefficient.

The presence of the Brillouin doublet in the scattered wave from the medium in which there is an acoustic wave may be seen as follows. For simplicity, we will ignore all tensor characteristics of the problem. Let the pressure wave be of the form

$$\Delta P = \Delta P_0 e^{i(qz - \omega_B t)} + \text{c.c.} \tag{5.77}$$

As a result of this change in pressure, a dielectric constant change occurs and is given by

$$\Delta\varepsilon = \left(\frac{\partial\varepsilon}{\partial\rho}\right)\Delta\rho$$

$$= \left(\frac{\partial\varepsilon}{\partial\rho}\right)\left(\frac{\partial\rho}{\partial P}\right)_s \Delta P. \tag{5.78}$$

Using the definition for the electrostrictive coefficient $\gamma^e = \rho(\partial\varepsilon/\partial\rho)$ and $C_s = 1/\rho(\partial\rho/\partial P)$, we thus have

$$\Delta\varepsilon = C_s\gamma^e\,\Delta P. \tag{5.79}$$

Writing the incident optical electric field as

$$\mathbf{E}_{inc} = \mathbf{E}_0\left[e^{i(k\cdot r - \omega t)} + \text{c.c.}\right], \tag{5.80}$$

we can see that the dielectric constant will give rise to an induced polarization

$$\mathbf{P} = \Delta\varepsilon\mathbf{E}_{inc}$$

$$= C_s\gamma^e\,\Delta P\mathbf{E}_{inc}. \tag{5.81}$$

From Equation (5.77) for ΔP and Equation (5.80) for $\mathbf{E}_{inc}$, the reader can easily verify that the scattered waves contain two components. One component possesses a wave vector $\mathbf{k}_- = \mathbf{k} - \mathbf{q}$ oscillating at a frequency $\omega - \omega_B$. This corresponds to the scattering of light from a retreating sound wave. The other scattered component is characterized by a wave vector $\mathbf{k}_+ = \mathbf{k} + \mathbf{q}$ oscillating at a frequency $\omega + \omega_B$. This corresponds to the scattering of light from an oncoming sound wave. These two components are very close to each other in magnitude.

The wave vectors of the incident $\mathbf{k}_1$ and scattered $\mathbf{k}_2$ light and the acoustic wave vector $\mathbf{q}$ are related by the wave vector addition rule shown in Figure 5.8 for both processes.

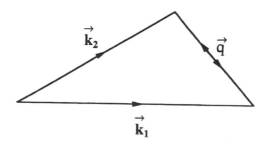

Figure 5.8. Wave vector addition rule for the Brillouin scattering processes.

For Stokes scattering, $\omega_2 = \omega_1 - \omega_B$ and $\mathbf{k}_2 = \mathbf{k}_1 - \mathbf{q}$; for anti-Stokes scattering, $\omega_2 = \omega_1 + \omega_B$ and $\mathbf{k}_2 = \mathbf{k}_1 + \mathbf{q}$.

Since $|\mathbf{k}_1| \approx |\mathbf{k}_2|$ and $|\mathbf{q}| = 2|k_1|\sin(\theta/2)$, the frequency ω_B of the acoustic wave is thus given by

$$\omega_B = |q|v_s = 2n\omega\frac{v_s}{c}\sin\left(\frac{\theta}{2}\right). \tag{5.82}$$

The spectral width of the scattered line is related to the acoustic damping constant Γ. This may be obtained by substituting ΔP in Equation (5.77) into the acoustic equation (5.73). This gives

$$\omega_B^2 = q^2(v_s^2 - i\omega_B\Gamma) \tag{5.83}$$

or

$$q^2 = \frac{\omega_B^2}{v_s^2 - i\omega_B\Gamma} \simeq \frac{\omega_B^2}{v_s^2}\left(1 + \frac{i\omega_B\Gamma}{v_s^2}\right), \tag{5.84}$$

$$q \sim \frac{\omega_B}{v} + \frac{i\Gamma q^2}{2v_s} \tag{5.85}$$

if $\omega_B\Gamma \ll v_s^2$ (which is usually the case). This shows that the acoustic wave is characterized by a damping constant

$$\Gamma_B = \frac{\Gamma q^2}{2}. \tag{5.86}$$

The inverse of this, $\tau_p = 1/\Gamma_B$, is often referred to as the phonon lifetime. As a result, the spectrum of the Brillouin doublet is broadened by an amount

$$\delta\omega = \frac{1}{\tau_p} = \Gamma_B. \tag{5.87}$$

Using $q = 2|k_1|\sin(\theta/2)$, we get

$$\delta\omega = 2n^2\Gamma\frac{\omega^2}{c^2}\sin^2\left(\frac{\theta}{2}\right). \tag{5.88}$$

5.8.2. Rayleigh Scattering

The entropy fluctuation ΔS obeys a diffusion equation similar to that for the temperature fluctuation:

$$\rho C_p\frac{\partial\Delta S}{\partial t} - \kappa\nabla^2 S = 0, \tag{5.89}$$

where C_p is the specific heat at constant pressure and K is the thermal conductivity. A solution of the diffusion equation is a *nonpropagative* function of the form

$$\Delta S = \Delta S_0 e^{-\Gamma_T t - i\mathbf{q} \cdot \mathbf{r}}, \tag{5.90}$$

where the thermal damping constant Γ_T is given by

$$\Gamma_T = \frac{k}{\rho C_p} q^2. \tag{5.91}$$

Following the preceding analysis, we can see that the scattering caused by the entropy fluctuation does not shift the frequency. Instead, because of the exponentially decaying dependence, $e^{-\Gamma_T t}$ it broadens the light by an amount $\delta\omega = \Gamma_T$. Again, since $\mathbf{q} = 2|k_1|\sin(\theta/2)$, we have

$$\delta\omega = \frac{4\kappa}{\rho C_p} |k_1|^2 \sin^2\left(\frac{\theta}{2}\right). \tag{5.92}$$

5.9. NONLINEAR LIGHT SCATTERING: SUPRAOPTICAL NONLINEARITY OF LIQUID CRYSTALS

Preceding sections deal mostly with what may be termed linear light-scattering processes by the liquid crystals, in which changes in the liquid crystalline properties give rise to the scattered light's temporal and spatial (directions) frequencies. The incident light or scattered lights do not create significant changes in the liquid crystal properties. On the other hand, there are so-called nonlinear light-scattering processes in which the light interacts sufficiently strongly with the liquid crystal to change its molecular or crystalline properties (e.g., director axis orientation) or molecular energy level populations (Raman scatterings). The reorientation of the director axis gives rise to a different effective refractive index experienced by the light and modifies its propagation, intensity, polarization states, etc., accordingly. These nonlinear optical processes will be discussed in detail in Chapters 9–12.

As a result of their extraordinarily large light-scattering abilities [cf. Eq. (5.33)], nematic liquid crystals are among the most nonlinear light-scattering medium and have been studied in this context over the past two decades. Recent studies[17] have shown that their extremely strong light-scattering properties can be utilized for self-action effect with unprecedented low optical power and for interaction length that are orders of magnitude shorter than before. Specifically, these authors have demonstrated an all-optical polarization switching effect based on stimulated orientational scattering with a 1.55 μm laser (as well as visible lasers) in which the milliwatt power incident beam is almost completely converted to an orthogonally polarized output after traversing a 200-μm-thick nematic liquid crystal (NLC) film. With optimal choice of materials, the operation power could easily be reduced to sub-milliwatt level.

Liquid crystals also ushered in the era of supranonlinearities characterized by an optical index change coefficient $n_2 \sim 1$ cm^2/W or larger. (n_2 is defined by $n_2 = \Delta n/I$, where Δn is the light-induced index change and I is the optical intensity.) By comparison, what used to be known as a highly nonlinear anisotropic liquid such as CS_2 is characterized by an n_2 about ten orders of magnitude lower. Such nonlinearities were first discovered in methyl red dye molecule-[18], C60-, and/or carbon nanotube-doped NLC.[19,20] These supraoptical nonlinearities have enabled many nonlinear processes such as optical limiting and coherent image processing[21] with even lower operational power threshold and make these NLC films competitive alternatives to their much more costly counterpart—optoelectronic devices (e.g., the optically addressed LCSLM, which will be discussed in Chapter 6).

What then is the limit on such supranonlinearity, as one may naturally ask? A glimpse on that can be gained by considering the basic light–liquid crystal (LC) interaction as depicted in Figure 5.9. The energy density involved in reorienting the LC axis by an angle θ is

$$U\left(\frac{\text{erg}}{\text{cm}^2}\right) \approx K\left(\frac{\partial\theta}{\partial x}\right)^2 L, \tag{5.93}$$

where L is the interaction length and K is the LC elastic constant. In a wave mixing type of interaction geometry as shown, the reorientation angle θ is of the form

$$0 = \theta_0 \sin(qx), \tag{5.94}$$

where $q = 2\pi/\Lambda$. Therefore, we have

$$U_{\text{LC}} \sim K\pi^2\theta^2 \frac{L}{\Lambda^2}. \tag{5.95}$$

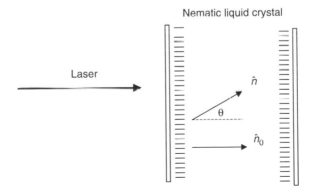

Nematic liquid crystal

Laser

$\hat{n}$

θ

$\hat{n}_0$

Figure 5.9. Typical laser–nematic director axis interaction geometry in a wave mixing experiment to measure the nonlinear refractive index change induced in the nematic liquid crystal.[18–20] Note that an external dc or ac voltage can be applied in conjunction with the optical field. Also the liquid crystal could be doped with photosensitive molecules or nanoparticulates to enhance their nonlinear scattering efficiency.

On the other hand, the energy provided by the light beam is

$$E_{\text{light}} = I\tau(1 - e^{-\alpha L}) \sim \alpha L I\tau, \tag{5.96}$$

where α is the absorption coefficient and τ is the response time. Equating E_{light} and U_{LC}, i.e., assuming complete conversion of absorbed light energy to reorientation, we get

$$\frac{K\pi^2\theta^2}{\alpha\Lambda^2} \approx I\tau. \tag{5.97}$$

For an interaction geometry as depicted in Figure 5.8, the change in index experienced by an incident electromagnetic wave

$$\Delta n \sim (n_e - n_o)\theta^2 \sim (n_e - n_o)\frac{I\tau\alpha\Lambda^2}{K}. \tag{5.98}$$

Writing $\Delta n = n_2 I$ yields the nonlinear index coefficient n_2:

$$n_2 \sim (n_e - n_o)\frac{\tau\alpha\Lambda^2}{K\pi^2}. \tag{5.99}$$

Depending on various parameters such as the birefringence, viscosity, and sample thickness, and other factors such as laser intensity, presence of other applied fields or photosensitive dopants, as well as the actual process involved, the value of n_2 and the response time can vary considerably. In NLC, typical τ is on the order of tens of milliseconds (10^{-2}s), for $\Lambda \sim 20$ μm. Using $K \sim 10^{-7}$ erg/cm, $(n_e - n_o) \sim 0.2$, and $\alpha \sim 100$ cm^{-1}, we have $n_2 \sim 1$ cm^2/W. Even larger n_2 values can be expected. For example, if $\Lambda \sim 100$ μm, $\tau \sim 1$ s, $K \sim 10^{-7}$, $(n_e - n_o) \sim 0.4$, and $\alpha \sim 200$ cm^{-1}, we have $n_2 \sim 1000$ cm^2/W! Such incredibly large optical nonlinearity was first observed by Khoo et al.[18] and more recently by Simoni et al.[22] One could indeed look forward to even more pleasant surprises and nonlinear optical wonders to emerge from studies of liquid crystals in the near future.

REFERENCES

1. deGennes, P. G. 1974. *The Physics of Liquid Crystals.* Clarendon: Oxford Press.
2. Landau, L. D., and E. M. Lifshitz. 1960. *Electrodynamics of Continuous Media.* London: Pergamon; see also Jackson, J. D. 1963. *Classical Electrodynamics.* New York: Wiley.
3. Chatelain, P. 1948. *Acta Crystallogr.* 1:315.
4. Khoo, I. C. 1983. *Phys. Rev. A.* 27:2747.
5. Filippini, J. C. 1977. *Phys. Rev. Lett.* 39:150; Malraison, B., Y. Poggi, and E. Guyon. 1980. *Phys. Rev. A.* 21:1012.

6. Landau, L. D., and E. M. Lifshitz. 1975. *Statistical Physics*. London: Pergamon.

7. Litster, J. D., and T. Stinson. 1973. *Phys. Rev. Lett.* 30:688; see also Stinson, T. W., and J. D. Litster. 1970. *ibid.* 25:503.

8. Liu, T. H. 1987. Ph.D. thesis, Pennsylvania State University, University Park; see also Khoo, I. C., and S. T. Wu. 1993. *Optics and Nonlinear Optics of Liquid Crystals*. Singapore: World Scientific.

9. Green, M., and S. J. Madden. 1989. *Appl. Opt.* 28:5202.

10. Zeldovich, B. Ya., and N. V. Tabiryan. 1981. *Sov. Phys. JETP.* 54:922.

11. Saupe, A. 1973. *Annu. Rev. Phys. Chem.* 24:441.

12. Orsay Liquid Crystal Group. 1969. *J. Chem. Phys.* 51:816.

13. See, for example, Jen, S., N. A. Clark, and P. S. Pershan. 1977. *J. Chem. Phys.* 66:4635.

14. Yariv, A. 1985. *Quantum Electronics*. New York: Wiley.

15. See, for example, Lalanne, J. R., B. Martin, and B. Pouligny. 1977. *Mol. Cryst. Liq. Cryst.* 42:153; see also Flytzanis, C., and Y. R. Shen. 1974. *Phys. Rev. Lett.* 33:14.

16. Fabelinskii, I. L. 1968. *Molecular Scattering of Light*. New York: Plenum.

17. Khoo, I. C., and J. Ding. 2002. All-optical cw laser polarization conversion at 1.55 micron by two beam coupling in nematic liquid crystal film. *Appl. Phys. Lett.* 81:2496–2498.

18. Khoo, I. C., S. Slussarenko, B. D. Guenther, and W. V. Wood. 1998. Optically induced space charge fields, dc voltage, and extraordinarily large nonlinearity in dye-doped nematic liquid crystals. *Opt. Lett.* 23:253–255.

19. Khoo, I. C. 1996. Holographic grating formation in dye- and fullerene C60-doped nematic liquid crystal film. *Opt. Lett.* 20:2137.

20. Khoo, I. C., J. Ding, Y. Zhang, K. Chen, and A. Diaz. 2003. Supra-nonlinear photorefractive response of single-wall carbon nanotube- and C60-doped nematic liquid crystal. *Appl. Phys. Lett.* 82:3587–3589.

21. Khoo, I. C., J. Ding, A. Diaz, Y. Zhang, and K. Chen. 2002. Recent studies of optical limiting, image processing and near-infrared nonlinear optics with nematic liquid crystals. *Mol. Cryst. Liq. Cryst.* 375:33–44.

22. Lucchetti, L., M. Di Fabrizio, O. Francescangeli, and F. Simoni. 2004. Colossal optical nolinearity in dye-doped liquid crystals. *Opt. Commun.* 233:417.

6

Liquid Crystal Optics and Electro-Optics

6.1. INTRODUCTION

Perhaps the most studied and applied property of liquid crystals is their light-scattering ability. With the aid of an externally applied (usually electric field) field, one can control or realign the anisotropic liquid crystal axis, thereby controlling the effective refractive index and phase shift experienced by the light traversing the liquid crystal. Such electro-optical processes form the basis for various optical transmission, reflection, switching, and modulation applications. Essentially, the liquid crystal cells are placed within a stack of phase-shifting (phase retardation) wave plates or polarizing elements to perform various electro-optical functions,[1,2] see Figure 6.1. These operations usually require a phase shift $\Delta\phi$ on the order of π (e.g., quarter wave plate requires $\pi/4$ and half-wave plate requires π, and so on). Depending on the actual configuration, the phase shift imparted by the liquid crystal cell $\Delta\phi \sim d(\Delta\mathrm{n})2\pi/\lambda$, where d is the path length of the light through the liquid crystal layer, Δn is the birefringence, and λ is the wavelength.

As seen in previous chapters, liquid crystals are noted for their large birefringence and easy susceptibility to external field perturbation. To create the required phase shift for optical application in the visible to near IR regime, only applied voltages of a few volts and a film thickness of a few microns are required. Since the interaction of the applied field (usually an ac field) with the nematic, for example, is essentially a dielectric one, the process of electro-optical control is essentially free of current flow and dissipation, i.e., very little power is consumed. As a result, liquid crystal has enjoyed wide spread and an ever increasing demand in various optical display, switching, information, and image processing industries.

The problem of polarized light propagation in liquid crystals is actually quite complex, and its quantitative description requires exact treatment of electromagnetic and anisotropic media. The problem is compounded by the fact that, in general, the director axis orientation within the cell is inhomogeneous and varies spatially in a nonuniform manner because of boundary anchoring effects in response to an applied field. Many sophisticated theoretical techniques and formalisms have been developed to quantitatively treat the detailed problem of light propagation in these highly

Liquid Crystals, Second Edition   By Iam-Choon Khoo
Copyright © 2007 John Wiley & Sons, Inc.

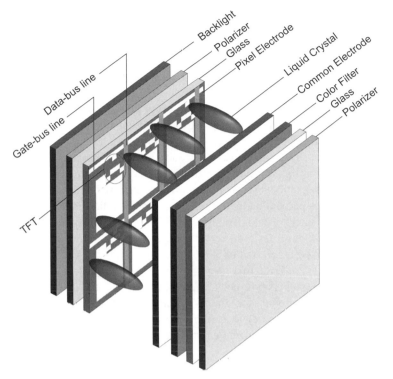

Figure 6.1. Schematic of a typical liquid crystal display pixel consisting of electronic driving circuit, polarizers, liquid crystal cell, color filter, and phase plate [not shown].

anisotropic media. These theoretical formalisms will be discussed in detail in the next chapter.

In this chapter, we focus on the basic principles and seek only some general understanding by dealing with analytically or conceptually solvable cases. We begin by reviewing the essentials of electro-optics and various optical modulations and switching processes. We provide also a general overview of some current liquid crystal display devices as well as other widely used or potential useful optical devices. Although our emphasis is on nematic liquid crystals, the basic optical principles described here are generally applicable to other liquid crystal phases and electro-optics crystals as well.

6.2. REVIEW OF ELECTRO-OPTICS OF ANISOTROPIC AND BIREFRINGENT CRYSTALS

6.2.1. Anisotropic, Uniaxial, and Biaxial Optical Crystals

All crystals are made up of many large constituent atoms and molecules assembled in various crystalline symmetries. As such, they are optically *anisotropic*. A polarized light incident on the crystal will experience different refractive indices depending on its states of polarization. If we denote the crystalline axes by 1, 2, and 3, the refractive indices are n_1, n_2, and n_3, for light polarized along x, y, and z, respectively.

For light of arbitrary polarizations, or equivalently, an electro-optics crystal of an arbitrary orientation relative to the polarized light propagation direction, the resulting propagation of polarized light in the crystal is rather complex as it involves all the indices. In practical situations, one therefore cuts the crystals or orients the light propagation direction such that the propagation direction and the polarization vectors coincide with one of the principal planes of the crystals.

Usually the z axis, sometimes also referred to as the crystal axis, is chosen as an axis of symmetry. For light propagating along this axis of symmetry, since light is a transverse electromagnetic wave, its polarization vector (defined by the direction of its electric field E) is perpendicular to the z axis, that is, E lies on the x-y plane. If n_1 and n_2 are unequal, such crystal is usually called *biaxial*. On the other hand, if $n_1 = n_2$ (for all intents and purposes), the crystal is called *uniaxial*. Conventionally, if $n_3 > n_1, n_2$, the crystal is referred to as *positive* uniaxial or biaxial, whereas if n_3 is $< n_1, n_2$, the crystal is referred to as *negative* uniaxial or biaxial.

In the case of uniaxial crystal such as nematic liquid crystal, that is, $n_1 = n_2$, usually these indices are denoted as n_o and n_3 is denoted as n_e. The meaning of the subscripts o and e will become clear in the following section when we discuss the propagation of light in a birefringent crystal as *ordinary* or *extraordinary* ray using the so-called index ellipsoid method,

$$\frac{x^2}{n_1^2} + \frac{y^2}{n_2^2} + \frac{z^2}{n_3^2} = 1. \tag{6.1}$$

Consider the case of light propagating along a direction k making an angle θ with the z axis, see Figure 6.2. The plane perpendicular to k going through the origin will define the polarization vectors allowed; that is, there will be two orthogonal polarization vectors, one lying in the x-z plane and one lying on the y-z plane. For example, if k lies on the z-y plane, then the two allowed polarization vectors lie on the x-z and y-z planes, with refractive indices n_o and $n_e(\theta)$, respectively.

From Figure 6.2, one can see that the so-called ordinary wave experiences an index that is independent of the propagation angle:

$$n_o(\theta) = n_o. \tag{6.2a}$$

On the other hand, the extraordinary wave experiences an index that is dependent on the propagation angle θ. From the index ellipsoid, the q dependent index can be simply calculated as

$$n_e(\theta) = n_e n_o / [n_e^2 \cos^2(\theta) + n_o^2 \sin^2(\theta)]^{1/2}. \tag{6.2b}$$

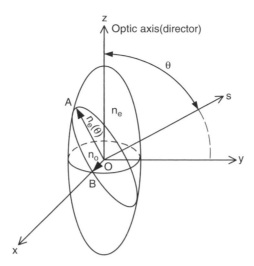

Figure 6.2. Index ellipsoid for a plane polarized optical wave propagating along **s** in a uniaxial crystals. $n_e(\theta)$ and n_o are the refractive indices for the extraordinary and ordinary components, respectively.

6.2.2. Index Ellipsoid in the Presence of an Electric Field: Linear Electro-Optics Effect

In general, application of an external field on a crystal will give rise to changes in the optical refractive indices. The basic mechanisms responsible for the changes vary widely. In typical electro-optical crystals such as lithium niobate, the change in refractive index is due to the Pockel effect, where the index change is proportional to the applied electric field vector. This is also the case in ferroelectric liquid crystals, in which the change depends on the magnitude as well as direction of the applied field. On the other hand, in nematic liquid crystals, the refractive index change is due to reorientation of the director axis, which is quadratic in the applied electric field— the so-called Kerr effect which will be treated in later sections in this chapter.

In this section, we treat the case of linear electro-optic effect. In the presence of an applied field, the index ellipsoid becomes[1]

$$\left(\frac{1}{n^2}\right)_1 x^2 + \left(\frac{1}{n^2}\right)_2 y^2 + \left(\frac{1}{n^2}\right)_3 z^2 + 2\left(\frac{1}{n^2}\right)_4 yz$$
$$+ 2\left(\frac{1}{n^2}\right)_5 xz + 2\left(\frac{1}{n^2}\right)_6 xy = 1, \tag{6.3}$$

where the various coefficients $(1/n^2)_i$ are dependent on the applied field E. When the applied field $E = 0$, the coefficients $(1/n^2)_i$ (for $i = 1, 2, 3$) correspond to $1/n_1^2$, $1/n_2^2$, and $1/n_3^2$, whereas $(1/n^2)_j = 0$ (for $j = 4, 5, 6$).

In a crystal exhibiting a linear optical effect, the changes in the coefficients $(1/n^2)$ on the applied field $E = E(E_1, E_2,$ and $E_3)$ are given by

$$\Delta\left(\frac{1}{n^2}\right)_i = \sum_{j=1}^{3} r_{ij}E_j \quad (i = 1,2,...,6 \text{ and } j = 1,2,3). \tag{6.4}$$

More explicitly, we have

$$\begin{vmatrix} \Delta(1/n^2)_1 \\ \Delta(1/n^2)_2 \\ \Delta(1/n^2)_3 \\ \Delta(1/n^2)_4 \\ \Delta(1/n^2)_5 \\ \Delta(1/n^2)_6 \end{vmatrix} = \begin{vmatrix} r_{11} & r_{12} & r_{13} \\ r_{21} & r_{22} & r_{23} \\ r_{31} & r_{32} & r_{33} \\ r_{41} & r_{42} & r_{43} \\ r_{51} & r_{52} & r_{53} \\ r_{61} & r_{62} & r_{63} \end{vmatrix} \begin{vmatrix} E_1 \\ E_2 \\ E_3 \end{vmatrix}. \tag{6.5}$$

For example, we have $\Delta(1/n^2)_3 = r_{31}E_1 + r_{32}E_2 + r_{33}E_3$. In other words, the application of an electric field along one direction generally will give rise to refractive index changes in all three directions. This is expected since the change in the refractive index is due to perturbation of the total electronic wave functions by the applied electric field. In crystals, owing to the close interactions between neighboring atoms/molecules, the response along some particular directions for a given applied field direction could be more or less intense than in other directions depending on crystalline structures and symmetry. Symmetry considerations could also allow one to specify which particular matrix elements r_{ij} should be or should not be nonzero (i.e., responds or do not respond) to the applied field.

For a widely used electro-optics crystal such as lithium niobate ($LiNbO_3$), we have the electro-optics coefficients $r_{33}=30.8$ (in units of 10^{-12} m/V), $r_{13}=8.6$, $r_{22}=3.4$, and $r_{42}=28$, with $n_e=2.29$ and $n_o=2.20$ (at $\lambda=550$ nm). It is important to note that these are typical values, and they could vary quite considerably depending on the presence of impurities (or dopants) and the method of growing these crystals, among other factors. For these values of electro-optics coefficients ($\sim10^{-11}$ m/V), an applied dc voltage of $\sim10,000$ V is needed to create a phase shift of $\sim\pi$ in a crystal of centimeter length. By comparison, in liquid crystal electro-optics devices, the typical ac voltage needed is around 1 V and the liquid crystal thickness is on the order of a few microns.

6.2.3. Polarizers and Retardation Plate

Most liquid crystal optical devices make use of the birefringence of the liquid crystal in combination with a multitude of optical elements to control, modulate, or switch

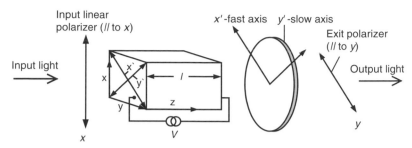

Figure 6.3. Typical electro-optic modulation scheme with polarizer–analyzer sandwiching an electro-optics crystals and a retardation plate.

the light. Figure 6.3 illustrates a standard electro-optical setup for modulating the intensity of light, where an electro-optics crystal is sandwiched between polarizers and retardation (phase) plate.

Linear polarizers are usually made of anisotropic absorbing materials[3] in which the absorption along a crystalline axis is much stronger than the orthogonal axis (which is usually called the transmission axis). Unpolarized light incident on the polarizer will emerge as light polarized along the transmission axis. Accordingly, in going through two crossed polarizers, the light will be extinguished, whereas maximal light transmission will occur if the two polarizers are oriented with their transmission parallel to each other.

Circular polarizers are usually made by putting in tandem a linear polarizer and a birefringent retardation plate, with the polarization vector bisecting the so-called fast and slow axes of the retardation plate (cf. Fig. 6.3 without the electro-optics crystal). Upon entering the retardation plate, the linearly polarized light will break up into two components along these axes. As a result of the difference in the refractive indices associated with the "fast" and "slow" axes, these two components will pick up different phase shifts and the light going through the retardation plate will acquire various polarization states depending on the phase shift. Figure 6.4 shows the resultant optical electric field upon traversing the linear polarizer + retardation plate for various exemplary values of phase difference between the fast and the slow axis components.

For $+\pi/2$ or $-\pi/2$ (i.e., quarter wave) phase shift, the two components add up to a circularly polarized light. For π or $-\pi$ phase shift, the light is again linearly polarized, but along a direction that is 90° with respect to the incident linear polarization direction. For other values of the phase difference, the light acquires elliptical polarizations.

Note on conventions: In this book, we use the convention typically used in optics texts—from the observer's point of view with light propagating towards the observer to denote the handedness of the circular or elliptical polarizations. Also, unless otherwise specifically stated, we express a plane propagating wave as $\sim e^{i(\omega t - kz)}$, where ω is the frequency and k is the wave vector.

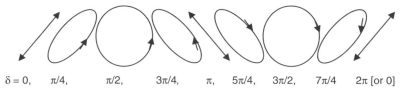

$$\delta = 0, \quad \pi/4, \quad \pi/2, \quad 3\pi/4, \quad \pi, \quad 5\pi/4, \quad 3\pi/2, \quad 7\pi/4 \quad 2\pi \text{ [or 0]}$$

Figure 6.4. Various states of polarization resulting from the addition of two orthogonal components of a polarized light with a relative phase shift.

6.2.4. Basic Electro-Optics Modulation

Consider the electro-optics switching/modulation scheme as depicted in Figure 6.3. Upon emergence from the input polarizer, the linearly polarized plane wave can be expressed as

$$\mathbf{E} = \mathbf{A}_i \exp i(\omega t - kz), \tag{6.6}$$

where $\mathbf{A}_i$ is its amplitude, $(\omega t - kz)$ is the phase, ω is the frequency of light, and k is the wave vector given by $k = n2\pi/\lambda$, with n the refractive index. The polarization vector is oriented at $45°$ with respect to the crystalline axes x' and y'. Upon entering the crystal, the light wave breaks down into two orthogonal components (this is equivalent to saying that the crystal allows the generation of these two waves), one along the x axis and one along the y axis.

$$E_x = (A_i \cos 45°) \exp i(\omega t - k_x z) = A \exp i\varphi_x, \tag{6.7}$$

$$E_y = (A_i \sin 45°) \exp i(\omega t - k_y z) = A \exp i\varphi_y, \tag{6.8}$$

where $k_x = n_x 2\pi/\lambda$ and $k_y = n_y 2\pi/\lambda$.

At the exit plane of the crystal of length l, these electric field components become

$$E_x(l) = A \cos(\omega t - k_x l), \tag{6.9}$$

$$\begin{aligned} E_y(l) &= A \cos(\omega t - k_y l) \\ &= A \cos(\omega t - k_x l - \Gamma_{\text{crystal}}), \end{aligned} \tag{6.10}$$

where $\Gamma_{\text{crystal}} = \varphi_x - \varphi_y$ is the phase difference between x and y,

$$\Gamma_{\text{crystal}} = (n_y - n_x)\frac{l2\pi}{z}. \tag{6.11}$$

The effect of the retardation plate on the $e-o$ crystal is to impart a further phase shift [e.g., $\pi/2$ phase shift for a quarter wave ($\lambda/4$) plate] on the y component. The total phase shift between the y and x components thus becomes

$$\Gamma = \Gamma_{\text{crystal}} + \text{phase shift by retardation plate.} \tag{6.12}$$

The electric field components after the retardation plate become

$$E_x(d) = A\cos(\omega t - k_x l), \tag{6.13}$$

$$E_y(d) = A\cos(\omega t - k_x l - \Gamma). \tag{6.14}$$

Depending on the value of Γ, the polarization state of the resulting electric field vector $E = E_x + E_y$ upon emergence from the retardation plate could be elliptical, linear, or circular, see Figure 6.4.

By summing the components of E_x and E_y on the transmission axis of the output polarizer (along y), a straightforward analysis gives the output intensity

$$I_o = A_i^2 \sin^2\left(\frac{\Gamma}{2}\right). \tag{6.15}$$

The intensity of the output beam from the output polarizer thus varies according to the value Γ, which is electrically controlled through the phase shift imparted by the electro-optic crystal, Γ_{crystal}.

6.3. ELECTRO-OPTICS OF NEMATIC LIQUID CRYSTALS

In general, the distortions on the electronic wave function of liquid crystal molecules caused by an applied field do not cause appreciable change to its contribution to the refractive indices (see Chapter 10). However, the orientation of the molecules can be dramatically altered by the applied field. This process alters the overall optical properties of the medium and is the principal mechanism used in liquid-crystal-based electro-optical devices. As noted in Section 6.2.2, the electrically induced orientational refractive index changes could be Pockel or Kerr effect. In this and the next sections, we shall focus on nematic liquid crystals in which the director axis reorientation is a Kerr-like effect; that is, the process is quadratic in the applied field.

6.3.1. Director Axis Reorientation in Homeotropic and Planar Cells: Dual-Frequency Liquid Crystals

As an illustration, consider the two typical nematic liquid crystal cells as depicted in Figure 1.18 in which the director axis is aligned in the homogeneous (or planar) and homeotropic states. For electro-optic application, the cell windows are typically coated with a transparent conductor such as indium tin oxide (ITO) to allow the application of an electric voltage (field) across the cell. In general, ac voltages are employed to avoid current flows, liquid crystal orientation instabilities, heating, and other electrochemical effects associated with a large dc field that will degrade the liquid crystals.[4]

The response of the nematic liquid director axis to the applied field depends on the dielectric anisotropy. For positive uniaxial liquid crystals (i.e., $n_e > n_o$), the director axis will tend to align along the electric field, whereas in negative uniaxial liquid crystals, the director axis will tilt away from the electric field. To produce field-induced director axis reorientation, one would employ positive uniaxial nematic liquid crystal in planar alignment and negative uniaxial nematic liquid crystal in homeotropic alignment. In the latter case, since the molecules are free to rotate in any (random in fact) direction away from the homeotropic state, usually some small pretilt is imparted on the cell surface to "guide" the molecular reorientation towards the pretilt direction when the field is applied.

It is important to bear in mind that the real liquid crystals in practical use are usually not single constituents or purely dielectric. Using special dopants or by mixing different liquid crystals, a final "compound" of more desirable physical, optical, and electro-optical properties could be obtained. For example, if the liquid crystal is doped with traces of materials that could modify the elastic constants K_{ii} or dielectric anisotropy $\Delta\varepsilon$ (usually in a manner that does not affect other desirable properties such as temperature range, stability against heat and moisture, etc.), then better response times and lower switching threshold could be realized. Figure 6.5 shows capacitance measurements of an undoped (5CB) planar aligned cell and a CdSe nanorod[5] doped 5CB cell. For a planar sample, the capacitance C is proportional to the effective dielectric constant of the liquid crystal sandwiched between two conducting electrodes (plates), that is, $C = \varepsilon_{LC}\varepsilon_0(A/d)$, where A is the area of the plate and d is the cell gap. For a planar sample at low voltage, the value at low applied voltage is $\varepsilon_\perp$, and at high voltage, the value approaches $\varepsilon_\parallel$. The figure clearly indicates that the CdSe-doped LC possesses larger dielectric anisotropy. This translates into a lower switching threshold in Freedericksz transition as well as orientational dc-optical field induced photorefractive effect.[6]

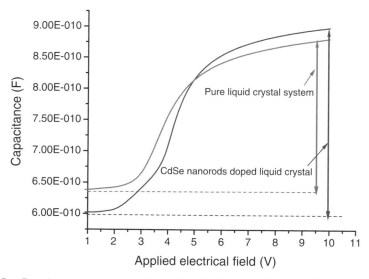

Figure 6.5. Capacitance measurements of an undoped 5CB and a CdSe-doped 5CB planar aligned cell as a function of the applied ac voltage. Cell thickness: mm, ac frequency: Hz. Note that the capacitance value corresponds to Cp and Cpara for low and high voltage, respectively.

Since the dielectric anisotropy is frequency dependent (cf. Fig. 3.5), one could create a mixture of liquid crystals with different dielectric dispersions such that the resulting so-called dual-frequency liquid crystal (DFLC) possesses an effective positive anisotropy at one frequency of the applied ac electric field, but possesses a negative anisotropy at another ac frequency.[7] In that case, a planar aligned cell, for example, could be switched to and fro the homeotropic alignment state by switching the frequency of the applied field. Since the response times of the director axis are dependent on the applied voltage, DFLC could allow one to switch between the two alignment states at a faster rate than the director axis's natural (no voltage applied) relaxation time.

6.3.2. Freedericksz Transition Revisited

Consider the three typical aligned nematic liquid crystal cells as depicted in Figures 6.6a–6.6c corresponding to planar, homeotropic, and twisted NLCs of positive anisotropy. With the applied electric field shown, they correspond to the splay, bend, and twist deformations in nematic liquid crystals. Strictly speaking, it is only in the third case that we have an example of pure twist deformation, so that only one elastic constant K_{22} enters into the free-energy calculation. In the first and second cases, in general, substantial director axis reorientation involves some combination of splay (S) and bend (B) deformations; pure S and B deformations, characterized by elastic constants K_{11} and K_{33} respectively, occur only for small reorientation.

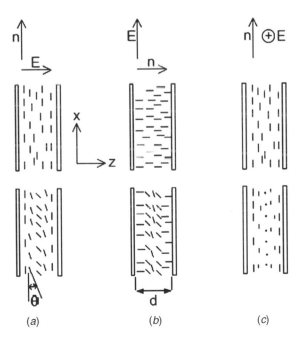

Figure 6.6. Geometry for observing (a) the S (splay) deformation, (b) the B (bend) deformation, and (c) the T (twist) deformation. The upper row shows the initial orientation, and the bottom row shows the deformation when the field exceeds the threshold values.

Case 1: One-elastic-constant approximation. For the simplest case under one-elastic-constant approximation and strong boundary anchoring condition, the free energy of the NLC layer can be obtained by integrating over the sample thickness d

$$G = \frac{1}{2} \int_0^d \left[K \left(\frac{d\theta}{dz} \right)^2 - \frac{\Delta\varepsilon}{4\pi} E^2 \sin^2\theta \right] dz, \tag{6.16}$$

where θ is the reorientation angle, K is the elastic constant, E is the applied field, and $\Delta\varepsilon = \varepsilon_\parallel - \varepsilon_\perp$. The solution of Equation (6.16) can be obtained using a standard variation method,

$$K \frac{\partial^2\theta}{\partial z^2} + \frac{\Delta\varepsilon}{4\pi} E^2 \sin\theta \cos\theta = 0. \tag{6.17}$$

The relation between the reorientation angle and external field can be derived by integrating Equation (6.17),

$$d = \int_0^{\theta_m} \left[c + \frac{\Delta\varepsilon E^2}{8\pi K} \cos(2\theta) \right]^{1/2} d\theta. \tag{6.18}$$

θ_m is the maximum reorientation angle. Constant c can be determined by minimizing the energy G. Then we derive the final expression relating θ and E,

$$\frac{Ed}{2} \left(\frac{\Delta\varepsilon}{4\pi K} \right)^{1/2} = \int_0^{\theta_m} \frac{d\theta}{\left(\sin^2\theta_m - \sin^2\theta \right)^{1/2}} = F(k). \tag{6.19}$$

The right-hand side of Equation (6.18) is an elliptic integral of the first kind, which is tabulated for values of $k = \sin\theta_m \leq 1$. For relatively small reorientation angles, the elliptic integral can be expanded as a series. Equation (6.18) can be approximated to the second-order,

$$E = \left(\frac{4\pi K}{\Delta\varepsilon} \right)^{1/2} \frac{\pi}{d} \left(1 + \frac{1}{4} \sin^2\theta_m + \cdots \right). \tag{6.20}$$

Equation (6.20) shows that a deformation with $\theta_m \neq 0$ occurs only if the external fields exceed the threshold $E > E_F$, where

$$E_F = \left(\frac{4\pi K}{\Delta\varepsilon} \right)^{1/2} \frac{\pi}{d}. \tag{6.21}$$

From Equation (6.21), the Freedericksz threshold voltage $V_F = E_F d = (4\pi K/\Delta\varepsilon)^{1/2} \pi$ is independent of sample thickness. For small director orientation, the maximum reorientation angle θ_m may be obtained from Equation (6.20):

$$\frac{V}{V_F} = 1 + \frac{1}{4}\sin^2\theta_m + \cdots. \tag{6.22}$$

Case 2: Freedericksz transition voltage including elastic anisotropies. Taking into account the anisotropic properties of the elastic constant (i.e., different K_{11}, K_{22}, and K_{33}), Equation (6.17) becomes

$$\frac{d}{dz}\left[(K_{11}\cos^2\theta + K_{33}\sin^2\theta)\frac{d\theta}{dz}\right] - (K_{33} - K_{11})\sin\theta\cos\theta\left(\frac{d\theta}{dz}\right)^2$$
$$= -\frac{\Delta\varepsilon}{4\pi}E^2\sin\theta\cos\theta. \tag{6.23}$$

Following similar steps as in case (i), we can obtain

$$\frac{V}{V_F} = 1 + \frac{1}{4}\left(1 + \kappa + \frac{\Delta\varepsilon}{\varepsilon_\perp}\right)\theta_m^2 + \cdots, \tag{6.24}$$

where

$$\kappa = \frac{K_{33} - K_{11}}{K_{11}} \tag{6.25}$$

and the Freedericksz transition voltage $V_F = (4\pi K/\Delta\varepsilon)^{1/2}\pi$.

Case 3: Freedericksz transition voltage including elastic conductivity. If the nematic liquid crystal layer is conductive, as a result of dopants or presence of impurities, Equation (6.23) changes slightly under the influence of electrical conductivity. The detailed lengthy derivation can be found in reference 8. We cite the final result as follows.

The maximum reorientation angle θ_m is described by

$$\frac{V}{V_F} = 1 + \frac{1}{4}(1 + \kappa + a)\theta_m^2 + \cdots, \tag{6.26}$$

where $V_F = (4\pi K/\Delta\varepsilon)^{1/2}\pi$, $\kappa = (K_{33} - K_{11}/K_{11})$ accounts for elastic anisotropy, and

$$a = \left[\frac{\sigma_\| - \sigma_\perp}{\sigma_\perp} + \left(\frac{\omega}{\omega_c}\right)^2\frac{\varepsilon_\| - \varepsilon_\perp}{\varepsilon_\perp}\right]\left[1 + \left(\frac{\omega}{\omega_c}\right)^2\right]^{-1} \tag{6.27}$$

with $\omega_c = 4\pi\sigma_\perp/\varepsilon_\perp$.

The Freedericksz transition voltage is identical to the case of nonconducting nematic liquid crystals. However, the presence of electrical conductivity does affect the director orientation for external voltage above threshold. For voltage above threshold V_F, the liquid crystal deformation depends on the ratio of elastic constants K_{33}/K_{11}, on the electric conductivity, and on the applied frequency. For very high frequency $\omega \to \infty$, the coefficient $a \to \Delta\varepsilon/\varepsilon_\perp$, the NLC becomes nonconductive. But for very low frequency $\omega \to 0$, the coefficient $a \to \Delta\sigma/\sigma_\perp$, the electric conductivity becomes dominant.

It is important to note that, in the above treatments, we have ignored flows and other possible conduction-induced effects and instabilities in the director axis alignments, which will occur if dc fields are used. For these and other practical reasons, ac fields are invariably the preferred choice in liquid crystal devices.

6.3.3. Field-Induced Refractive Index Change and Phase Shift

As a result of the ac field-induced director axis reorientation from the initial value, a polarized light going through the nematic liquid crystal cell as an extraordinary wave will experience refractive index changes and phase shifts, following the discussions in preceding sections. As shown in the preceding section, owing to the hard-boundary conditions, that is, $\theta = 0$ at the cell surfaces, the director axis reorientation angle θ varies as a function of the propagation distance into the cell, see Figure 6.7, with the maximum orientation angle at the center of the cell. Accordingly, the phase shift

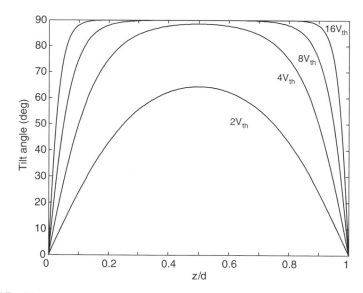

Figure 6.7. Director axis reorientation profile in the cell at various applied voltage above the Freedericksz transition.

experienced by the light after its passage through the liquid crystal cell is given by an integral of the form

$$\Delta\phi = \int_0^d \frac{2\pi}{\lambda}[n_e(z,\theta) - n_o]dz. \tag{6.28}$$

In actual devices where the precise value of the phase shift is needed, the actual phase shift has to be calculated with quantitative numerical simulation of the director axis reorientation profile. For analysis purposes (e.g., estimation of required voltage, thickness dependence, etc.), the director axis reorientation profile that obeys the hard-boundary condition ($\theta = 0$ at the boundary surfaces) can be approximated by a sinusoidal function of the form

$$\theta = \theta_0 \sin\left(\frac{\pi z}{d}\right) \tag{6.29}$$

for the planar and homeotropic cells discussed previously.

In many current liquid crystal display devices, the liquid crystal director axis is in the twisted configuration. The 90° twisted cell is shown in Figure 6.6c. It consists of two cell window surfaces that are treated to favor planar alignment of the director axis, with the director axis rotated by an angle $\theta = 90°$ from one cell window to the other. In the supertwist nematic (STN), the rotation of the director axis from one surface to the other can vary from 180° to 270°.[9] STN director axis arrangement is achieved by having large pretilt angles with the aid of chiral dopants.

Clearly, because of these more complex director axis initial alignments, their subsequent reorientation by an applied field (usually applied perpendicular to the cell surfaces) is more complicated, although it is still governed by the basic physical principle that the director axis will tend to assume a new configuration to minimize the total free energy. Consider, for example, the simpler case of a 90° twisted cell. Under an applied field perpendicular to the cell surfaces, the director axis will begin to tilt towards the cell normal, see Figure 6.8. Studies of the phase modulation[10] and transmission of a polarized light by such cell indicate that there are two thresholds corresponding to the initiation of significant changes in the tilt and rotation angles ϕ and θ, respectively. Just above a first threshold V_{tilt} given by

$$V_{\text{tilt}} = \pi \left(\frac{[K_{11} + (K_{33} - 2K_{22})/4]}{\varepsilon_0 \Delta\varepsilon}\right)^{1/2}, \tag{6.30}$$

the director axis will tilt along the applied field direction, while collectively preserving the original twist profile (θ remains unchanged). In this case, the main effect on the incident light is a phase-only modulation for the optical setup used in Figure 6.3. At much higher applied voltage, the rotational angle θ will also be modified by the

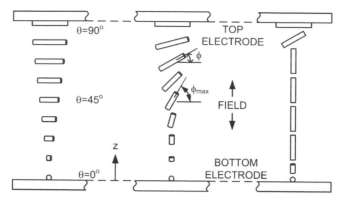

Figure 6.8. Tilting and unwinding of the director axis of a 90° twisted nematic liquid crystal cell under the action of an applied field.

applied field. Eventually, for applied voltage $>> V_{tilt}$, the nematic axis unwinds and $\theta \rightarrow 0$ in the bulk of the cell.

For STN cells, similar studies have identified the corresponding threshold voltages for various pretilt angles and initial twist angles.[11]

6.4. NEMATIC LIQUID CRYSTAL SWITCHES AND DISPLAYS

It is important to note here that liquid crystal display technology is advancing by leaps and bounds, very much like other optoelectronic and micro- or nanoelectronics technologies, as one strives to obtain higher resolution, faster response, wider field of view, larger display area, and more functions in each display pixel. It would require a (continuously updated) treatise to explore the details of the various display designs. Therefore, our focus in this chapter, just as in the rest of this book, is on the fundamental principles governing electro-optics of liquid crystals in exemplary configuration and cell designs. These and the computation methods described in the next chapter will enable one to apply to specific design or configuration independent of advancing time.

There are two types of LC switches for display application: transmissive and reflective. Both types of display make use of the polarizing and birefringent properties of liquid crystals, in conjunction with polarizers and phase (retardation) plates, see Figures 6.1 and 6.3. The unique physical properties of liquid crystals such as their broadband (from near UV to far infrared) birefringence, and transparency make them the preferred materials for display and many other optical devices. Because of the large birefringence, it requires only a very thin cell to create the necessary phase shift for optical devices. For example, with a Δn of 0.2 operating at the visible region (e.g., $\lambda = 500$ nm), the cell thickness d needed to produce a phase shift of π is about 1.5 μm, while the voltage required is $\sim$ 1 V.

6.4.1. Liquid Crystal Switch: On-Axis Consideration for Twist, Planar, and Homeotropic Aligned Cells

In Section 6.2.4, we discuss a polarizers+electro-optics crystal setup (Fig. 6.3) commonly used for electro-optical intensity modulation purposes.

Consider, for example, a 90° twisted aligned cell, cf. Figure 6.6c in place of the electro-optics crystal. In the normally black (NB) operation, the cell is placed between two parallel polarizers (i.e., without the quarter wave plate). In passing the cell, the polarization vector of the light, initially pointing along the x direction, will follow the director axis orientation and exit the cell with its polarization along the y direction, and therefore it will be extinguished by the output polarizer whose transmission axis is along the y direction. On the other hand, if the output polarizer is orthogonal to the input, then the y-polarized light will be maximally transmitted— the so-alled normally white (NW) mode. For both types of polarizer orientations (crossed or parallel), the transmission states can be switched from one to the other if the director axis of the twisted nematic (TN) cell is "unwound" by a sufficiently strong applied field to the homeotropic state, see Section 6.3.3, Equation (6.30). The homeotropic state preserves the polarization of the on-axis light, and therefore the exit light will remain x polarized. As a result the NB pixel will be switched to the bright state and the NW pixel will be switched to the dark state.

In the case of homeotropic or planar aligned cell, it functions very much as the electro-optics crystal, namely, as an electrically tunable phase shifter placed within the polarizer/retardation plates setup as shown in Figure 6.3. Initially, if placed between two crossed polarizer, the e and o waves see the same index (for on-axis-propagation), and thus, the output is dark—the NB mode. When the director axis is reoriented by the applied field, it will impart a phase shift between the e and o waves, and the transmission will be "modulated" accordingly, see Equation (6.15).

6.4.2. Off-Axis Transmission, Viewing Angle, and Birefringence Compensation

The preceding discussions on liquid crystal electro-optics switch are for on-axis light propagation. These considerations would suffice to modulate the laser beam made to propagate along the axis. However, for display applications, the backplane illumination is diffuse, and thus, off-axis propagation of light through the cell has to be taken into account. With references to all the modulation setups discussed in the preceding section, it is clear that for off-axis light, the transmission function T is now a function of many variables.

Consider the TN cell as depicted in Figure 6.6c in the NB operation. For on-axis light, the initial transmission is 0. When the voltage is on, the transmission is at a maximum for the on-axis light, see preceding section. However, for the off-axis light, the e and o waves will pick up an extra phase shift because of the extra optical path length through the cell, see Section 6.2.4. The transmission is therefore now dependent on the angle of incidence and wavelength of the light, as well as the director axis reorientation angles θ and ϕ, which are in turn functions of the applied voltage V, such that, $T = T(\theta, \phi, \lambda, V)$, and is not necessarily at the maximum value.

This off-axis consideration clearly also applies to the simpler case of a homeotropically aligned cell, Figure 6.6b; that is, off-axis light will "pick up" polarization vectors not exactly aligned with (or perpendicular to) the output analyzer, so that what is supposed to be a bright (or dark) output could now have some finite transmission. This is often referred to as light leakage effect, and is simply due to the extra phase shift experienced by the off-axis light.

Since the problem of light leakage is caused by the phase shift experienced by light propagating through the birefringent liquid crystals, various methods of correcting for this effect based on introducing a birefringent compensation film[12] have been introduced. One simple but effective means is to place a birefringent film (of opposite anisotropy to that of the liquid crystal) adjacent to the LC film, so that when polarized light traverses the liquid crystal layer and the birefringent, so-called compensation, film, it effectively experiences net zero phase shift.

To illustrate the above statements, consider, for example, the arrangement as depicted in Figure 6.9. For a given angle of incidence (defined by θ) and a director axis orientation (defined by ϕ), the effective refractive index as seen by the extraordinary ray is given by Equation (6.2b),

$$n_e(\theta, \phi) = n_e n_o / [n_e^2 \cos^2(\phi + \theta) + n_o^2 \sin^2(\phi + \theta)]^{1/2}. \tag{6.31}$$

This gives a phase shift $\Delta\phi$ between the extraordinary and ordinary waves (which sees an index of n_o),

$$\Delta\phi = \left(\frac{2\pi}{\lambda}\right)\left(\frac{d}{\cos\theta}\right)[n_e(\theta, \phi) - n_o]. \tag{6.32}$$

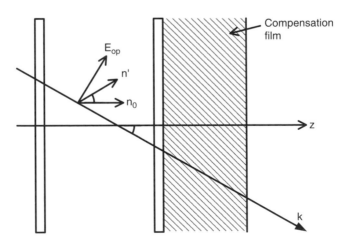

Figure 6.9. Light propagating through a initially homeotropic LC cell. The director axis is tilted by an ϕ with respect to the initial alignment and the propagation direction makes an θ with respect to the z axis.

Consider the limiting case of $\phi = 0$. This corresponds, for example, to the director axis reorientation of a TN cell at high applied voltage, or a homeotropic cell at $V = 0$. In that case, we have

$$\Delta\phi = \left(\frac{2\pi}{\lambda}\right)\left(\frac{d}{\cos\theta}\right)\left[\left(\frac{n_e n_o}{[n_e^2 \cos^2(\theta) + n_o^2 \sin^2(\theta)]^{1/2}}\right) - n_o\right]. \tag{6.33}$$

Clearly, the phase shift is dependent on the film thickness d, the wavelength λ, and the angle of incidence θ. Since θ can be positive or negative depending on whether the light comes from either side of the z axis, the phase shift is thus asymmetrical with respect to the angle of incidence. Accordingly, the compensation film should possess the corresponding asymmetric angular dependent phase retardation property. Also, the compensation film should have birefringence $[n_e(\theta) - n_o]$ of opposite sign to that of the liquid crystal.

For arbitrary angle ϕ or director axis angular and spatial distributions, and more complicated cell structure, the phase shift, and therefore the transmission of light through the cell and other accompanying polarization selective elements, is not amenable to simple analytical treatment. More sophisticated Jones matrix methods or numerical technique such as the finite difference time domain (FDTD) numerical methods discussed in the next chapter are needed to solve such a complex propagation problem.

6.4.3. Liquid Crystal Display Electronics

Figure 6.1 depicts a typical display panel comprising rows and columns of pixels. Each pixel typically measures several microns and consists of an aligned nematic (or ferroelectric) liquid crystal between polarizers, phase plates, color filter, mirror, etc., in conjunction with an electronic thin film transistor (TFT) circuitry. Information such as images are transmitted via the electronic circuitry using either direct or matrix addressing scheme.[13]

In the direct addressing scheme, the external electrode of the segment substrate is directly connected to each element of the display area, whereas the other substrate has the common ground electrode. This method is only suitable for the low-information-content displays due to the very high cost associated with many-pixel operation.

There are two types of matrix addressing schemes—passive and active. The passive matrix (PM) addressing scheme requires the row and column electrodes to address each individual pixel. This scheme still promises well in the area of bistable device such as ferroelectric liquid crystal (FLC) display and bistable twisted nematic (BTN) display because they do not need a control unit for gray-scale capability. The active matrix (AM) addressing scheme is the most developed and widely adopted one in current LC displays. In this scheme, each pixel is connected to a small electronic switch or TFT made with a-Si, poly-Si, or CdSe. This switch not only enables the pixel to hold the video information until it can be refreshed, but also prevents cross talk among neighboring addressed pixels.

6.5.   ELECTRO-OPTICAL EFFECTS IN OTHER PHASES OF LIQUID CRYSTALS

Although by far nematics are the most extensively used ones, other phases (smectic, cholesteric, etc.) of liquid crystals and "mixed systems" such as polymer-dispersed liquid crystals capable of field-induced reorientation have also been employed for electro-optical studies and applications. They are basically based on the same basic mechanism of field-induced director axis reorientation similar to nematic liquid crystals; i.e., the response is Kerr like in that it is independent of the direction of the electric field. In general, nematic liquid crystal electro-optics devices switch at a rate of several tens of hertz, corresponding to response times from a few to tens of microseconds.

On the other hand, ferroelectric liquid crystals operate very much like Pockel cells; their response depends linearly on the applied electric field vector, that is, they depends on both the magnitude and direction of the electric field. They generally switch faster than nematic cells. To date ferroelectric liquid crystals possess a switching speed of ~ 1 μs or less.

6.5.1.   Surface Stabilized FLC

Recall that in the one-elastic-constant approximation $(K_{11}=K_{22}=K_{33}=K)$, the dynamical Equation (4.92) for a ferroelectric liquid crystal under an applied field, see Figure 4.10, is given by

$$K \sin^2 \theta \frac{\partial^2 \phi}{\partial y^2} + \frac{e}{2} \varepsilon_\perp E^2 \sin 2\phi + P_s E \sin \phi = \gamma_1 \frac{\partial \phi}{\partial t}. \tag{6.34}$$

For a typical FLC material and device: $K \sim 10^{-11}$ N, $\theta \sim 20°$, $\varepsilon_\perp \sim 10^{-11}$ F/m, $E \sim 10^6$ V/m, $P_s \sim 10^{-5}$ C/m², and $e = \cos^2 \theta + (\varepsilon_\parallel/\varepsilon_\perp)\sin^2 \theta - 1$. If e is appreciable so that the magnitude of the dielectric term [second term on left-hand-side of Eq. (6.34)] may be comparable to that of the spontaneous polarization (third) term, then the elastic term becomes the smallest among the three. Under this assumption, Equation (6.34) becomes

$$\frac{1}{2} \Delta\varepsilon E^2 \sin 2\phi + P_s E \sin \phi = \gamma_1 \frac{\partial \phi}{\partial t}. \tag{6.35}$$

A solution of Equation (6.35) is given by[14]

$$\frac{t}{\tau} = \frac{1}{1-a^2} \left\{ \ln \frac{\tan(\phi/2)}{\tan(\phi_0/2)} + a \ln \frac{(1+a\cos\phi)\sin\phi_0}{(1+a\cos\phi_0)\sin\phi} \right\}, \tag{6.36}$$

where $a = \Delta\varepsilon E \sin^2 \theta / 2P_s$, $\phi_0 = \phi(t=0)$, and τ is the response time

$$\tau = \frac{\gamma_1}{P_s E}.$$ (6.37)

In practical application, a SSFLC film is arranged such that the input light polarization is parallel to the director and maintained at one of the bistable states, for example, the "up" state, as shown in Figure 6.10a. The input and output polarizers are crossed, that is, a dark state. When the sign of the electric field is reversed, the director is rotated from the up to the down state, leading to some transmitted light, that is, bright state (see Fig. 6.10b). The normalized transmission is proportional to the tilt angle (θ) and phase retardation $\delta(=2\pi d\Delta n/\lambda)$:

$$T_\perp = \sin^2(4\theta)\sin^2\left(\frac{\delta}{2}\right).$$ (6.38)

The optimal transmission occurs at $\theta=22.5°$ and $\delta=\pi$. In order to achieve a uniform rotation, the SSFLC layer thickness is often limited to about 2μm. As mentioned before, FLC cells, in general, switch faster than nematic cells; the response time $\gamma_1/P_s E$ ranges from 1μs to several tens of microseconds, owed largely to the lower rotational viscosity, larger spontaneous polarization, and higher switching fields.

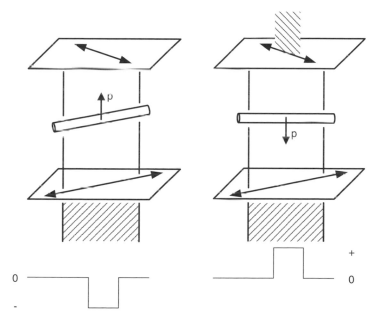

Figure 6.10. Schematic of a ferroelectric liquid crystal light switch with polarized light and crossed polarizer sandwiching the cell for two applied fields in opposite directions.

6.5.2. Soft-Mode FLCs

Soft-mode FLC (SMFLC) was introduced[15] as an alternative way to switch FLC. In this case, instead of varying the azimuthal angle (ϕ) around the tilt cone, i.e., θ is constant as in SSFLC, SMFLCs use changes in the tilt (θ) while ϕ remains constant. As a result, SSFLCs exhibit bistability, but SMFLCs are capable of continuous intensity change. The SMFLCs employ smectic-A* phase, and uniform alignment is much easier to obtain for SMFLCs than for SSFLCs, which employ smectic-C*.

The experimental setup for realizing the soft-mode ferroelectric effect is depicted in Figure 6.11. For this geometry, the free-energy density is given by[15]

$$F = \tfrac{1}{2} a\theta^2 + \tfrac{1}{4} b\theta^4 - \mu E\theta \cos\phi - \tfrac{1}{2}\varepsilon_0 \Delta\varepsilon E^2, \qquad (6.39)$$

where $a = \mu/a(T - T_c)$, μ is the structure coefficient which is equivalent to the dipole moment per unit volume for unit tilt angle, a and b are expansion coefficients (both a and b are positive for a second-order transition), and ϕ is the angle between the electric field and the direction of the ferroelectric polarization. The last term is the dielectric free energy.

In the small angle approximation, the θ^4 term can be neglected. Choosing $\phi = 0$, a solution of θ is as follows:

$$\theta = e_c E, \qquad (6.40)$$

$$e_c = \frac{\mu}{\alpha(T - T_c)}, \qquad (6.41)$$

where e_c is called the electroclinic coefficient. At a given temperature, the field-induced tilt angle is proportional to the applied electric field E.

The dynamic behavior of the SMFLC is governed by equating the viscous torque:

$$\Gamma^v = -\gamma_\theta \frac{\partial\theta}{\partial t} \qquad (6.42)$$

to the elastic torque:

$$\Gamma^\theta = -\frac{\partial F}{\partial \theta} = -\alpha(T - T_c)\theta - b\theta^3 + \mu E \qquad (6.43)$$

to yield

$$\Gamma_\theta \frac{\partial\theta}{\partial t} = \mu E - \alpha(T - T_c)\theta - b\theta^3. \qquad (6.44)$$

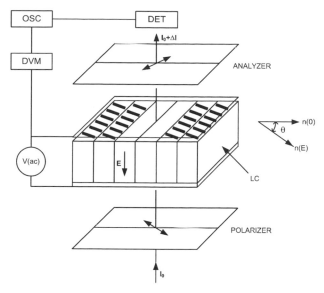

Figure 6.11. A SMFLC switching set-up.

In the low field limit, the $b\theta^3$ term can be neglected and Equation (6.44) has the following solutions:

$$\theta(t) = \frac{\mu E}{\alpha(T - T_c)}\left\{1 - 2\exp\left[-\frac{\alpha(T - T_c)t}{\gamma_\theta}\right]\right\},\tag{6.45}$$

where the response time τ is given by

$$\tau = \frac{\gamma_0}{\alpha(T - T_c)}.\tag{6.46}$$

For high field, the $b\theta^3$ term has to be included and the field-dependent response time becomes more apparent:

$$\tau = \frac{\gamma_\theta}{\alpha(T - T_c)} + O(E^2).\tag{6.47}$$

Studies[16] have shown that near the smectic A–smectic C* transition, the response time is dependent on temperature and the applied field: $\tau \sim E^{-2x}$ with $1/3 \leq x \leq 1$. Response times as fast as 1 μs have been demonstrated.

To date, many more "modes" of switching operations for FLC cells have been developed, such as the deformable helix effect analog switching,[17] the "Sony-mode"

V-shaped switching,[18] "continuous director rotation (CDR) half-V switching,"[19] and various other ways of making a SSFLC bistable cell behave V-shaped, each with its own special characteristics.[20–23]

6.6.   NONDISPLAY APPLICATIONS OF LIQUID CRYSTALS

Besides display-type devices, liquid crystals find increasing uses in other electro-optical devices throughout an extremely broad spectral range (from near UV to far infrared and into the microwave regime). Their fluid nature and compatibility with most optoelectronic materials allow them to be easily incorporated into other device elements in various configurations, forms, shapes, and sizes. It is not surprising that a whole host of tunable lens, filters, switches, and beam/image processing devices have emerged.[24–27] Because of their organic nature, liquid crystals have also recently emerged as good candidates for biochemical sensing applications.[28] As electronic materials, liquid crystals have also been employed in the development of light emitting diodes and electroluminescence devices.[29] A complete exposition of these applications will require a treatise. In the following sections, we will discuss exemplary cases corresponding to distinct mechanisms or device configurations involved.

6.6.1.   Liquid Crystal Spatial Light Modulator

As we have briefly alluded to in Chapter 5, liquid crystal is capable of very low threshold reorientation if the light energy is efficiently transferred to reorient the liquid crystals. One of the approaches is to use liquid crystal in conjunction with highly photosensitive materials such as photoconducting semiconductors such as CdS, ZnS, CdSe, and hydrogenated amorphous silicon (α-Si:H) to construct image sensing and display devices such as spatial light modulators.[30]

Figure 6.12 shows the schematic construction of a typical optically addressed liquid crystal spatial light modulator (OALCSLM) operating in the reflective mode. It consists of an aligned liquid crystal layer sandwiched between transparent electrodes and adjacent to a photoconducting semiconductor layer (α-Si:H) sensitive to the writing beams and a dielectric mirror coated to block the reading beam.

In the absence of light, the semiconductor layer is highly resistive, and thus the bias voltage applied across the electrodes suffers the largest voltage drop across this layer, and less in the LC layer. When illuminated by light, the semiconductor layer becomes conducting, and the voltage drop now takes place across the LC layer. This causes director axis reorientation and all the resulting optical property changes associated with it. Accordingly, the spatial distribution in the incident optical intensity (e.g., an image) will be recorded as an orientation and therefore an optical phase shift spatial profile. A probing beam will sense this phase shift and reproduce the image on the other side.

Figure 6.12 shows how the OALCSLM is used as a four-wave-mixing-based holographic imaging device. The incident signal is coherently mixed with a reference

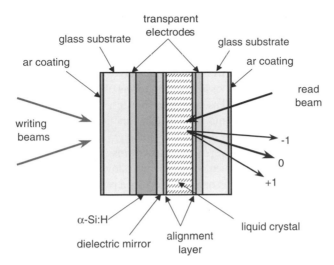

Figure 6.12. Schematic of the various components of a typical optically addressed liquid crystal spatial light modulator (OALCSLM) operating in the reflective mode.

beam and records an intensity interference pattern on the LC from the left side. In the language of holography, this interference pattern may be called a hologram. The holographic interference pattern is transferred to the liquid crystal layer via the photoconducting layer and the bias field. When the hologram is illuminated with a viewing light (the probe beam), a signal or reconstructed image is obtained.

As will be explained in detail in Chapters 11 and 12, in the language of nonlinear optics, this may also be called optical wave front conjugation, where the signal is actually a time-reversed replica of the input image and will emerge after passing through the distorting medium with the phase aberration corrected [30]

From the point of view of nonlinear optics, OALCSLM is actually one of the most nonlinear optical devices ever made as one can see from the following estimate. Typically OALCSLM has a threshold sensitivity of about 50 μW/cm^2. In typical SLM two-beam interference operation, a writing beam optical intensity I on the order of 500 μW/cm^2 is used and the induced diffraction grating generates a first-order diffraction efficiency of 10%. This corresponds to a phase shift experienced by the probe beam $\delta\phi = \Delta n 2\pi d/\lambda = n_2 I 2\pi d/\lambda \sim 0.1\pi$. For a SLM liquid crystal layer thickness d of 5 μm, these values for $\delta\phi$, I, and d mean that the nonlinear coefficient n_2 is on the order of 10 cm^2/W at the optical wavelength $\lambda = 0.5$ μm. Although it is electrically assisted (by the bias voltage) and optically enhanced (by the photoconducting semiconductor layer), OALCSLM is by far the most nonlinear practical device for image/information processing application.

Current thrusts in SLM developments are focused on making faster, higher resolution, higher sensitivity, and broader spectral response devices. These expectations call for developing higher birefringent material, low viscosity, and high speed and high information content micro- or nanoelectronics circuit platforms.

6.6.2.   Tunable Photonic Crystals with Liquid Crystal Infiltrated Nanostructures

We turn our attention here to photonic crystals. Photonic crystals in 1-, 2- and 3D forms made of various optoelectronic materials have received intense research interest owing to the rich variety of possibilities in terms of material compositions, lattice structures, and electronic as well as optical properties.[31–33] By using an active tunable material as a constituent, photonic crystals can function as tunable filters, switches, and lasing devices. In particular, liquid crystals have been employed in many studies[34–36] involving opals and inverse opal structures. However, most of the tuning mechanisms are via thermal means, which are slow and unsuitable for practical devices. Also, the tuning range is quite limited, as a result of the relatively small volume fraction of the tunable material.

Recent studies[37] of non-close-packed inverse opal photonic crystal structure, see Figure 6.13, impregnated with nematic liquid crystal have shown that, due to the higher volume fraction for NLC infiltration, a much larger tuning range (> 20 nm) of the Bragg reflection peak can be achieved, see Figure 6.14.

As one can see from literature dealing with the fabrication of photonic crystals by self-assembly methods, the process is rather tedious. As mentioned in the previous chapter, optical holography[38] offers a quick one-step process, but the method applies only to polymer-dispersed liquid crystal (PDLC) or systems where the structures can be optically written.[39,40]

6.6.3.   Tunable Frequency Selective Planar Structures

A completely different approach to making a 2D patterned structure has been practiced in the field of microwave and electromagnetic antenna designs. These planar

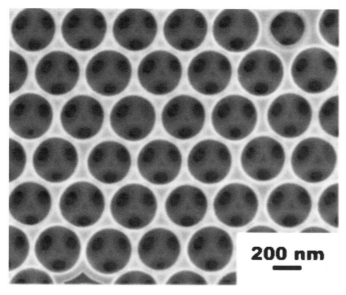

Figure 6.13.   Non-close-packed inverse opal photonic crystal structure.

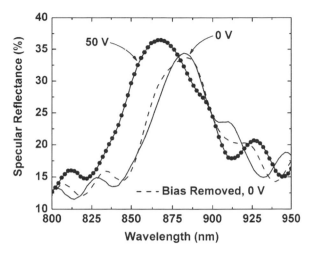

Figure 6.14. Electrical tuning of Bragg reflection peak of nematic liquid crystal filled non-close packed inverse opal photonic crystal.

frequency selective structures[41] are often also referred to as frequency selective surfaces (FSS). Basically they are two-dimensional periodic arrays of metallic patches or aperture elements that possess low-pass, high-pass, bandpass, or multi-band filtering properties for incident electromagnetic waves. Total reflection or total transmission will occur when the wavelength corresponds to the "resonant length" of the PFSS screen elements. The pattern of metallic patches on the dielectric substrate is designed by a genetic algorithm (GA) technique. Details of the GA may be found in previous studies.[42]

To realize low loss visible/IR FSS, the metallic portion of the FSS screen is replaced by suitable dielectric materials which have lesser loss in the visible/IR. An exemplary structure design[39] is given in Figure 6.15a, especially tailored for operation as an optical filter at 600 THz (0.5 µm). The unit cell size of the FSS is 0.393 µm × 0.393 µm, with a thickness of 0.172 µm, and consists of two different dielectric materials in the same layer with $\varepsilon_{r1}=1.82$ and $\varepsilon_{r2}=2.82$. The response for this polarization-independent all-dielectric PFSS is shown in Figure 6.15b, which shows a sharp transmission dip or stop band centered at the desired wavelength of 0.5 µm.

The planar geometry of the FSS allows easy incorporation of a liquid crystal overlayer. The dimensions of the FSS unit cell as shown in the inset of Figure 6.16 are 3.04 µm × 3.04 µm. The FSS is composed of polyimide ($\varepsilon_{r1}=2.6$) blocks (1.14 µm × 1.52 µm) embedded in a layer of silica ($\varepsilon_{r2}=3.9$) with a 1 µm overlayer of liquid crystal. Both the polyimide blocks and silica have a thickness of 1.0 µm. By changing the dielectric constants of the LC from 2 ($n_o=1.414$) to 4 ($n_e=2$), the stop band in the transmission response can be shifted from 92 THz (~ 3.26 µm) to 83 THz (3.64 µm), that is, a tuning range of 380 nm for a birefringence Δn of 0.6. Accordingly, since the birefringence of nematic liquid crystals spans the entire visible to far infrared spectrum, LC-FSS will allow one to design and fabricate very

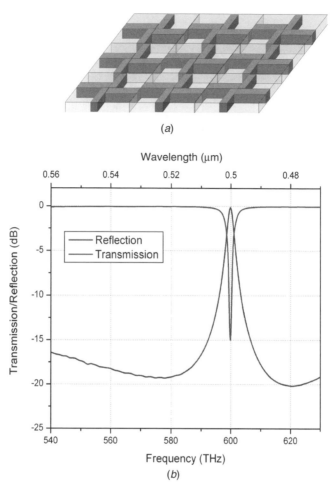

Figure 6.15. (a) Unit cell of an all-dielectric polarization independent FSS for operation in the visible region (λ=0.5 μm; 600 THz) as a stop-band filter. (b) Transmission (upper curve) and reflection (lower curve) spectra for the all-dielectric FSS.

broad band high-extinction-ratio tunable filters/switches, and also negative index materials/structures.[39,43]

6.6.4. Liquid Crystals for Molecular Sensing and Detection

The remarkable physical properties of liquid crystals also make them uniquely suitable for many other types of applications. For example, the color changes in cholesteric liquid crystal as a function of temperature are now widely used in temperature sensing/display.

Sensing and detection can also be done on the molecular level.[28,44–46] These approaches are based on the fact that the macroscopic properties such as alignment,

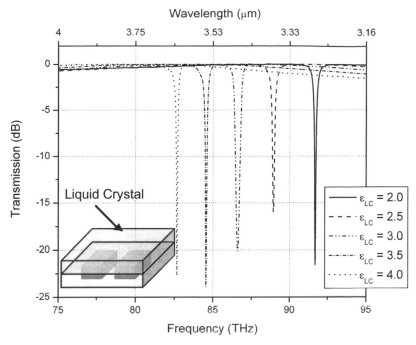

Figure 6.16. Transmission spectra of an all-dielectric FSS with a liquid crystal superstrate. As the dielectric constant of the liquid crystal is tuned from 2 to 4, the structure exhibits broadly tunable filter action.

order parameter, etc., of liquid crystals depend critically on the constituent molecules and their electronic structures and properties, which in turn are extremely sensitive to the presence of other trace molecules.

For example, by exposing a liquid crystal cell to the sensing targets, the effect of trace amounts of adsorbed molecules on the alignment surfaces will give rise to a macroscopic change in the different director axis orientation profiles (cf. Figures 6.17a–6.17d).[44] The director axis reorientation is in turn manifested in an overall change in the optical properties[28] of the liquid crystal cell, for example, color changes for LC cells situated within an interferometer setup. Similar approaches have been employed for detecting ligand–receptor and organoamines–carboxylic acids bindings.

In another work,[44] phase change of homogeneously aligned nematic liquid crystals resulting from absorbed molecules near the phase transition temperature is used in a surface plasmon resonance setup[45] to monitor small changes in the concentration of the analyte. At a certain analyte concentration, the liquid crystal will undergo phase transition, and two surface plasmon resonances will appear. The wavelength shift can be correlated to the relative concentration and the structure/shape of the analyte. Another approach[46] uses a dual-waveguide interferometric technique to monitor the density and thickness of absorbed layers at a solid–liquid interface. A ferroelectric liquid crystal is used to switch the polarization of the input beam without causing appreciable beam movement.

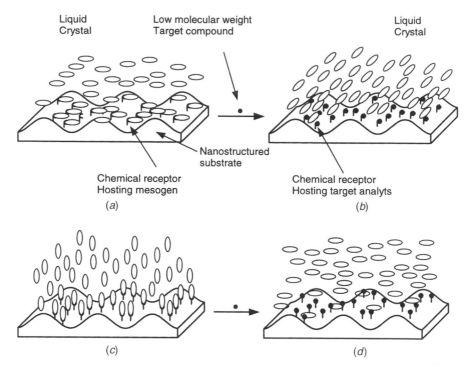

Figure 6.17. Schematic depiction of aligned liquid crystal for chemical sensing. (a) Planar alignment of liquid crystal on nanostructured substrate with receptor hosting mesogen. (b) Target compound creates in-plane rotation of planar alignment. (c) Homeotropic alignment of liquid crystal on nanostructured substrate with receptor hosting mesogen. (d) Target compound creates rotation of homeotropic to planar alignment (after Ref. 44).

6.6.5. Beam Steering, Routing, and Optical Switching and Laser Hardened Optics

We close this chapter with further statements on liquid crystals as a preferred material for optical and electro-optical applications. To date, liquid crystals and related optical technologies have been incorporated in the design and fabrication of filters, lens, waveguides, diffractive and reflective elements, routers and interconnects, etc., of various forms, shapes, and functions used in optical communication system[47] as well as in free-space beam steering systems.[26] Their compatibility with almost all optoelectronic materials as well as polymers and organic materials allows even more possibilities and flexibility in the emerging field of flexible displays[48] and polymer cholesteric liquid crystal flake/fluid display.[49]

Although most optical elements involve low level light, liquid crystals are actually excellent laser-hardened materials capable of handling very intense pulsed lasers or high power continuous wave cw lasers.[50] These studies have shown that common liquid crystals such as 5CB and E7 can withstand a nanosecond laser pulse of ~ 10 J/cm^2 (corresponding to an intensity $\sim 10^{10}$ W/cm^2), thus making them particularly useful to construct high power laser optics such as polarization rotators, wave plates, optical isolators, and laser blocking notch filters. In Chapter 12, we will discuss some current developments of ultrafast self-activated laser blocking and attenuating devices based on nonlinear multiphoton absorption properties of liquid crystals. Under extended illumination of the liquid crystals (E7 or 5CB) with cw laser intensity of several kW/cm^2, for example, in stimulated orientation scatterings with focused lasers (cf. Chapter 12), liquid crystals also do not suffer any structural/chemical damages. These laser-hardened properties of liquid crystals are likely to find an ever increasing usage along with rapid progress in laser/optical technologies.

REFERENCES

1. Yariv, A. 1997. *Optical Electronics in Modern Communications.* 5th ed. New York: Oxford University Press.

2. Born, M., and E. Wolf. 1964. *Principles of Optics.* New York: Macmillan.

3. Shurcliff, W. A. 1962. *Polarized Light.* Cambridge, MA: Harvard University Press.; see also Yeh, P., and C. Gu. 1999. *Optics of Liquid Crystal Displays.* New York: Wiley Interscience.

4. Helfrich, W. 1969. *J. Chem. Phys.* 51:4092.

5. Khoo, I. C., Y. Z. Williams, B. Lewis, and T. Mallouk. 2005. Photorefractive CdSe and gold nanowire-doped liquid crystals and polymer-dispersed–liquid-crystal photonic crystals. *Mol. Cryst. Liq. Cryst.* 446:233–244.

6. Khoo, I. C., Kan Chen, and Y. Zhang Williams. 2006. Orientational photorefractive effect in undoped and CdSe nano-rods doped nematic liquid crystal: Bulk and interface contributions. *IEEE J. Sel. Top. Quantum Electron.* 12(3): 443–450.

7. Kirby, Andrew K., and Gordon D. Love. 2004. Fast, large and controllable phase modulation using dual frequency liquid crystals. *Opt. Express.* 12:1470–1475.

8. Gruler, H., and L. Cheung. 1975. Dielectric alignment in an electrically conducting nematic liquid crystal. *J. Appl. Phys.* 46:5097.

9. See, for example, Scheffer, T., and J. Nehring. 1993. Twisted nematic and supertwisted nematic mode LCDs. In *Liquid Crystals: Applications and Uses*, B. Bahadur (ed.), Singapore: World Scientific.; see also Kataoh, K., Y. Endo, M. Akatsuka, M. Ohgawara, and K. Sawada. 1987. Application of retardation compensation: A new highly multiplexable black-white liquid crystal display with two supertwisted nematic layers. *Jpn. J. Appl. Phys., Part 2* 26:L1784.

10. Konforti, N., E. Marom, and S. T. Wu. 1988. Phase-only modulation with twisted nematic liquid crystal spatial light modulators. *Opt. Lett.* 13:251.

11. See, for example, Raynes, E. P. 1996. The theory of supertwist transitions. *Mol. Cryst. Liq. Cryst. Lett.* 4:1.

12. See, for example, Fergason, J. L. 1983. Liquid crystal display with improved angle of view and response times for TN cell. US Patent 4,385,806, issued May 31,1983. Note that in the onstate, the TN cell resembles a homeotropically aligned cell and so the compensation technique could also apply there.

13. Lueder, E. 2001. *Liquid Crystal Displays: Addressing Schemes and Electro-optical Effects.* New York: Wiley.

14. Xue, J. Z., M. A. Handschy, and N. A. Clark. 1987. Electrooptical switching properties of uniform layer tilted surface stabilized ferroelectric liquid crystal devices. *Liq. Cryst.* 2:707.

15. See, for example, Anderssen, G., I. Dahl, W. Kuczynski, S. T. Lagerwall, and K. Skarp. 1988. The soft mode ferroelectric effect. *Ferroelectrics.* 84:285.

16. Lee, S. D., and J. S. Patel. 1989. Temperature and field dependence of the switching behaviour of induced molecular tilt near the smectic-A-C* transition. *Appl. Phys. Lett.* 55:122–124.

17. Beresnev, L. A., V. G. Chigrinov, D. I. Dergachev, E. P. Poshidaev, J. Funfschilling, and M. Schadt. 1989. Deformed helix ferroelectric liquid-crystal display: A new electrooptic mode in ferroelectric chiral smectic-C liquid crystals. *Liq. Cryst.* 5 (4):1171–1177.

18. Nito, K., H. Takanashi, and A. Yasuda. 1995. Dynamics of FLCs with a tilted bookshelf structure using time-resolved FTIR spectroscopy. *Liq. Cryst.* 19 (5):653–658.

19. Nonaka, T. J. Li, A. Ogawa, B. Hornung, W. Schmidt, R. Wingen, and H.-R. Dübal. 1999. Material characteristics of an active matrix LCD based upon chiral smectics. *Liq. Cryst.* 26(11):1599–1602.

20. Inui, S. N. Iimuro, T. Suzuki, H. Iwane, K. Miyachi, Y. Takanishi, and A. Fukuda. 1996. Thresholdless antiferroelectricity in liquid crystals and its application to displays. *J. Mater. Chem.* 6(4):671–973.

21. Maclennan, J. E. P. Rudquist, R. F. Shao, D. R. Link, D. M. Walba, N. A. Clark, and S. T. Lagerwall. 1999. V-shaped switching in ferroelectric liquid crystals. In *Liquid Crystals III.* Denver, Colorado; Khoo, I. C., ed. 1999. *SPIE,* Opt. Eng 136–139. *Proceedings of* Denver, Colorado: Int. Soc.

22. Clark, N. A. D. Coleman, and J. E. Maclennan. 2000. Electrostatics and the electro-optic behaviour of chiral smectics C: "Block" polarization screening of applied voltage and "V-shaped" switching. *Liq. Cryst.* 27(7):985–990.

23. Rudquist, P. D. Krüerke, J. E. Lagerwall, J. E. Maclennan, N. A. Clark, and D. M. Walba. 2000. The hysteretic behavior of 'V-shaped switching' smectic materials. *Ferroelectrics.* 246:21–33.

24. See, for example, Hands, Philip J. W., Andrew K. Kiby, and Gordon D. Love. 2004. Adaptive modally addressed liquid crystal lenses. *Proc. SPIE.* 5518:136–143, and references therein.

25. See, for example, Gat, N. 2000. Imaging spectroscopy using tunable filters: A review. *Proc. SPIE.* 4056:50–64, and references therein.

26. See, for example, Mcmanamon, P. F., T. A. Dorschner, D. L. Corkum, L. J. Friedman, D. S. Hobbs, M. Holz, S. Liberman, H. Q. Nguyen, D. P. Resler, R. C. Sharp, and E. A. Watson. 1996. *Proc. IEEE.* 84:268.

27. Park, J.–H., I. C. Khoo, C.-J. Yu, M.–S. Jung, and S.–D. Lee. 2005. *Appl. Phys. Lett.* 86:021906, and references therein.

28. Ramssy, G., ed. 1998. *Commercial Biosensors: Applications to Clinical, Bioprocess, and Environmental Samples.* New York: Wiley.

29. See, for example, Suzuki, Masayoshi, Hiromoto Sato, Peer Kirsch, Atsushi Sawada, and Shohei Naemura. 2004. Light-emitting materials based on liquid crystals. *Proc. SPIE.* 5518:51–65, and references therein.

30. See, for example, Gruneisen, Mark T., and James M. Wilkes. 1997. Compensated Imaging by real-time holography with optically addresses spatial light modulators. In G. Burdge and S. C. Esener (eds.) *Spatial Light Modulators,* OSA TOPS Vol. 14.

31. Mach, P., P. Wiltzius, M. Megens, D. A. Weitz, K.-H. Lin, T. C. Lubensky, and A. G. Yodh. 2002. *Phys. Rev. E.* 65:031720.

32. Mertens, G., T. Röder, R. Schweins, K. Huber, and H.-S. Kitzerow. 2002. *Appl. Phys. Lett.* 80:1885–1887.

33. Larsen, Thomas Tanggaard, Anders Bjarklev, David Sparre Hermann, and Jes Broeng. 2003. *Opt. Express.* 11:2589–2596.

34. Yoshino, K., Y. Shimoda, Y. Kawagishi, K. Nakayama, and M. Ozaki. 1999. *Appl. Phys. Lett.* 75:932.

35. Kubo, S., Z.-Z. Gu, K. Takahashi, A. Fujishima, H. Segawa, and O. Sato. 2005. *Chem. Mater.* 17:2298.

36. Kang, D., J. E. Maclennan, N. A. Clark, A. A. Zakhidov, and R. H. Baughman. 2001. *Phys. Rev. Lett.* 86:4052.

37. Graugnard, E., J. S. King, S. Jain, C. J. Summers, Y. Zhang-Williams, and I. C. Khoo. 2005. Electric field tuning of the Bragg peak in large-pore TiO_2 inverse shell opals. *Phys. Rev. B.* 72:233105.

38. Tondiglia, V. P., L. V. Natarajan, R. L. Sutherland, D. Tomlin, and T. J. Bunning. 2002. Holographic formation of electro-optical polymer-liquid crystal photonic crystals. *Adv. Mater.* 14:187–191.

39. Khoo, I. C., Yana Williams, Andres Diaz, Kan Chen, J. Bossard, D. Werner, E. Graugnard, and C. J. Summers. Forthcoming. Liquid-crystals for optical filters, switches and tunable negative index material development. *Mol. Cryst. Liq. Cryst.*

40. Divliansky, Ivan, and Theresa S. Mayer. 2006. Three-dimensional low-index-contrast photonic crystals fabricated using a tunable beam splitter. *Nanotechnology.* 17:1241–1244.

41. Wu, T. K., ed., 1995. *Frequency Selective Surface and Grid Array.* New York: Wiley.

42. See, for example, Bossard, J. A., D. H. Werner, T. S. Mayer, and R. P. Drupp. 2005. *IEEE. Trans. Antennas Propag.* 53 (4): 1390–1400, and references therein by D. H. Werner et al.

43. Li, L., and D. H. Werner. 2005. *Proceedings of the 2005 IEEE Antennas and Propagation Society International Symposium and USNC/URSI National Radio Science Meeting,* Washington DC, Vol. 4A, pp. 376–379, July 3-8, 2005.

44. Shah, R. R., and N. L. Abbott. 2001. Principles for measurement of chemical exposure based on recognition-driven anchoring transitions in liquid crystals. *Science.* 293:1296–1298; Shah, R. R., and N. L. Abbott. 2003. Orientational transitions of liquid crystals driven by binding of organoamines to carboxylic acids presented at surfaces with nanometer-scale topology. *Langmuir.* 19(2):275–284, and references therein; see also Brake, J. M., M. K. Daschner, Y.-Y Luk, and N. L. Abbott. 2003. Biomolecular interactions at phospholipid-decorated surfaces of liquid crystals. *Science.* 302:2094–2097.

45. Kiser, B., D. Pauluth, and G. Gauglitz. 2001. Nematic liquid crystals as sensitive layers for surface plasmon resonance sensors. *Anal. Chim. Acta.* 434:231–237.

46. Cross, G. H., A. Reeves, S. Brand, M. J. Swann, L. L. Peel, N. J. Freeman, and J. R. Lu. 2004. The metrics of surface adsorbed small molecules on the Young's fringe dual-slab waveguide interferometer. *J. Phys. D: Appl. Phys.* 37: 74–80.

47. See, for example, De La Tocnaye, J. L. D. 2004. Engineering liquid crystals for optimal uses in optical communication systems. *Liq. Cryst.* 31:241–269, and the numerous references quoted therein.

48. Sato, H., et al. 2005. Rollable ferroelectric liquid crystal devices monostabilized with molecular aligned polymer walls and networks. *Liq. Cryst.* 32:221, and references therein; see also Fujisaki, Y., et al. 2005. Liquid crystal cells fabricated on plastic substrate driven

by low-voltage organic thin film transistor with improved gate insulator and passivation layer. *Jpn. J. Appl. Phys. Part 1*. 44:3728.

49. See, for example, Marshall, K. L., E. Kimball, S. Mcnamara, T. Z. Kosc, A. Trajkovska-Petkoska, and S. D. Jacobs. 2004. Electro-optical behaviour of polymer cholesteric liquid crystal flake/fluid suspensions in a microencapsulated matrix, in *Liquid Crystals VIII*, edited by I. C. Khoo, SPIE Proceedings Vol. 5518 (SPIE Bellingham, WA 2004) pp. 170–181 and references therein; see also, Marshall, K. L., T. Z. Kosc, S. D. Jacobs, S. M. Faris, and L. Li. 2003. Electrically switchable polymer liquid crystal and polymer birefringent flake/fluid host systems and optical devices utilizing same, U.S. Patent No. 6,665,042 B1 (Dec 2003).

50. Jacobs, S. D., K. A. Cerqua, K. L. Marshall, A. Schmid, M. J. Guardalben, and K. J. Skerrett. 1988. Liquid-crystal laser optics: Design, fabrication and performance. *J. Opt. Soc. Am. B*. 9:1962–1979.

7

Electromagnetic Formalisms
for Optical Propagation

7.1. INTRODUCTION

In general, the methods for solving the problem of light propagation through a medium depend on whether the light–matter interaction is linear or nonlinear. In linear optics, the light fields (electric or magnetic) are not intense enough to create appreciable changes in the optical properties of the medium (e.g., refractive index, absorption or scattering cross sections, etc.) and so its propagation through the medium is dictated primarily by its properties. In nonlinear optics, the light–matter interactions are sufficiently intense so that the optical properties of the medium are affected by the optical fields. For example, the optical field is intense enough to cause director axis reorientations and so it experiences an ever changing refractive index as it propagates in the medium. Such coupled interactions give rise to many so-called self-action effects such as self-focusing, self-phase modulations, stimulated scattering, etc. These nonlinear optical phenomena and other related processes will be discussed in Chapters 11 and 12.

In this chapter, we discuss the case of linear optics of liquid crystal structures extensively used in optical display applications. Such structures usually consist of various optical phase shifting or polarizing elements in tandem and an aligned liquid crystal cell serving as an electrically controlled phase-shifting element. In the previous chapter, we discuss some aspects of the optics of an anisotropic medium and illustrate mostly on-axis type of propagation and simple director axis reorientation geometry that allow analytical solutions or conceptual understanding. These limiting-case studies would suffice if we desired only to grasp the "physics" of the processes. However, exact quantitative determination of the polarization state of light and the director axis reorientation profile in a liquid crystal cell under an applied ac field is mandated by the stringent requirements placed on practical display devices. More sophisticated models and techniques are needed to calculate both on-axis and off-axis propagations, inhomogeneous director axis distribution in

Liquid Crystals, Second Edition   By Iam-Choon Khoo
Copyright © 2007 John Wiley & Sons, Inc.

conjunction with a multitude of phase plates, filters, retarders, etc. Accordingly, a detailed exposition of some of the fundamental theoretical formalisms and techniques currently being used or developed for LC devices will be presented. For completeness, we shall first review the current understanding of electromagnetic theories of complex anisotropic media.

7.2. ELECTROMAGNETISM OF ANISOTROPIC MEDIA REVISITED

7.2.1. Maxwell Equations and Wave Equations

The first important step is to revisit the electromagnetism of an anisotropic medium, which has been briefly alluded to in Chapters 3 and 6. For optical propagation in an anisotropic medium such as a liquid crystal, the physical parameters of interest are the field vectors, the material polarization induced by these fields, and the power flow. The electromagnetic nature of light and the optical responses of the liquid crystals (assuming source- and current-free $\rho = 0, J = 0$) are described by Maxwell's equations:

$$\frac{\partial \vec{D}}{\partial t} = \nabla \times \vec{H}, \tag{7.1}$$

$$\mu_o \frac{\partial \vec{H}}{\partial t} = -\nabla \times \vec{E}, \tag{7.2}$$

$$\nabla \cdot \boldsymbol{D} = 0, \tag{7.3}$$

$$\nabla \cdot \boldsymbol{B} = 0, \tag{7.4}$$

supplemented by the constitutive equations:

$$\boldsymbol{D} = \varepsilon_0 \boldsymbol{E} + \boldsymbol{P} = \varepsilon \boldsymbol{E}, \tag{7.5}$$

$$\boldsymbol{B} = \mu_0 (\boldsymbol{H} + \boldsymbol{M}) = \mu \boldsymbol{H}, \tag{7.6}$$

where E (V/m) and H (A/m) are the electric and magnetic field, and D (Coul/m^2) and B (Wb/m^2) are the electric and magnetic displacement vectors, respectively. P and M are the corresponding electric and magnetic polarizations induced by the electric and magnetic fields in the materials. ε_0 (8.854 $\times$ 10^{-12} F/m) is the permittivity and μ_0 (4π $\times$ 10^{-7} H/m) is the permeability of vacuum.

In a source- and current-free environment, i.e., $\rho = 0$, $\boldsymbol{J} = 0$, Equations (7.1)–(7.6) can be combined by successive differentiation and substitution to yield the wave equation:

$$\nabla^2 E - \mu\varepsilon \frac{\partial^2 E}{\partial t^2} = \mu_0 \frac{\partial^2 P}{\partial t^2}, \tag{7.7}$$

7.2.2. Complex Refractive Index

Among the most important optical properties of a material is its refractive index, or more correctly speaking, the refractive index experienced by a particular polarized light traversing the medium. For simplicity, we shall focus our attention in this chapter to crystalline materials such as liquid crystals in which there are three well-defined principal crystalline axes. For such crystals, the dielectric and permeability tensors are diagonal. Furthermore, we also limit our attention to *linear* responses, i.e., the induced polarization $\boldsymbol{P}$ is proportional to the optical electric field $\boldsymbol{E}$:

$$\boldsymbol{P} = \varepsilon_0 \chi_e \boldsymbol{E}, \tag{7.8}$$

where χ_e is the linear electric susceptibility.

In the principal axes coordinate, χ_e is diagonal. Accordingly, the optical dielectric tensor ε, or permittivity, of the medium is diagonal:

$$\varepsilon = \varepsilon_0 \begin{pmatrix} n_1^2 & 0 & 0 \\ 0 & n_2^2 & 0 \\ 0 & 0 & n_3^2 \end{pmatrix} - \begin{pmatrix} \varepsilon_1 & 0 & 0 \\ 0 & \varepsilon_2 & 0 \\ 0 & 0 & \varepsilon_3 \end{pmatrix}. \tag{7.9}$$

The permittivity ε (and its magnetic counterpart, the permeability μ) are complex in general, with an imaginary component accounting for losses (e.g., the finite electric conductivity). The relative permittivity ε_r may be expressed as

$$\varepsilon_r = \frac{\varepsilon}{\varepsilon_0} = 1 + \chi_e = 1 + \chi_e' + i\chi_e'' = \varepsilon_r' + i\varepsilon_r''. \tag{7.10}$$

Similarly, the linear magnetic polarization is expressed in terms of the magnetic field intensity and the magnetic susceptibility:

$$\boldsymbol{M} = \chi_m \boldsymbol{H} \tag{7.11}$$

and

$$\frac{\mu}{\mu_0} = 1 + \chi_m = 1 + \chi_m' + i\chi_m'' = \mu_r' + i\mu_r'' . \tag{7.12}$$

If the plane- polarized wave propagating in the positive z direction is expressed in the form $e^{i(k_0 nz - \omega t)}$, then the imaginary parts of both the permittivity and the permeability need to be positive in order for the system to be causal.

Substituting the expressions obtained above for the permittivity and permeability of the medium, we get the refractive index,

$$n_+ = \sqrt{\frac{\mu\varepsilon}{\mu_0\varepsilon_0}} = \sqrt{(\varepsilon_r' + i\varepsilon_r'')(\mu_r' + i\mu_r'')}$$

$$= \sqrt{(\varepsilon_r'\mu_r' - \varepsilon_r''\mu_r'') + i(\varepsilon_r'\mu_r'' + \varepsilon_r''\mu_r')} = n' + in'', \tag{7.13}$$

where n_+ indicates that we choose the square root for which the imaginary part n'' is positive. This corresponds to power flow in the $+z$ direction (as can be demonstrated by calculating the Poynting vector), and gives an exponential attenuation of the fields in the direction of propagation of the wave when it propagates in a lossy medium. If the wave propagates in the $-z$ direction, the opposite root should be taken (n_-).

7.2.3. Negative Index Material

A negative index will occur whenever $\text{Re}\{n_+\} < 0$ (or $\text{Re}\{n_-\} > 0$ for a wave propagating in the $-z$ direction). From Equation (7.13), one can see that a sufficient (but not necessary) condition to achieve this is to have $\varepsilon_r' < 0$ and $\mu_r' < 0$. But even if the real part of the permittivity or the permeability is positive, $\text{Re}\{n_+\}$ may still be negative if

$$\varepsilon_r'\mu_r'' + \varepsilon_r''\mu_r' < 0. \tag{7.14}$$

As an example, consider a material with positive real relative permeability μ_r'. Condition (7.14) above still holds as long as

$$\varepsilon_r' < -\frac{\varepsilon_r''\mu_r'}{\mu_r''}. \tag{7.15}$$

Hence, the existence of a negative index material as defined will depend critically on the ratio of the imaginary parts of the relative permittivity and permeability. For the specific case of a material with a very small magnetic response, for which $\mu_r' \approx 1$ and, $\mu_r'' \ll 1$, a large negative real permittivity together with a small loss tangent (for the imaginary permittivity) may be required in order to satisfy inequality.[1]

The idea was first proposed almost three decades ago,[1] and was first experimentally demonstrated in the microwave regime. In view of their application in subwavelength resolution imaging and fabricating planar lensless optical elements, there have been intense research efforts in developing negative index materials in the visible and near infrared spectra. The main problem in the development of actual negative index material (NIM) lies in reducing the high loss associated with the large imaginary part of ε.

A negative ε' occurs in metals at IR/visible frequencies due to plasmonic resonance, but there is no naturally occurring material that exhibits negative μ' at these frequencies. A simple example of a metallic nanoresonator suitable for NIM development is a pair of closely spaced parallel metallic nanorods as shown in Figure 7.1. When the parallel rods are illuminated with a high-frequency electromagnetic (EM) wave, a current is induced along each rod forming a "closed" loop via displacement currents between the rods. The parallel-nanorods structure exhibits electromagnetic resonance at frequencies that depend on the plasmonic resonance of the metal, and the dimensions and spacing of the rods.[2-5] The dimensions (L, w, d) can be tuned so that $\varepsilon'(\mu)$ and $\mu'(\omega)$ are simultaneously negative at the desired frequencies, thereby producing an effective negative refractive index. For optical application, these spacings/dimensions are on the order of submicrons to 1 μm.

In many ways, such structures are analogous to studies[6,7] conducted in the far IR and microwave regimes with planar structures that consist of an array of nanopatterned metallo-dielectric or all-dielectric films, the so-called frequency selective surfaces (FSS) discussed in Chapter 6, Section 6.6.3. These structures exhibit both frequency selectivity and negative index refraction properties. More recently, Khoo et al.[7] have demonstrated that nematic liquid crystals containing nano-core-shell spheres made of polaritonic and drude materials could possess tunable refractive indices ranging from negative through zero and positive values (Fig. 7.2). In Chapter 4, we also discuss the possibility of realizing negative index properties in photonic crystals such as the 1D cholesteric

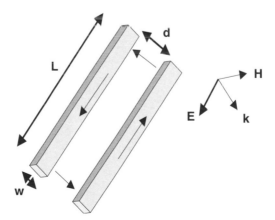

Figure 7.1.   Paired metallic nanorods that exhibits negative index of refraction in the visible spectrum.

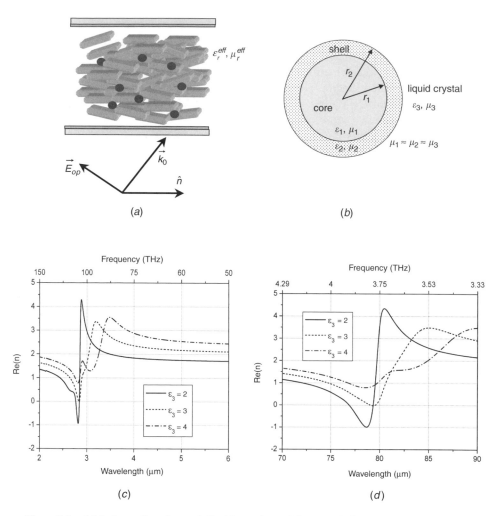

Figure 7.2. (a) A planar aligned nematic liquid crystal containing core-shell nanospheres. (b) Cross section of the core [μ_1, ε_1]-shell [μ_2, ε_2] nanosphere with the liquid crystal host [μ_3, ε_3]. (c) Tunable refractive index of the nanosphere dispersed liquid crystal as a function of liquid crystal host permittivity in the optical region. (d) Tunable refractive index as a function of liquid crystal host permittivity in the terahertz region.

liquid crystal that has electromagnetic coupling. Recently, negative index refraction in general chiral materials[8] has been demonstrated.

7.2.4. Normal Modes, Power Flow, and Propagation Vectors in a Lossless Isotropic Medium

Consider again a monochromatic plane wave of frequency ω with electric and magnetic wave vectors E and H, respectively:

$$E \exp[i(kr - \omega t)], \tag{7.16a}$$

$$H \exp[i(kr - \omega t)], \tag{7.16b}$$

where k is the propagation wave vector with a magnitude given by $k = (\omega/c)n$, where n is the refractive index of the medium. We also assume that the medium is lossless, that is, n is real and positive.

Substituting Equations (7.16) in the Maxwell Equations (7.1–7.4) yields

$$k \times E = \omega\mu H, \tag{7.17}$$

$$k \times H = -\omega\varepsilon E = -\omega D. \tag{7.18}$$

Eliminating E or H from these equations will yield decoupled equations for H and E:

$$k \times (k \times H) + \omega^2\,\mu\varepsilon H = 0, \tag{7.19}$$

$$k \times (k \times E) + \omega^2\,\mu\varepsilon E = 0. \tag{7.20}$$

Equations (7.17) and (7.18) show that the propagation wave vector k, the electric field E, and the magnetic field H are orthogonal to one another, see Figure 7.3. From consideration of the electromagnetic energy, one can deduce that the power flow associated with the electromagnetic wave is described by the Poynting vector S:

$$S = E \times H. \tag{7.21}$$

Equations (7.17) and (7.18) can be written as

$$H = \left(\frac{n}{\mu c}\right) k \times E, \tag{7.22}$$

$$D = -\left(\frac{n}{c}\right) k \times H. \tag{7.23}$$

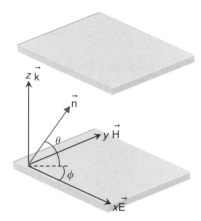

Figure 7.3. Illustration of a plane light wave propagating along z-axis through a nematic liquid crystal confined within two flat windows. The electric field E is along the x-axis, H is along the y-axis and k is along z-axis. D is in the x-z plane.

Combining these equations yields

$$D = -\left(\frac{n^2}{c^2\mu}\right)k \times (k \times E) = \left(\frac{n^2}{c^2\mu}\right)[E - k(k \cdot E)]. \tag{7.24}$$

Note that $D^2 = (n^2/c^2\mu)\,[E \cdot D]$ since $D \cdot k = 0$.

These equations show that D, K, and E all lie in the same plane, that is, in the x-z plane (see Fig. 7.3), and D, E, and S are all orthogonal to H. These points are important to keep in mind whenever one considers light propagation in an anisotropic medium. In particular, when considering obliquely incident polarized light into a birefringent medium, the normal and tangential components of various field vectors and their continuity across the interfaces and resulting different directions of propagation will come into play.

7.2.5. Normal Modes and Propagation Vectors in a Lossless Anisotropic Medium

We now discuss the case where the medium is anisotropic. The general forms of the electromagnetic (light) waves, that is, its polarization states, power flows, propagation modes, and so on, in an anisotropic medium are understandably more complicated than in an isotropic medium. It is more instructive to discuss the power flow for specific propagation directions. Consider again Equation (7.19). Writing out K and E explicitly as

$$K = k_1\mathbf{i} + k_2\mathbf{j} + k_3\mathbf{k},$$

$$E = E_1\mathbf{i} + E_2\mathbf{j} + E_3\mathbf{k},$$

Equation (7.19) yields

$$
\begin{pmatrix}
\omega^2\mu\varepsilon_1 - k_2^2 - k_3^2 & k_1 k_2 & k_1 k_3 \\
k_2 k_1 & \omega^2\mu\varepsilon_2 - k_1^2 - k_3^2 & k_2 k_3 \\
k_3 k_1 & k_3 k_2 & \omega^2\mu\varepsilon_2 - k_1^2 - k_2^2
\end{pmatrix}
\begin{pmatrix}
E_1 \\
E_2 \\
E_3
\end{pmatrix} = 0, \quad (7.25)
$$

where we have expressed the dielectric tensor in terms of the principal axes of the crystals:

$$
\varepsilon = \varepsilon_0
\begin{pmatrix}
n_1^2 & 0 & 0 \\
0 & n_2^2 & 0 \\
0 & 0 & n_3^2
\end{pmatrix}
=
\begin{pmatrix}
\varepsilon_1 & 0 & 0 \\
0 & \varepsilon_2 & 0 \\
0 & 0 & \varepsilon_3
\end{pmatrix}.
\tag{7.26}
$$

Nontrivial solutions for E's, which are termed eigenmodes of the wave in the medium, exist if the determinant of the matrix multiplying $\mathbf{E}$ is zero, that is,

$$
\det
\begin{vmatrix}
\omega^2\mu\varepsilon_1 - k_2^2 - k_3^2 & k_1 k_2 & k_1 k_3 \\
k_2 k_1 & \omega^2\mu\varepsilon_2 - k_1^2 - k_3^2 & k_2 k_3 \\
k_3 k_1 & k_3 k_2 & \omega^2\mu\varepsilon_2 - k_1^2 - k_2^2
\end{vmatrix} = 0. \quad (7.27)
$$

For a given frequency ω, Equation (7.27) is a three-dimensional surface (*normal surface*) in the $\mathbf{k}$ (k_1, k_2, k_3) space. In general, for a given propagation direction $\mathbf{s}$ there will be two solutions for the $\mathbf{k}$ values corresponding to the intersections of $\mathbf{s}$ with the normal surface.

These considerations give rise to the index ellipsoid method for determining the eigenmodes and their polarization states and refractive indices. For propagation in the 1-2 (x-y) plane, the two eigenmodes of the matrices (one polarized along z and one in the x-y plane) are given by:[9]

$$
E_1 =
\begin{pmatrix}
0 \\
0 \\
1
\end{pmatrix}, \quad
k_1 = \frac{n_3 \omega}{c}
\tag{7.28a}
$$

and

$$E_2 = \begin{pmatrix} \dfrac{s_1}{n^2 - n_1^2} \\[2mm] \dfrac{s_2}{n^2 - n_2^2} \\[2mm] 0 \end{pmatrix}, \quad k_2 = \frac{\omega}{c}\left(\frac{n_1^2 n_2^2}{n_1^2 \cos^2\theta + n_2^2 \sin^2\theta} \right)^{1/2} = n\frac{\omega}{c}. \qquad (7.28b)$$

For propagation in the 2-3 (y-z) plane (i.e., $k_1 = 0$), we have

$$E_1 = \begin{pmatrix} 1 \\ 0 \\ 0 \end{pmatrix}, \quad k_1 = \frac{n_1 \omega}{c}, \qquad (7.29a)$$

$$E_2 = \begin{pmatrix} 0 \\[2mm] \dfrac{s_2}{n^2 - n_2^2} \\[2mm] \dfrac{s_3}{n^2 - n_3^2} \end{pmatrix}, \quad k_2 = \frac{\omega}{c}\left(\frac{n_2^2 n_3^2}{n_2^2 \cos^2\theta + n_3^2 \sin^2\theta} \right)^{1/2}. \qquad (7.29b)$$

For propagation in the 3-1 (z-x) plane ($k_2 = 0$), we have

$$E_1 = \begin{pmatrix} 0 \\ 1 \\ 0 \end{pmatrix}, \quad k_1 = \frac{n_2 \omega}{c}, \qquad (7.30a)$$

$$E_2 = \begin{pmatrix} \dfrac{s_1}{n^2 - n_1^2} \\[2mm] 0 \\[2mm] \dfrac{s_3}{n^2 - n_3^2} \end{pmatrix}, \quad k_2 = \frac{\omega}{c}\left(\frac{n_1^2 n_3^2}{n_1^2 \cos^2\theta + n_3^2 \sin^2\theta} \right)^{1/2}. \qquad (7.30b)$$

For uniaxial materials such as liquid crystals characterized by n_o ($\varepsilon_1 = \varepsilon_2 = \varepsilon_0 n_o^2$) and n_e ($\varepsilon_3 = \varepsilon_0 n_e^2$), one can deduce from the normal surfaces that they consist of two parts. The sphere gives the relation between ω and k of the ordinary (O) wave; the ellipsoid of revolution gives a similar relation for the extraordinary (E) wave:

$$O \text{ wave} : n = n_o, \tag{7.31a}$$

$$E \text{ Wave} : \frac{1}{n^2} = \frac{\cos^2\theta}{n_o^2} + \frac{\sin^2\theta}{n_e^2}, \tag{7.31b}$$

where θ is the angle between the direction of propagation and the optic axis (the crystal axis) (see Fig. 7.4).

Using the equations discussed above, one can deduce the propagation wave vectors and electric fields for the following typical cases. For propagation perpendicular to the z axis (i.e., in the x-y plane), the unit vector s becomes

$$s = \begin{pmatrix} \cos\phi \\ \sin\phi \\ 0 \end{pmatrix}, \tag{7.32a}$$

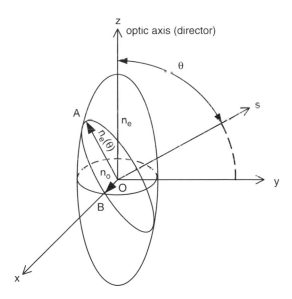

Figure 7.4. Index ellipsoid for a uniaxial medium such as nematic liquid crystal. S is the propagation direction [in the y-z plane].

where ϕ is the angle of the direction of propagation in the x-y plane. From Equation (7.28), the directions of polarization for the ordinary and extraordinary waves are given respectively by

$$O \text{ wave} : E_o = \begin{pmatrix} \sin\phi \\ -\cos\phi \\ 0 \end{pmatrix}, \quad E \text{ wave} : E_e = \begin{pmatrix} 0 \\ 0 \\ 1 \end{pmatrix}. \tag{7.32b}$$

For propagation in the x-z plane (e.g., Section 7.4), s is given by

$$s = \begin{pmatrix} \sin\theta \\ 0 \\ \cos\theta \end{pmatrix}, \tag{7.33a}$$

where θ is the angle of s measured from the z axis. The refractive index of the ordinary wave is $n = n_o$ and the refractive index of the extraordinary wave is

$$n_e(\theta) = \left(\frac{\cos^2\theta}{n_o^2} + \frac{\sin^2\theta}{n_e^2} \right)^{-1/2}, \tag{7.33b}$$

and their polarization vectors are given by Equation (7.30), respectively,

$$O \text{ wave} : E_o = \begin{pmatrix} 0 \\ 1 \\ 0 \end{pmatrix}, \quad E \text{ wave} : E_e = \begin{pmatrix} n_e^2 \cos\theta \\ 0 \\ -n_o^2 \sin\theta \end{pmatrix}. \tag{7.33c}$$

For the simplest case of propagation along the z axis, the axis of symmetry for the uniaxial liquid crystal, two modes of propagation characterized by equal refractive indices $n_1 = n_2 = n_o$ along the x and y directions exist.

7.3. GENERAL FORMALISMS FOR POLARIZED LIGHT PROPAGATION THROUGH LIQUID CRYSTAL DEVICES

As discussed previously, liquid crystal display devices/pixels are made up of a multitude of optical polarizing, phase shifting, and polarization rotation and absorption/transmission elements. Furthermore, starting with a uniform director axis of an aligned liquid crystal cell, the reorientation profile of the director axis under the action of an applied field is, in general, inhomogeneous. The problem becomes even more complex when one considers off-axis propagation. Such complicated "polarized" optical systems render it impossible to have analytical solutions or "pictures" as presented in the previous chapter, and call for more powerful and sophisticated analytical method. In this and the following section, we discuss various theoretical formalisms that have been developed to tackle these problems.

7.3.1. Plane-Polarized Wave and Jones Vectors

We shall begin with a discussion of the basic formalism for treating polarized light. Consider a plane-polarized monochromatic light wave, the electric field vector of which can be written as[10]

$$\mathbf{E} = \mathbf{A}\cos(\omega t - \mathbf{k\cdot r}), \tag{7.34}$$

where $k = n\,\omega/c = n2\pi/\lambda$ and $\mathbf{k\cdot E} = 0$.

For propagation along the z direction, the two transverse components may be written as

$$\begin{aligned} E_x &= A_x\cos(\omega t + \delta_x), \\ E_y &= A_y\cos(\omega t + \delta_y). \end{aligned} \tag{7.35}$$

As discussed previously, the electric field vector can assume various polarization states. Jones calculus is a method to treat propagation and evolution of these polarization states in an anisotropic crystal, which will impart various phase shifts to the principal axes components of the electric field. We begin by defining the Jones vector:

$$\mathbf{J} = \begin{pmatrix} A_x e^{i\delta_x} \\ A_y e^{i\delta_y} \end{pmatrix}. \tag{7.36}$$

Jones vector is a mathematical representation of the x and y components at any given time t, such that, $\mathbf{E}_x(t) = \mathrm{Re}\,(J_x e^{i\omega t}) - \mathrm{Re}\,[A_x e^{i(\omega t + kx)}]$.

In a lossless medium, we assume an amplitude of unity for the electric field (i.e., $A = 1$),

$$\mathbf{J}^*\cdot\mathbf{J} = 1. \tag{7.37}$$

For a linearly polarized light ($\delta_x = \delta_y$) making an angle ψ, $A_x = \cos\psi$ and $A_y = \sin\psi$:

$$\begin{pmatrix} \cos\psi \\ \sin\psi \end{pmatrix}. \tag{7.38}$$

For example, the Jones vectors for a linearly x-polarized light (corresponding to $\psi = 0$) and a y-polarized light ($\psi = \pi/2$) are described respectively:

$$\mathbf{x} = \begin{pmatrix} 1 \\ 0 \end{pmatrix}, \quad \mathbf{y} = \begin{pmatrix} 0 \\ 1 \end{pmatrix}. \tag{7.39}$$

For a linearly polarized light whose plane of polarization is orthogonal to the above, such that, $\psi \to \psi + \pi/2$, we have

$$\begin{pmatrix} -\sin\psi \\ \cos\psi \end{pmatrix}. \tag{7.40}$$

Similarly, the Jones vectors for the right and left circularly polarized lights are given by

$$\mathbf{R} = \frac{1}{\sqrt{2}} \begin{pmatrix} 1 \\ +i \end{pmatrix}, \quad \mathbf{L} = \frac{1}{\sqrt{2}} \begin{pmatrix} 1 \\ -i \end{pmatrix}. \tag{7.41}$$

Note that $\mathbf{R}^* \cdot \mathbf{L} = 0$:

$$\mathbf{R} = \frac{1}{\sqrt{2}}(\mathbf{x} + i\mathbf{y}), \quad \mathbf{L} = \frac{1}{\sqrt{2}}(\mathbf{x} - i\mathbf{y}), \quad \mathbf{x} = \frac{1}{\sqrt{2}}(\mathbf{R} + \mathbf{L}), \quad \mathbf{y} = \frac{-i}{\sqrt{2}}(\mathbf{R} - \mathbf{L}). \tag{7.42}$$

Note: In the above definition of right and left circularly polarized light, we adopt the convention frequently used in optics, that is, from the point of view of an observer looking at the light head on. For $\mathbf{R}$, the observer will see a clockwise rotation of the electric field vector, while for $\mathbf{L}$, the observer will see a counterclockwise rotation of the electric field vector. This is also the convention used in Chapter 4 when we discussed circularly polarized light in the context of the optical properties of cholesteric liquid crystals.

To describe a general state of polarization, we ascribe a phase shift δ between the x and y components (see Chapter 6). Accordingly, the Jones vector for a general elliptical polarization state is of the form .[9,10]

$$\mathbf{J}(\psi, \delta) = \begin{pmatrix} \cos\psi \\ e^{i\delta}\sin\psi \end{pmatrix}. \tag{7.43}$$

We now consider the application of the Jones matrix method to the simple problem of polarized light through a birefringent phase plate or retardation plate (see Fig. 7.5). As illustrated in the previous chapter, depending on the phase shift between the fast and slow axes components imparted by the retardation plates, the resulting emergent light will have different states of polarization. To apply the Jones matrix method, we represent the incident polarization state by a Jones column vector:

$$V = \begin{pmatrix} V_x \\ V_y \end{pmatrix}, \tag{7.44}$$

where V_x and V_y are two complex numbers representing the complex field amplitudes along x and y. To determine how the light propagates in the retardation plate, we need to resolve it into components along the fast and slow axes of the crystal, that is, a rotation around the z axis by an angle ψ:

$$\begin{pmatrix} V_s \\ V_f \end{pmatrix} = \begin{pmatrix} \cos\psi & \sin\psi \\ -\sin\psi & \cos\psi \end{pmatrix}\begin{pmatrix} V_x \\ V_y \end{pmatrix} = R(\psi)\begin{pmatrix} V_x \\ V_y \end{pmatrix}, \tag{7.45}$$

where V_s and V_f are the slow and fast components, respectively. Let n_s and n_f be the refractive indices of the slow and fast components, respectively. The polarization state of the emerging beam in the crystal coordinate system is thus given by

$$\begin{pmatrix} V_s' \\ V_f' \end{pmatrix} = \begin{pmatrix} \exp\left(-in_s\dfrac{\omega}{c}l\right) & 0 \\ 0 & \exp\left(-in_f\dfrac{\omega}{c}l\right) \end{pmatrix}\begin{pmatrix} V_s \\ V_f \end{pmatrix}, \tag{7.46}$$

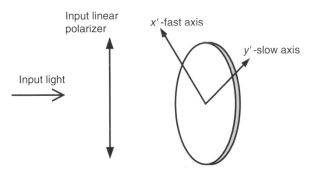

Figure 7.5. Propagation of a polarized light through a birefringent phase plate: fast and slow axes.

where l is the thickness of the plate and ω is the radian frequency of the light beam. Because of the index difference, the two components experience a phase delay after their passage through the crystal respectively:

$$\Gamma = (n_s - n_f)\frac{\omega l}{c}. \tag{7.47}$$

Writing $\phi = \frac{1}{2}(n_s + n_f)\omega l/c$, Equation 7.46 becomes

$$\begin{pmatrix} V'_s \\ V'_f \end{pmatrix} = e^{-i\phi}\begin{pmatrix} e^{-i\Gamma/2} & 0 \\ 0 & e^{i\Gamma/2} \end{pmatrix}\begin{pmatrix} V_s \\ V_f \end{pmatrix}. \tag{7.48}$$

The Jones vector of the polarization state of the emerging beam in the laboratory frame (i.e., xy coordinate system) is given by transforming back from the crystal to the laboratory coordinate system, that is, applying a rotation matrix with ψ replaced by $-\psi$:

$$\begin{pmatrix} V'_x \\ V'_y \end{pmatrix} = \begin{pmatrix} \cos\psi & -\sin\psi \\ \sin\psi & \cos\psi \end{pmatrix}\begin{pmatrix} V'_s \\ V'_f \end{pmatrix}. \tag{7.49}$$

This yields

$$\begin{pmatrix} V'_x \\ V'_y \end{pmatrix} = R(-\psi)W_0 R(\psi)\begin{pmatrix} V_x \\ V_y \end{pmatrix}, \tag{7.50}$$

where

$$R(\psi) = \begin{pmatrix} \cos\psi & \sin\psi \\ -\sin\psi & \cos\psi \end{pmatrix} \tag{7.51}$$

and

$$W_0 = e^{-i\phi}\begin{pmatrix} e^{-i\Gamma/2} & 0 \\ 0 & e^{i\Gamma/2} \end{pmatrix}. \tag{7.52}$$

In the Jones matrix formalism, therefore, a retardation plate is described by a matrix $W(\psi,\Gamma)$ characterized by its phase retardation Γ and its azimuth angle ψ:

$$W(\psi,\Gamma) \equiv W = R(-\psi)W_0 R(\psi)$$

$$= \begin{vmatrix} e^{-i(\Gamma/2)}\cos^2\psi + e^{i(\Gamma/2)}\sin^2\psi & -i\sin\dfrac{\Gamma}{2}\sin(2\psi) \\ -i\sin\dfrac{\Gamma}{2}\sin(2\psi) & e^{-i(\Gamma/2)}\sin^2\psi + e^{i(\Gamma/2)}\cos^2\psi \end{vmatrix}. \quad (7.53)$$

Note that the Jones matrix of a wave plate is a unitary matrix, that is,

$$W^+ W = 1, \quad (7.54)$$

where $W_{ij}^* = (W^+)_{ji}$.

Using these transformation matrices, one can derive the polarization vectors for light propagating (along the z axis) through various polarizers and phase retardation elements such as those described in the previous chapter.

7.3.2. Jones Matrix Method for Propagation Through a Nematic Liquid Crystal Cell

Consider the example of propagation through a general twisted NLC cell in a typical liquid crystal display pixel as depicted schematically in Figure 7.6. The pixel consists of an input (or entrance) polarizer, a liquid crystal cell with the director axis oriented along the x direction at the input end. At the output end, the director axis is oriented at an angle θ with the x axis before merging with the exit

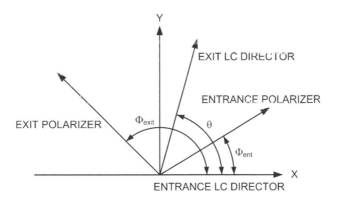

Figure 7.6. Schematic depiction of a pixel element that consists of an input and an output polarizer sandwiching a LC cell with the director axis oriented along x-direction at the input end. At the output end, the director axis is oriented at an angle θ with the x axis before merging from the exit polarizer as shown.

polarizer as shown. The transmission of the light through such a pixel has been calculated to be of the form[11]

$$T = \cos^2(\theta - \phi_{exit} + \phi_{ent}) + \sin^2(\theta\sqrt{1+u^2})\sin 2(\theta - \theta_{exit})\sin 2\phi_{ent}$$
$$+ \frac{1}{2\sqrt{1+u^2}}\sin(2\theta\sqrt{1+u^2})\sin 2(\theta - \theta_{exit} + \phi_{ent})$$
$$- \frac{1}{1+u^2}\sin^2(\theta\sqrt{1+u^2})\cos 2(\theta - \theta_{exit})\cos 2\phi_{ent}, \tag{7.55}$$

where

$$u = \left(\frac{\pi d}{\theta\lambda}\right)\left(\frac{n_e}{\sqrt{1+v\sin^2\theta_\delta}} - n_0\right), \quad v = \left(\frac{n_e}{n_0}\right)^2 - 1, \tag{7.56}$$

and θ_δ is the pretilt angle.

For the case of a planar aligned sample, that is, $\theta = 0$, we have

$$T = \cos^2(\phi_{ent} - \phi_{exit}) - \sin 2\phi_{ent}\sin 2\phi_{exit}$$
$$\times \sin^2\left[\left(\frac{\pi d}{\lambda}\right)\left(\frac{n_e}{\sqrt{1+\eta\sin^2\theta_\delta}} - n_0\right)\right]. \tag{7.57}$$

For crossed entrance–exit polarizers oriented such that $\phi_{ent} = 45°$ and $\phi_{exit} = 135°$,
this expression reduces to simply

$$T_\perp = \sin^2\left(\frac{\pi d \Delta n}{\lambda}\right). \tag{7.58}$$

For a 90° TN cell ($\theta = 90°$), we have

$$T_\perp = 1 - \frac{\sin^2\left(\frac{\pi}{2}\sqrt{1+u^2}\right)}{1+u^2} \quad \text{(for crossed polarizers),} \tag{7.59}$$

$$T_\parallel = 1 - T_\perp \quad \text{(for parallel polarizers).} \tag{7.60}$$

For a 180 STN cell, we have

$$T_\perp = \frac{\sin^2\left(\pi\sqrt{1+u^2}\right)}{1+u^2}. \tag{7.61}$$

For a 270 STN cell

$$T_\perp = 1 - \frac{\sin^2\left(\dfrac{3\pi}{2}\sqrt{1+u^2}\right)}{1+u^2}, \quad \text{where } u = \frac{3d\,\Delta n}{\lambda}. \tag{7.62}$$

7.3.3. Oblique Incidence: 4×4 Matrix Methods

The discussion above applies in the case when the light is incident normally on the plane of the optical elements (see Fig. 7.5). In this case, the Jones matrices are two-component vectors. However, for off-axis or obliquely incident light, this cannot account for various physical effects, such as Fresnel refraction and reflection of light at the interface, breakup of the light into ordinary and extraordinary components, and so on, and most importantly, multiple reflections at the interfaces where there is discontinuity in the refractive indices (e.g., between air and glass, between glass and polarizers). At each interface, four electric and magnetic tangential field components (E_x, E_y, H_x, and H_y) are required to fully describe the transmission and reflection.[12]

The 4×4 matrix method relates the tangential components of these field vectors at the exit (Z_2) to the inputs at the entrance plane (Z_1):

$$\Psi(Z_2) = \mathbf{M}(Z_2, Z_1)\Psi(Z_1), \tag{7.63}$$

where $\Psi = (E_x, H_y, E_y, -H_x)$ and $\mathbf{M}$ is the so-called transfer matrix. $\mathbf{M}$ depends on the dielectric anisotropy ($\varepsilon_\parallel - \varepsilon_\perp$), the layer thickness ($Z_2 - Z_1$), and the propagation wave vector of the incident light. For the case where the light is incident from the glass (index n_g) to the LC interface at an angle θ_{in}, and the director axis is oriented with respect to the z axis by the Euler angles $0(z)$ and $\phi(z)$, Wohler et al.[12] gives

$$\mathbf{M} = \beta_0 I + \beta_1 \Delta + \beta_2 \Delta^2 + \beta_3 \Delta^3, \tag{7.64}$$

where I is the identity matrix and Δ is a 4×4 matrix:

$$\Delta = \begin{bmatrix} \Delta_{11} & \Delta_{12} & \Delta_{13} & 0 \\ \Delta_{21} & \Delta_{22} & \Delta_{23} & 0 \\ 0 & 0 & 0 & \Delta_{34} \\ \Delta_{23} & \Delta_{13} & \Delta_{43} & 0 \end{bmatrix}. \tag{7.65}$$

In these equations,

$$\Delta_{11} = \frac{-(\Delta\varepsilon \sin\theta \cos\theta \sin\phi)X}{\varepsilon_{33}},$$

$$\Delta_{12} = \frac{1 - X^2}{\varepsilon_{33}},$$

$$\Delta_{13} = \frac{(\Delta\varepsilon \sin\theta \cos\theta \cos\phi)X}{\varepsilon_{33}},$$

$$\Delta_{21} = \left(\frac{\varepsilon_\perp}{\varepsilon_{33}}\right)(\varepsilon_\parallel - \sin^2\theta \cos^2\phi), \tag{7.66}$$

$$\Delta_{23} = -\left(\frac{\varepsilon_\perp}{\varepsilon_{33}}\right)(\Delta\varepsilon \sin^2\theta \sin\phi \cos\phi),$$

$$\Delta_{34} = 1,$$

$$\Delta_{43} = \left(\frac{\varepsilon_\perp}{\varepsilon_{33}}\right)(\varepsilon_\parallel - \Delta\varepsilon \sin^2\theta \sin^2\phi) - X^2,$$

$$\varepsilon_{33} = \varepsilon_\perp + \Delta\varepsilon \cos^2\theta,$$

$$X = n_g \sin\theta_{\text{in}},$$

and

$$\beta_0 = -\sum_{i=1}^4 \lambda_j \lambda_k \lambda_l \frac{f_i}{\lambda_{ij}\lambda_{ik}\lambda_{il}},$$

$$\beta_1 = -\sum_{i=1}^4 (\lambda_j\lambda_k + \lambda_j\lambda_l + \lambda_k\lambda_l)\frac{f_i}{\lambda_{ij}\lambda_{ik}\lambda_{il}},$$

$$\beta_2 = -\sum_{i=1}^4 (\lambda_j + \lambda_k + \lambda_l)\frac{f_i}{\lambda_{ij}\lambda_{ik}\lambda_{il}}, \tag{7.67}$$

$$\beta_3 = \sum_{i=1}^4 \frac{f_i}{\lambda_{ij}\lambda_{ik}\lambda_{il}},$$

where, $\lambda_{ij} = \lambda_i - \lambda_j$, $f_i = \exp(-ik_o\lambda_i d)$ and j, k, $l = 1, 2, 3, 4$, with

$$\lambda_{1,2} = \pm \sqrt{\Delta\varepsilon - X^2} \tag{7.68}$$

and

$$\lambda_{3,4} = -\sqrt{\frac{\varepsilon_{13}}{\varepsilon_{33}}} X \pm \sqrt{\frac{\varepsilon_{\parallel}\varepsilon_{\perp}}{\varepsilon_{33}}} \left[\varepsilon_{33} - \left(1 - \frac{\Delta\varepsilon}{\varepsilon_{\parallel}} \sin^2\theta\cos^2\phi \right) X^2 \right]^{1/2}. \tag{7.69}$$

Once **M** is found, the transmitted field components can be obtained from the equation.[12,13]

7.4. EXTENDED JONES MATRIX METHOD

Instead of the 4×4 matrix method described in the preceding section, oblique incidence can also be treated using the so-called extended Jones matrix method[14,15] involving 2×2 matrices, if the effects of multiple reflections can be neglected. To avoid confusion, we employ the *cgs units* adopted by these authors. Maxwell equations become

$$\nabla \cdot \mathbf{D} = 0,$$

$$\nabla \times \mathbf{H} = \left(\frac{1}{c}\right)\left(\frac{\partial D}{\partial t}\right),$$

$$\nabla \times \mathbf{E} = -\left(\frac{1}{c}\right)\left(\frac{\partial B}{\partial t}\right), \tag{7.70}$$

$$\nabla \cdot \mathbf{B} = 0,$$

where $\mathbf{D} = \varepsilon\mathbf{E}$,

$$\varepsilon = \begin{bmatrix} \varepsilon_{xx} & \varepsilon_{xy} & \varepsilon_{xz} \\ \varepsilon_{yx} & \varepsilon_{yy} & \varepsilon_{yz} \\ \varepsilon_{zx} & \varepsilon_{zy} & \varepsilon_{zz} \end{bmatrix}, \tag{7.71}$$

and

$$\varepsilon_{xx} = n_0^2 + \left(n_e^2 - n_0^2\right)\cos^2\theta\cos^2\phi,$$

$$\varepsilon_{xy} = \varepsilon_{yx} = \left(n_e^2 - n_0^2\right)\cos^2\theta \sin\phi \cos\phi,$$

$$\varepsilon_{xz} = \varepsilon_{zx} = \left(n_e^2 - n_0^2\right)\sin\theta \cos\theta \cos\phi,$$

$$\varepsilon_{yy} = n_0^2 + \left(n_e^2 - n_0^2\right)\cos^2\theta \sin^2\phi, \tag{7.72}$$

$$\varepsilon_{yz} = \varepsilon_{zy} = \left(n_e^2 - n_0^2\right)\sin\theta \cos\theta \sin\phi,$$

$$\varepsilon_{zz} = n_0^2 + \left(n_e^2 - n_0^2\right)\sin^2\theta.$$

Consider a plane wave of the form $\exp[i(\mathbf{k}\cdot\mathbf{r} - \omega t)]$ incident on a liquid crystal cell as depicted in Figure 7.7. For propagation in the x-z plane at an angle θ_k with the z axis, we have $k=(k_0 \sin\theta_k, 0, k_0 \cos\theta_k)$.

In the extended Jones matrix method, the liquid crystal cell is divided into N (usually several hundreds for accurate computation) layers, with the dielectric tensor of each layer differing from that of its adjacent layer. The input and output polarizers are considered as two separate layers characterized by dielectric tensors of the form of Equation (7.71). In each layer, there are four eigenwaves: two transmitted and two reflected waves. At each interface, the boundary condition that applies is that the tangential components of the electric field are continuous.

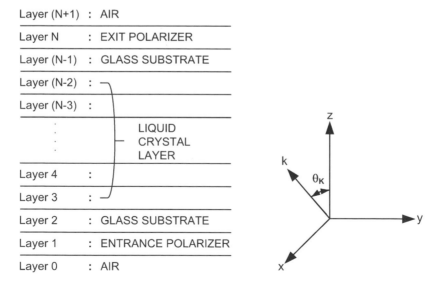

Figure 7.7. $N+1$ layers representation of a LC cell between polarizers for transfer matrix calculation. Insert shows the incident beam direction.

For the liquid crystal pixel, as shown in Figure 7.7, the extended Jones matrix becomes

$$J = J_N J_{N-1} \cdots J_1.$$ (7.73)

The Jones matrix at each layer is given by

$$J_n = (SGS^{-1})_n,$$ (7.74)

where

$$S = \begin{bmatrix} 1 & C_2 \\ C_1 & 1 \end{bmatrix} \text{ and } G = \begin{bmatrix} \exp(ik_{z1}d) & 0 \\ 0 & \exp(ik_{z1}d) \end{bmatrix},$$

where d is the corresponding layer thickness and

$$C_1 = \frac{\left[\left(\dfrac{k_x}{k_0}\right)^2 - \varepsilon_{zz}\right]\varepsilon_{yx} + \left(\dfrac{k_z k_{z1}}{k_0 k_0} + \varepsilon_{zx}\right)\varepsilon_{yz}}{\left[\left(\dfrac{k_{z1}}{k_0}\right)^2 + \left(\dfrac{k_x}{k_0}\right)^2 - \varepsilon_{yy}\right]\left[\left(\dfrac{k_x}{k_0}\right)^2 - \varepsilon_{zz}\right] - \varepsilon_{yz}\varepsilon_{zy}},$$ (7.75)

$$C_2 = \frac{\left[\left(\dfrac{k_x}{k_0}\right)^2 - \varepsilon_{zz}\right]\varepsilon_{xy} + \left(\dfrac{k_x k_{z2}}{k_0 k_0} + \varepsilon_{xz}\right)\varepsilon_{zy}}{\left[\left(\dfrac{k_{z2}}{k_0}\right)^2 - \varepsilon_{xx}\right]\left[\left(\dfrac{k_x}{k_0}\right)^2 - \varepsilon_{zz}\right] - \left(\dfrac{k_x k_{z2}}{k_0 k_0} + \varepsilon_{zx}\right)\left(\dfrac{k_x k_{z2}}{k_0 k_0} + \varepsilon_{xz}\right)},$$ (7.76)

$$\frac{k_{z1}}{k_0} = \sqrt{n_0^2 - \left(\frac{k_x}{k_0}\right)^2},$$ (7.77)

$$\frac{k_{z2}}{k_0} = -\frac{\varepsilon_{xz}}{\varepsilon_{zz}}\frac{k_x}{k_0} + \frac{n_0 n_e}{\varepsilon_{zz}}\sqrt{\varepsilon_{zz} - \left(1 - \frac{n_e^2 - n_0^2}{n_e^2}\cos^2\theta\sin^2\phi\right)\left(\frac{k_x}{k_0}\right)^2}.$$ (7.78)

Let $E_x^{(0)}$ and $E_y^{(0)}$ be the incident field. The transmitted field $E_x^{(N+1)}, E_y^{(N+1)}$ is then given by

$$\begin{bmatrix} E_x^{(N+1)} \\ E_y^{(N+1)} \end{bmatrix} = J \begin{bmatrix} E_x^{(0)} \\ E_y^{(0)} \end{bmatrix}. \tag{7.79}$$

For an obliquely incident light, as shown in Figure 7.7, the transmission T is

$$T_{\text{opt}} = \frac{\left|E_x^{(N+1)}\right|^2 + \cos^2(\theta_p)\left|E_y^{(N+1)}\right|^2}{\left|E_x^{(0)}\right|^2 + \cos^2(\theta_p)\left|E_y^{(0)}\right|^2}. \tag{7.80}$$

where $\theta_p = \sin^{-1}[\sin(\theta_k)/\text{Re}(n_p)]$ is the exit angle of the light leaving the polarizer, and $\text{Re}(n_p)$ is the average real part of n_o and n_e of the polarizer.

With the inclusion of the transmission at the air–glass interfaces at the entrance and exit, the net optical transmission of the "pixel" becomes

$$T'_{\text{opt}} = T_{\text{ent}} T_{\text{opt}} T_{\text{exit}}. \tag{7.81}$$

For normal incidence ($\theta_k = 0$)

$$T_{\text{ent}} = T_{\text{exit}} = \frac{4\,\text{Re}(n_p)}{\left[1 + \text{Re}(n_p)\right]^2}, \tag{7.82}$$

whereas for $\theta_k \neq 0$, we have

$$T_{\text{ent}} = T_p \cos^2(\alpha_{\text{ent}}) + T_v \sin^2(\alpha_{\text{ent}}), \tag{7.83}$$

$$T_{\text{exit}} = T_p \cos^2(\alpha_{\text{exit}}) + T_v \sin^2(\alpha_{\text{exit}}), \tag{7.84}$$

where α_{ent} and α_{exit} are the entrance and exit angles at the glass–air interfaces, respectively.

Finally, the transmission for parallel (p) and vertical (s) components of the transmission are given by

$$T_p = \frac{\sin(2\theta_k)\sin(2\theta_p)}{\sin^2(\theta_k + \theta_p)\cos^2(\theta_k - \theta_p)}, \tag{7.85}$$

$$T_v = \frac{\sin(2\theta_k)\sin(2\theta_p)}{\sin^2(\theta_k + \theta_p)}, \tag{7.86}$$

where

$$\cos(\alpha_{ent}) = \frac{E_x^{(0)}}{\sqrt{|E_x^{(0)}|^2 + \cos^2(\theta_k)|E_y^{(0)}|^2}}, \tag{7.87}$$

$$\cos(\alpha_{exit}) = \frac{E_x^{(N+1)}}{\sqrt{|E_x^{(N+1)}|^2 + \cos^2(\theta_k)|E_y^{(N+1)}|^2}}. \tag{7.88}$$

The transmission for various cell geometries, director axis orientations, pretilts, and so on have been calculated.[14]

7.5 FINITE-DIFFERENCE TIME-DOMAIN TECHNIQUE

With the advance of computing power and speed, many other numerical methods traditionally used in radio and long wavelength regimes that were too cumbersome and time consuming for applications involving short wavelength electromagnetic waves (light) have become feasible. One numerical technique that is attracting increased use is the so-called finite-difference time-domain technique.

The finite-difference time-domain (FDTD) method provides direct numerical solutions of Maxwell's equations in both space and time domains,[16,17] and is widely used to study electromagnetic radiation and scattering, especially when complex, inhomogeneous geometries are involved. The FDTD method can also be used to analyze optical wave propagation through liquid crystal (LC) structures.[18,19] In simple stratified structures (i.e., structures in which the dielectric tensor varies along the direction normal to the display surface) matrix-type methods, such as the extended Jones method[14] and the Berreman matrix method,[20] can be satisfactorily used to analyze optical propagation. However, these methods cannot give the accurate optical information required for advanced displays possessing smaller pixels, in-plane electrodes, or multidomains because of the underlying multidimensional director deformations.[21] By contrast, the FDTD approach is well suited to tackle these complex propagation problems, and has been successfully implemented for applications such as optics in textured LCs,[22] diffraction gratings,[23] and light interaction with LCs.[24]

7.5.1. The Implementation of FDTD Methods

Yee first devised a set of finite-difference equations for the time-dependent Maxwell equations for isotropic media.[16] In Yee's algorithm, space and time are discretized and the finite central-difference expressions for their derivatives are used. Figure 7.8 shows the grid positions of the electric and magnetic field components in three-dimensional space. At all grid points, the electric and magnetic components of the coupled Maxwell equations are iteratively computed until the steady state is obtained.[17]

To describe the application of the FDTD method to LCs, let us start from Maxwell's equations:

$$\frac{\partial \vec{D}}{\partial t} = \nabla \times \vec{H}, \tag{7.89}$$

$$\mu_o \frac{\partial \vec{H}}{\partial t} = -\nabla \times \vec{E}, \tag{7.90}$$

$$\vec{D} = \tilde{\varepsilon}\vec{E}. \tag{7.91}$$

The optical dielectric tensor $\tilde{\varepsilon}$ is represented as

$$\tilde{\varepsilon} = \begin{bmatrix} \varepsilon_{xx} & \varepsilon_{xy} & \varepsilon_{xz} \\ \varepsilon_{yx} & \varepsilon_{yy} & \varepsilon_{yz} \\ \varepsilon_{zx} & \varepsilon_{zy} & \varepsilon_{zz} \end{bmatrix} \tag{7.92}$$

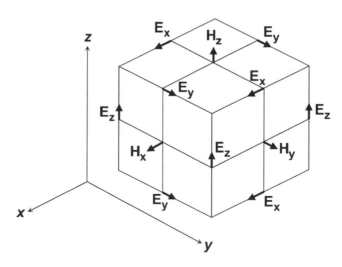

Figure 7.8. Positions of the electric and magnetic field components in Yee grids.

where each component of $\tilde{\varepsilon}$ is related to the refractive indices n_o (ordinary ray) and n_e (extraordinary ray) by

$$\varepsilon_{ij} = n_o^2 \delta_{ij} + \left(n_e^2 - n_o^2\right) n_i n_j \qquad (i, j = x, y, z), \tag{7.93}$$

where δ_{ij} is Kronecker's delta, and n_i and n_j are LC director components.

For leapfrog time stepping, each field component is discretized and every $\vec{D}$ $(\vec{H})$ component is updated by a circulating $\vec{H}$ $(\vec{D})$ component. As an example, the x components of the fields are updated using the following equations:

$$
\begin{aligned}
D_x^{n+1/2}(i+1/2, j, k) = D_x^{n-1/2}(i+1/2, j, k) \\
+ \Delta t \left\{
\begin{array}{c}
\dfrac{H_z^n(i+1/2, j+1/2, k) - H_z^n(i+1/2, j-1/2, k)}{\Delta y} \\[2mm]
- \dfrac{H_y^n(i+1/2, j, k+1/2) - H_y^n(i+1/2, j, k-1/2)}{\Delta z}
\end{array}
\right\},
\end{aligned} \tag{7.94}
$$

$$
\begin{aligned}
H_x^{n+1}(i, j+1/2, k+1/2) = H_x^n(i, j+1/2, k+1/2) \\
+ \dfrac{\Delta t}{\mu_o} \left\{
\begin{array}{c}
\dfrac{D_y^{n+1/2}(i, j+1/2, k+1) - D_y^{n+1/2}(i, j+1/2, k)}{\Delta z} \\[2mm]
- \dfrac{D_z^{n+1/2}(i, j+1, k+1/2) - D_z^{n+1/2}(i, j, k+1/2)}{\Delta y}
\end{array}
\right\},
\end{aligned} \tag{7.95}
$$

where n is the time step number, $\{i, j, k\}$ are the indices of the grid points, Δy and Δz are the unit cell dimensions (distances between consecutive grid points in the y or z directions), and Δt is the time step. If the dimensions of the unit cell are the same, that is, $\Delta x = \Delta y = \Delta z = \Delta$, the time step interval should satisfy the bounding condition known as "Courant condition" in order to obtain numerical stability:[17]

$$\Delta t \leq \frac{\Delta}{\sqrt{n} \cdot c}, \tag{7.96}$$

where c is the velocity of light in free space and n is the dimension of the FDTD space.

From Equation (7.94), the dielectric displacement components are first obtained and they are used to calculate the corresponding electric field components using the inverted constitutive relation of Equation (7.91). The obtained electric field components are then used for to update the magnetic components. This procedure is iterated until timestepping is concluded.

In order to determine the solution of the wave propagation in infinite (unbounded) regions, the computational space should be truncated with nonphysical perfectly absorbing media. This can be achieved by surrounding the computational domain with a perfectly matched layer (PML) introduced by Berenger.[25] In the PML regions, each field component is split into the additive subcomponents according to the coordinate axis and Maxwell's equations are modified with the split field components having electric conductivities and magnetic losses which are designed to be matched well with each other. This PML provides a reflectionless boundary for plane waves of arbitrary incidence angle, frequency, and polarization. Figure 7.9 shows how the main computational domain (i.e., the physical domain) is surrounded by PML regions. In this simple one-dimensional case, Maxwell's equations are expressed as

$$\frac{\partial}{\partial t}D_y + \sigma_z^D D_y = \frac{\partial}{\partial z}H_x, \qquad (7.97)$$

$$\mu_0 \frac{\partial}{\partial t}H_x + \sigma_z^H H_x = \frac{\partial}{\partial z}E_y. \qquad (7.98)$$

The matching condition[19] of the PML conductivities σ_z^D and σ_z^H given by

$$\sigma_z^D = \frac{\sigma_z^H}{\mu_0}. \qquad (7.99)$$

In addition, the PML conductivity increases as the thickness of the slab increases as follows:

$$\sigma_z^D(z) = \left(\frac{z}{\delta}\right)^m \cdot \sigma_{max}, \qquad (7.100)$$

where δ is the PML thickness and m is a small number which is usually determined in the range from 3 to 4.[17] The constant σ_{max} can be equal to

$$\sigma_{max} = -\frac{c(m+1)\ln[R(0)]}{2\delta}, \qquad (7.101)$$

where c is the velocity of light in free space and $R(0)$ is the desired reflection error at normal incidence.

The incident plane wave onto the LC structure can be realized using a total/scattered (TF/SF) field formulation.[17] This technique is based on the linearity of Maxwell's equations. According to this technique, the total electric and magnetic fields $\vec{E}_{tot}$ and $\vec{H}_{tot}$ are separated into two subcomponents—one is the component of the incident field ($\vec{E}_{inc}$, $\vec{H}_{inc}$) and the other is that of the scattered field ($\vec{E}_{scat}$, $\vec{H}_{scat}$) which results from the interaction between the incident field and any material in the main domain:

$$\vec{E}_{tot} = \vec{E}_{inc} + \vec{E}_{scat}, \qquad (7.102)$$

$$\vec{H}_{\text{tot}} = \vec{H}_{\text{inc}} + \vec{H}_{\text{scat}}. \tag{7.103}$$

The main computational space is then separated by a virtual surface into an inner and an outer space, as shown in Figure 7.9. In the inner space, the TF region, the FDTD procedure is performed on both the incident and scattered waves, whereas in the outer space, the SF region, only the scattered fields are considered. By using this TF/SF technique, one can obtain the propagating plane wave not interacting with the absorbing boundary conditions.[26] In implementing this technique, the incident field should be added at the TF/SF boundary. For the simple one-dimensional case of Figure 7.9, if the incident field is known in the whole main domain, each field component at the two TF/SF boundaries is corrected by adding the incident field component as follows:

$$E_y^{n+1/2}(K_L) = \{E_y^{n+1/2}(K_L)\} - H_x^n(K_L - 1/2)\big|_{\text{inc}}, \tag{7.104}$$

$$H_x^{n+1}(K_L - 1/2) = \{H_y^{n+1}(K_L - 1/2)\} - E_y^{n+1/2}(K_L - 1)\big|_{\text{inc}}, \tag{7.105}$$

$$E_y^{n+1/2}(K_R) = \{E_y^{n+1/2}(K_R)\} + H_x^n(K_R + 1/2)\big|_{\text{inc}}, \tag{7.106}$$

$$H_x^{n+1}(K_R + 1/2) = \{H_x^{n+1}(K_R + 1/2)\} + E_y^{n+1/2}(K_R + 1)\big|_{\text{inc}}. \tag{7.107}$$

The curly brackets in Equations (7.104)−(7.107) indicate that all field components are updated by a time-stepping procedure before the incident field correction term is added.

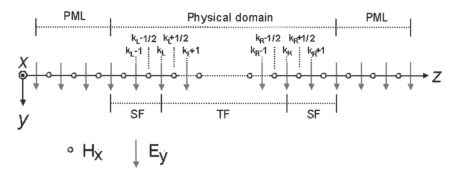

Figure 7.9. One-dimensional Yee grids: k is the index number of grids. TF and SF indicates total field and scattered field regions, respectively.

7.5.2. Example: FDTD Computations of the Twisted Nematic Cell in One Dimension

As a simple illustration[27] of the FDTD technique, consider a twisted nematic liquid crystal (E7) cell in one dimension (in z axis). The cell thickness is set equal to 5 μm. The pretilt angles of the top and bottom glasses are assumed to be nearly zero (0.01°) The director of the bottom glass is parallel to the x axis and the total twist angle is set equal to 90°.

In this case, Maxwell's equations in the main computational region are

$$\frac{\partial}{\partial t} D_x = -\frac{\partial}{\partial z} H_y, \tag{7.108}$$

$$\frac{\partial}{\partial t} D_y = \frac{\partial}{\partial z} H_x, \tag{7.109}$$

$$\frac{\partial}{\partial t} H_x = \frac{1}{\mu_o} \frac{\partial}{\partial z} E_y, \tag{7.110}$$

$$\frac{\partial}{\partial t} H_y = -\frac{1}{\mu_o} \frac{\partial}{\partial z} E_x. \tag{7.111}$$

Within the PML regions, Maxwell's equations can be rewritten as

$$\frac{\partial}{\partial t} D_x + \sigma_z D_x = -\frac{\partial}{\partial z} H_y, \tag{7.112}$$

$$\frac{\partial}{\partial t} D_y + \sigma_z D_y = \frac{\partial}{\partial z} H_x, \tag{7.113}$$

$$\mu_o \frac{\partial}{\partial t} H_x + \sigma_z^* H_x = \frac{\partial}{\partial z} E_y, \tag{7.114}$$

$$\mu_o \frac{\partial}{\partial t} H_y + \sigma_z^* H_y = -\frac{\partial}{\partial z} E_x. \tag{7.115}$$

The material constants of the nematic liquid crystal E7 used in this calculation are elastic constants $K_1 = 11.2 \times 10^{-12}$ N, $K_2 = 6.8 \times 10^{-12}$ N, and $K_3 = 18.6 \times 10^{-12}$ N; and the dielectric constants $\varepsilon_\perp = 5.15$ and $\varepsilon_\parallel = 18.96$. The ordinary and extraordinary refractive indices are $n_o = 1.5185$ and $n_e = 1.737$, respectively. The dielectric tensor is determined from the director profiles, which are obtained by the minimization of the total free-energy density.

Consider an x-polarized input plane wave ($\lambda = 633 \times 10^{-9}$ m) from the bottom glass. It is assumed that the index of the surrounding medium is 1. For Yee grids, the unit cell space is $\Delta z = \lambda/40 = 1.5825 \times 10^{-8}$ m and the time step is $\Delta t = 2.639 \times 10^{-17}$s. Figures 7.10 and 7.11 show the electric field distributions after 15,000 time steps. As shown in Figure 7.10, in the off state (0 V), as the incident wave

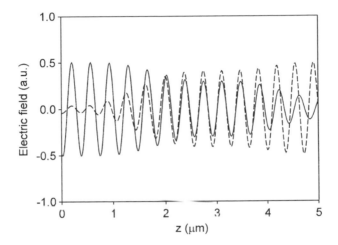

Figure 7.10. Electric field distribution through the LC layer at the off state in a TN cell. The solid and dotted line denotes E_x and E_y field, respectively.

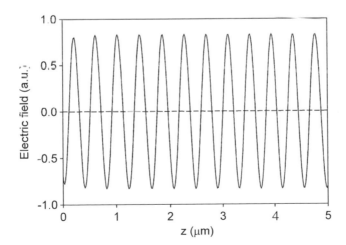

Figure 7.11. Electric field distribution through the LC layer at the on state in a TN cell. The solid and dotted line denotes E_x and E_y field, respectively.

propagates through the LC layer, the polarization state is changed from E_x to E_y (the magnitude of the x component progressively diminishes, while the magnitude of the y component grows), confirming the well-known "waveguide" property of the twisted nematic cell. However, when the voltage is applied, the molecules become untwisted and consequently the x-polarized input remains unchanged in magnitude, while the y-polarized input remains at vanishing value, as shown in Figure 7.11; that is, the polarization plane of the light is no longer rotated in the on state (5 V).

Although we have demonstrated a basic one-dimensional propagation case, FDTD is a very powerful and accurate method to describe almost every optical response including refraction, reflection, diffraction, and scattering phenomena, since the underlying principles are governed by Maxwell's equations. It is likely to see increased usage in analyzing optical propagation through inhomogeneous anisotropic media such as liquid crystals in complex geometries.[21–24]

REFERENCES

1. Veselago, V. G., 1968. The electrodynamics of substances with simultaneously negative values of ε and μ. *Sov. Phys. Usp.* 10 (4):509–514.

2. Shalaev, V. M., W. S. Cai, U. K. Chettiar, H. K. Yuan, A. K. Sarychev, V. P. Drachev, and A. V Kildishev. 2005. *Opt. Lett.* 30:3356–3358.

3. Smith, D. R., et al. 2000. Composite medium with simultaneously negative permeability and permittivity. *Phys. Rev. Lett.* 84:4184.

4. Shelby, R. A., D. R. Smith, and S. Schultz. 2001. Experimental verification of a negative index of refraction. *Science.* 292:77.

5. Pendry, J. B., 2000. Negative refraction makes a perfect lens. *Phys. Rev. Lett.* 85:3966.

6. Gingrich, M. A., D. H. Werner, and A. Monorchio. The synthesis of planar left-handed metamaterials from frequency selective surfaces using genetic algorithms. In *2004 IEEE AP-S International Symposium on Antennas and Propagation and USNC/URSI National Radio Science Meeting.*

7. Khoo, I. C., D. H. Werner, X. Liang, A. Diaz, and B. Weiner. 2006. Nano-sphere dispersed liquid crystals for tunable negative-zero-positive index of refraction in the optical and terahertz regimes, *Optics Letts.* 31:2592.

8. Eritsyan, S., 2000. Diffraction reflection of light in a cholesteric liquid crystal in the presence of wave irreversibility and Bragg formula for media with non-identical forward and return wavelengths. *J. Exp. Theor. Phys.* 90:102–108; see also Bita, I., and E. L. Thomas. 2005. Structureally chiral photonic crystals with magneto-optic activity: Indirect photonic bandgaps, negative refraction, and superprism effect. *J. Opt. Soc. Am. B.* 22:1199–1210.

9. Yeh, P., and C. Gu. 1999. *Optics of Liquid Crystal Displays.* New York: Wiley Interscience.

10. Yariv, A., 1997. *Optical Electronics in Modern Communications.* 5th ed. New York: Oxford University Press.

11. See, for example, Ong, H. L., 1988. Origin and characteristics of the optical properties of general twisted nematic liquid crystal displays. *J. Appl. Phys.* 64:614–628.

12. Wohler, H., G. Haas, M. Fritsch, and D. A. Mlynski. 1988. Faster 4×4 matrix method for uniaxial inhomogeneous media. *J. Opt. Soc. Am. A.* 5:1554.

13. See also Yang, K. H., 1990. Elimination of the Fabry–Perot effect in the 4×4 matrix method for inhomogeneous uniaxial media. *J. Appl. Phys.* 68:1550.

14. Lien, A., 1990. Extended Jones matrix representation for the twisted nematic liquid crystal display at the oblique incidence. *Appl. Phys. Lett.* 57:2767.

15. Ong, H. L. 1991. Electro-optics of electrically controlled birefringence liquid crystal display by 2×2 propagation matrix and analytical expression at oblique angle. *Appl. Phys. Lett.* 59:155–157.

16. Yee, K. S. 1966. Numerical solution of initial boundary value problems involving Maxwell's equations in isotropic media. *IEEE Trans. Antennas Propag.* 14(3):302.

17. Taflove, A., and S. C. Hagness. 2000. *Computational Electrodynamics: The Finite-Difference Time-Domain Method.* 2nd ed. Massachusetts: Artech House.

18. Witzigmann, B., P. Regli, and W. Fuchtner. 1998. Rigorous electromagnetic simulation of liquid crystal displays. *J. Opt. Soc. Am. A.* 15(3):753.

19. Kriezis, E. E., and S. J. Elston. 1999. Finite-differece time domain method for light wave propagation within liquid crystal devices. *Opt. Commun.* 165:99.

20. Berreman, D. W. 1972. Optics in stratified and anisotropic media matrix formulation. *J. Opt. Soc. Am.* 62:502.

21. Titus, C. M., P. J. Bos, J. R. Kelly, and E. C. Gartland. 1999. Comparison of analytical calculations to finite-difference time-domain simulations of one-dimensional spatially varying anisotropic liquid crystal structures. *Jpn. J. Appl. Phys., Part 1.* 38:1488.

22. Hwang, D. K., and A. D. Rey. 2006. Computational studies of optical textures of twist disclination loops in liquid-crystal films by using the finite-difference time-domain method. *J. Opt. Soc. Am. A.* 23(2):483.

23. Wang, Bin, D. B. Chung, and P. J. Bos. 2004. Finite-difference time-domain optical calculations of polymer-liquid crystal composite electrodiffractive device. *Jpn. J. Appl. Phys. Part 1.* 43(1):176.

24. Ilyina, V., S. J. Cox, and T. J. Sluckin. 2004. FDTD method for light interaction with liquid crystals. *Mol. Cryst. Liq. Cryst.* 422(1):271.

25. Berenger, J.-P. 1994. A perfectly matched layer for the absorption of electromagnetic waves. *J. Comput. Phys.* 114:185.

26. Sullivan, D. M. 2000. *Electromagnetic Simulation Using the FDTD Method.* New Jersey: IEEE Press.

27. Park, J. H. (unpublished).

8

Laser-Induced Orientational Optical Nonlinearities in Liquid Crystals

8.1. GENERAL OVERVIEW OF LIQUID CRYSTAL NONLINEARITIES

In linear optical processes the physical properties of the liquid crystal, such as its molecular structure, individual or collective molecular orientation, temperature, density, population of electronic levels, and so forth, are not affected by the optical fields. The direction, amplitude, intensity, and phase of the optical fields are affected in a unidirectional way (i.e., by the physical parameters of the liquid crystal). The optical properties of liquid crystals may, of course, be controlled by some externally applied dc or low-frequency fields; this gives rise to a variety of electro-optical effects which are widely used in many electro-optical display and image-processing applications as discussed in previous chapters.

Liquid crystals are also optically highly nonlinear materials in that their physical properties (temperature, molecular orientation, density, electronic structure, etc.) are easily perturbed by an applied optical field.[1–5] Nonlinear optical processes associated with electronic mechanisms will be discussed in Chapter 10. In this and the next chapter, we discuss the principal nonelectronic mechanisms for the nonlinear optical responses of liquid crystals.

Since liquid crystalline molecules are anisotropic, a polarized light from a laser source can induce an alignment or ordering in the isotropic phase, or a realignment of the molecules in the ordered phase. These result in a change in the refractive index.

Other commonly occurring mechanisms that give rise to refractive index changes are laser-induced changes in the temperature, ΔT, and the density, $\Delta\rho$. These changes could arise from several mechanisms. A rise in temperature is a natural consequence of photoabsorptions and the subsequent inter- and intramolecular thermalization or nonradiative energy relaxation processes. In the isotropic phase the change in the refractive index is due to the density change following a rise in temperature. In the nematic phase the refractive indices are highly dependent on the temperature through their dependence on the order parameters, as well as on the density, as discussed in Chapter 3.

Liquid Crystals, Second Edition By Iam-Choon Khoo
Copyright © 2007 John Wiley & Sons, Inc.

Temperature changes inevitably lead to density changes via the thermoelastic coupling (i.e., thermal expansion). However, density changes can also be due to the electrostrictive effect (i.e., the movement of the liquid crystal molecules toward a region of high laser field). To gain a first-order understanding of the electrostrictive effect,[6,7] let us ignore for the moment the anisotropy of the liquid crystalline parameters. Consider a molecule situated in a region illuminated by an optical field E. The field induces a polarization $P = \alpha E$, where α is the molecular polarizability. The electromagnetic energy expended on the molecule is thus

$$u_E = - \int_0^E P \cdot dE = - \tfrac{1}{2}\alpha(E \cdot E). \tag{8.1}$$

If the electric field is spatially varying (i.e., there are regions of high and low energy densities E^2), there will be a force acting on the molecule given by

$$F_{\text{molecule}} = - \nabla u_E = \tfrac{1}{2}\alpha\nabla(E \cdot E), \tag{8.2}$$

that is, the molecule is pulled into the region of increasing field strength (see Fig. 8.1). As a result, the density in the high field region is increased by an amount $\Delta\rho$. This increase in density gives rise to an increase in the dielectric constant (and therefore the refractive index) by an amount

$$\Delta\varepsilon = \rho\frac{\partial\varepsilon}{\partial\rho}\left(\frac{\Delta\rho}{\rho}\right)$$
$$= \gamma^e\left(\frac{\Delta\rho}{\rho}\right), \tag{8.3}$$

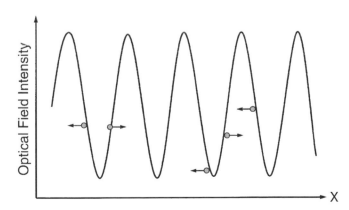

Figure 8.1. Molecules are pulled toward regions of higher field strength by electrostrictive forces.

where

$$\gamma^e = \rho\left(\frac{d\varepsilon}{d\rho}\right)$$

(8.4)

is the electrostrictive constant.

Large density and temperature changes could also give rise to flows and director axis reorientations. An intense laser field induces flows in liquid crystals via the pressure it exerts on the system.[4,6] The pressure $p(r, t)$ creating the flow may originate from the thermoelastic or electrostrictive effects mentioned previously in conjunction with thermal and density changes. In nonabsorbing liquid crystals the flow is due mainly to electrostrictive forces of the type shown in Equation (8.2), which are derived generally from the so-called Maxwell stress:[9]

$$\boldsymbol{F} = (\boldsymbol{D} \cdot \boldsymbol{\nabla})\boldsymbol{E}^* - \tfrac{1}{2}\boldsymbol{\nabla}(\boldsymbol{E} \cdot \boldsymbol{D}^*).$$

(8.5)

Electrostrictive effects are highly dependent on the gradient of the electromagnetic fields, which are naturally present in tightly focused or spatially highly modulated pulsed lasers.

Flows also give rise to director axis realignment (see Fig. 8.2). In the extreme case of flow, the liquid crystal is forced to vacate the site it occupied (i.e., an empty space is left). Laser-induced flow effects thus give rise to large index changes, as observed in several studies involving nanosecond or picosecond laser pulses.[6-9]

In this and the next chapters, laser-induced changes in the director axis orientation $\theta(r, t)$, density $\rho(r, t)$, temperature $T(r, t)$, and flows are separately discussed for all the principal mesophases of liquid crystals. An intense laser pulse can also generate electronic nonlinearities in liquid crystals. This is treated separately in Chapter 10.

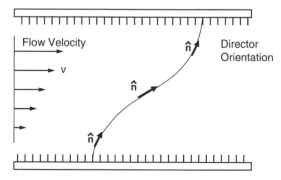

Figure 8.2.   An example of the flow-induced director axis reorientation effect in nematics. v is the flow velocity.

8.2. LASER-INDUCED MOLECULAR REORIENTATIONS IN THE ISOTROPIC PHASE

8.2.1. Individual Molecular Reorientations in Anisotropic Liquids

In the isotropic phase the liquid crystal molecules are randomly oriented owing to thermal motion, just as in conventionally anisotropic liquids. An intense laser field will force the anisotropic molecules to align themselves in the direction of the optical field through the dipolar interaction (see Fig. 8.3), in order to minimize the energy. Such a process is often called laser-induced ordering; that is, the laser induced some degree of preferred orientation in an otherwise random system. Because the molecules are birefringent, this partial alignment gives rise to a change in the effective optical dielectric constant (i.e., an optical field intensity-dependent refractive index change).

If the laser is polarized in the x direction, as shown in Figure 8.4, the induced polarization in the x direction is given by

$$P_x = \varepsilon_0 \, \Delta \chi_{xx}^{\mathrm{op}} E_x, \tag{8.6}$$

where $\Delta \chi^{\mathrm{op}}$ is the optically induced change in the susceptibility. In terms of the principal axes 1 and 2,

$$P_x = P_2 \cos \theta + P_1 \sin \theta, \tag{8.7}$$

where

$$P_1 = \varepsilon_0 \chi_{11} E_1 = \varepsilon_0 \chi_{11} E_x \sin \theta, \tag{8.8}$$

$$P_2 = \varepsilon_0 \chi_{22} E_2 = \varepsilon_0 \chi_{22} E_x \cos \theta. \tag{8.9}$$

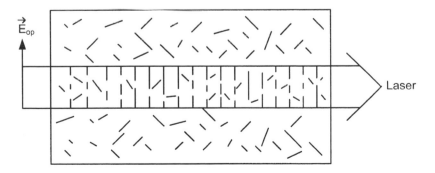

Figure 8.3.   Laser-induced ordering in an anisotropic liquid.

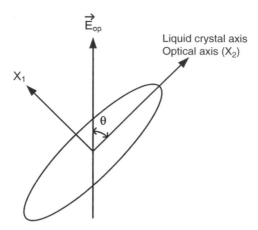

Figure 8.4.   Interaction of a linearly polarized light with an anisotropic (birefringent) molecule.

We therefore have

$$\Delta\chi_{xx}^{op} = \chi_{22}\cos^2\theta + \chi_{11}\sin^2\theta. \tag{8.10}$$

This may be expressed in terms of the average susceptibility:

$$\bar{\chi} = \tfrac{1}{3}(\chi_{22} + 2\chi_{11}) \tag{8.11}$$

and the susceptibility anisotropy:

$$\Delta\chi = \chi_{22} - \chi_{11} \tag{8.12}$$

by

$$\Delta\chi^{op} = \bar{\chi} + \Delta\chi\langle\!\langle\cos^2\theta - \tfrac{1}{3}\rangle\!\rangle. \tag{8.13}$$

In Equation (8.13) the angle brackets containing the factor $(\cos^2\theta - \tfrac{1}{3})$ signify that $\Delta\chi^{op}$ is a macroscopic parameter and an ensemble average; it can be expressed in terms of the (induced) order parameter $Q \equiv \langle\tfrac{3}{2}\cos^2\theta - \tfrac{1}{2}\rangle$ by

$$\Delta\chi_{xx}^{op} = \bar{\chi} + \tfrac{2}{3}\Delta\chi Q. \tag{8.14}$$

The total polarization P_x therefore becomes

$$\begin{aligned} P_x &= [\varepsilon_0\bar{\chi} + \varepsilon_0\tfrac{2}{3}\Delta\chi Q]E_x \\ &= P_x^{L} + P_x^{NL}, \end{aligned} \tag{8.15}$$

where the linear polarization $P_x^L = \varepsilon_0 \bar{\chi} E$ is the contribution from the unperturbed system, and the nonlinear polarization

$$P_x^{NL} = \tfrac{2}{3} \varepsilon_0 \Delta\chi QE \qquad (8.16)$$

arises from the laser-induced molecular orientation, or ordering.

From Equation (8.16) one can see that the molecular orientational nonlinearity in the isotropic phase of a liquid crystal is directly proportional to the laser-induced order parameter Q. In typical anisotropic liquids (e.g., CS_2 or liquid crystals at temperatures far above T_C), the value of Q may be obtained by a statistical mechanics approach. In the completely random system, the average orientation is described by a distribution function $f(\theta)$:

$$Q = \langle \tfrac{3}{2}\cos^2\theta - \tfrac{1}{2} \rangle = \int_0^\pi f(\theta)\left(\tfrac{3}{2}^2\theta - \tfrac{1}{2}\right) d \qquad (8.17)$$

In a steady state the equilibrium value for $f(\theta)$ is given by

$$f(\theta) = \frac{e^{-\varepsilon/K_BT}}{\int_0^\pi f(\theta)d\cos\theta}, \qquad (8.18)$$

where ε is the interaction energy of a molecule [$\varepsilon = \varepsilon_0(\Delta\alpha/2)|E|^2$] and $\Delta\alpha$ is the molecular polarization anisotropy ($\Delta\alpha = \Delta\chi/N$). K_B is the Boltzmann constant.

Laser-induced individual molecular orientations in liquid crystalline systems have been studied by several groups.[10–13] In the study by Lalanne et al.,[13] both the uncorrelated individual molecular reorientations and the correlated reorientation effects, described in the next section, induced by picosecond laser pulses have been measured. Typically, the individual molecular motion is characterized by a response time on the order of a few picoseconds, and fluctuations in these individual molecular motions give rise to a broad central peak in the Rayleigh scattering measurement; this is usually referred to as the Rayleigh wing scattering component, and it always exists in ordinary liquids (see Chapter 5). On the other hand, the correlated molecular reorientational effect discussed in the next section is characterized by a response time on the order of 10^1–10^2 ns. This gives rise to a narrow central component in the Rayleigh scattering spectrum and could also be called a Rayleigh wing component because it also originates from orientational fluctuations.

Besides these molecular motions, an optical field could also induce other types of orientation effects in liquid crystalline systems, for example, nuclear reorientation caused by the field-induced nuclear orientational anisotropy. This process is sometimes referred to as the nuclear optical Kerr effect[10] as it results in an optical intensity-dependent change in the optical dielectric constant. Such effects in liquid crystals have been investigated by Deeg and Fayer.[10] In general, these nuclear motions are characterized by a rise time of a few picoseconds and a decay time of about 10^2 ps in experiments involving subpicosecond laser pulses. The dynamics of these nuclear motions are also more complex than the individual molecular reorientational effects discussed previously.[10]

8.2.2. Correlated Molecular Reorientation Dynamics

For liquid crystals, owing to pretransitional effects near T_C, the induced ordering Q exhibits interesting correlated dynamics and temperature-dependent effects.

In general, short intense laser pulses are required to create appreciable molecular alignment in liquids. To quantitatively describe the pulsed laser-induced effect, a time-dependent approach is needed. In this regime $f(\theta)$ obeys a Debye rotational diffusion equation:[10]

$$\eta \frac{\partial f}{\partial t} = \frac{1}{\sin\theta} \frac{\partial}{\partial\theta} \left[\sin\theta \left(\frac{\partial f}{\partial\theta} + 2\Delta\alpha \frac{|E|^2}{K_B T} \sin\theta \cos\theta\, f \right) \right], \qquad (8.19)$$

where η is the viscosity coefficient. Substituting Equation (8.17) in Equation (8.19), one obtains a dynamical equation for Q:

$$\frac{\partial Q}{\partial t} = -\frac{Q}{\tau_D} + \frac{2\Delta\,\alpha\,|E|^2}{3\eta}, \qquad (8.20)$$

where the relaxation time τ_D is given by

$$\tau_D = \frac{\eta}{5KT}. \qquad (8.21)$$

Equation (8.20) shows that $Q \sim |E|^2$, and thus the nonlinear polarization from Equation (8.16) becomes

$$P_x^{\mathrm{NL}} \sim |E|^2\, E. \qquad (8.22)$$

In nonlinear optics terminology, see Chapter 11, this is a third-order nonlinear polarization which is related to the electric field E by a third-order nonlinear susceptibility that is proportional to $|E|^2$. A nonlinear susceptibility of this form is equivalent to an intensity-dependent refractive index change.

In the vicinity of the phase transition temperature T_C, molecular correlations in liquid crystals give rise to interesting so-called pretransitional phenomena. This is manifested in the critical dependences of the laser-induced index change and the response time on the temperature. These critical dependences are described by

Landau's[14] theory of second-order phase transition advanced by deGennes,[15] as explained in Chapter 2.

The free energy per unit volume in the isotropic phase of a liquid crystal, in terms of a general order parameter tensor Q_{ij}, is given by

$$F = F_0 + \tfrac{1}{2} A Q_{ij} Q_{ji} - \tfrac{1}{4} \chi_{ij} E_i^* E_j,$$ (8.23)

$$A = a(T - T^*),$$ (8.24)

where A and T^* are constants defined in Chapter 2.

From Equation (8.23) the dynamical equation for Q_{ij} becomes[11]

$$\eta \frac{\partial Q_{ij}}{\partial t} + A Q_{ij} = f_{ij},$$ (8.25)

where

$$f_{ij} = \tfrac{1}{6} \Delta \chi \left(E_i^* E_j - \tfrac{1}{3} | E |^2 \, \delta_{ij} \right).$$ (8.26)

The solution for Q_{ij} is

$$Q_{ij}(t) = \int_{-\infty}^{t} \left[\frac{f_{ij}(t')}{\eta} e^{-(t-t')/\tau} \right] dt',$$ (8.27)

where

$$\tau = \frac{\eta}{A} = \frac{\eta}{a(T - T^*)}.$$ (8.28)

The exact form of $Q_{ij}(t)$, of course, depends on the temporal characteristics of the laser field $E(t')$.

For simplicity, we assume that the incident laser pulse is polarized in the i direction, for example. Furthermore, we assume that the laser is a square pulse of duration τ_p. We thus have $f_{ij} = f_{ii} = \tfrac{1}{9} \Delta \chi E^2$. For $0 < t < \tau_p$, we have

$$Q_{ii} = \frac{1}{9} \left(\frac{\Delta \chi}{\eta} \right) \int_0^t E^2 e^{-(t-t')/\tau} dt'$$

$$= \frac{\tau}{9} \left(\frac{\Delta \chi}{\eta} \right) E^2 (1 - e^{-t/\tau}).$$ (8.29)

For time after the laser pulse, that is, $t > \tau_p$, the order parameter freely relaxes and is described by an exponential function:

$$Q_{ii}(t) = \frac{\tau}{9} \frac{\Delta\chi}{\eta} E^2 [1 - e^{-\tau_p/\tau}] e^{-t/\tau}. \tag{8.30}$$

The subscripts ii on Q denote that we are evaluating the ith component of the polarization. Clearly, Q_{ii} relaxes with a time constant τ, which is given in Equation (8.28). The dependence of τ on $(T - T^*)^{-1}$ shows that there is a critical slowing down as the system approaches T^*.

Since the linear polarization P^{NL} is proportional to Q, and Q is proportional to τ [cf. Eq. (8.29)], P^{NL} is therefore proportional to τ [i.e., proportional to $(T - T^*)^{-1}$]. This dependence on the temperature shows that the nonlinearity of the isotropic phase will be greatly enhanced as one approaches T^*, just as its response time is greatly lengthened.

These phenomena, the critical slowing down of the relaxation and the enhancement of the optical nonlinearity near T_C, have been experimentally observed by Wong and Shen.[10] In MBBA, for example, the observed relaxation times vary from about 100 ns at $T - T^* > 10°$ to 900 ns at $T - T^* < 1°$ (see Fig. 8.5); the nonlinearity χ_{1122}, for example, varies as 2.2×10^{-10} esu/$(T - T^*)$ (see Fig. 8.6).

8.2.3. Influence of Molecular Structure on Isotropic Phase Reorientational Nonlinearities

As explained in Chapter 1, molecular structures dictate the inter- and intramolecular fields, which in turn influence all the physical properties of the liquid crystals.

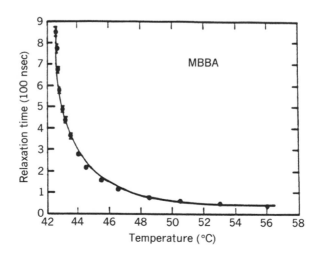

Figure 8.5.   Observed independence of the orientation relaxation time as a function of the temperature above T_C (after Wong and Shen[10]).

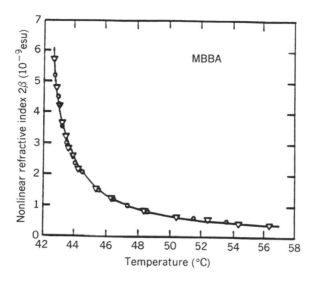

Figure 8.6. Measured optical nonlinear susceptibility as a function of the temperature above T_C (see Wong and Shen[10]).

Molecular structures therefore are expected to influence the reorientational optical nonlinearities in both their magnitude and response time. The study by Madden et al.[16] has shed some light on this topic.

In the transient degenerate four-wave mixing study (see Chapter 11) conducted by these workers, the observed optical nonlinearity associated with molecular reorientation can be expressed[16] in terms of the parameter $C(t,k)$:

$$C(t,k) = \left\langle \left(\frac{\mathscr{L}}{K_B T} \right) \left(\sum_{\substack{i \\ \text{over all} \\ \text{molecules}}} \alpha_{xz}^i(t) e^{i\mathbf{k}\cdot\mathbf{r}^i(t)} \right) \left(\sum_j \alpha_{xz}^j(0) e^{i\mathbf{k}\cdot\mathbf{r}^j(0)} \right)^* \right\rangle, \quad (8.31)$$

where α^i and $\mathbf{r}^i$ are the polarizability and position of molecule i and $\mathscr{L}$ is a local field correction factor. K_B is the Boltzmann constant and T is the temperature. $C(t,k)$ is a measure of the time correlation function of a Fourier component of the polarizability of the liquid crystals.

Note that in Equation (8.31) the $i \neq j$ terms contribute only when there is a correlation between the orientation of different molecules. Molecular orientational correlation in liquid crystals affects both the amplitude of $C(0, k)$ and its relaxation behavior.

In the case of weak coupling between molecular orientations and the transverse components of the momentum density of the liquid crystal,[17,18] the preceding expression reduces to

$$C(t,k) \equiv C(t) = \rho \mathscr{L} \frac{2\gamma^2}{15 K_B T} g e^{-t/\tau}, \quad (8.32)$$

where γ is the molecular polarizability anisotropy, ρ is the number density, τ is the relaxation time, and g is the static orientational correlation:

$$g \equiv \left\langle \alpha_{xz}^i(0) \sum_{\substack{j \\ \text{over all} \\ \text{molecules}}} \alpha_{xz}^j(0) \right\rangle \Big/ \langle \alpha_{xz}^i(0)\alpha_{xz}^j(0)\rangle. \qquad (8.33)$$

In an ordinary isotropic fluid, $g \approx 1$. On the other hand, for liquid crystal molecules near T_c, $g \approx K_B T/A(T - T^*)$, where A comes from the Landau expansion [see Eq. (8.24)] of the free energy discussed in the preceding section. Using g for a liquid crystal and Equation (8.32), we have $C(t) = 2/15\rho e^{t/\tau}(\ell\gamma^2/A)$.

From Equation (8.32) for $C(t)$, one may conclude that the most important factor influencing the optical nonlinearity is γ, the molecular polarizability anisotropy. If everything else about the molecule remains constant, obviously a larger γ means a larger optical nonlinearity. Consider, for example, cyanobiphenyls (nCBs), which are stable liquid crystals well known for their large polarizability anisotropy. In general, going to heavier members of a homologous series (i.e., larger-number n) increases γ by increasing the size and anisotropy of the molecule.[18] One would expect therefore a corresponding trend in the observed nonlinearity $C(0)$.

However, this is contradicted by the experimental measurements.[17] Table 8.1 shows the observed results. As n is increased from 3 to 10, where γ increases, the factor $\ell\gamma^2/A$ and the optical nonlinearity actually drop.

This "deviation" from the preceding notion of how molecular structures should influence optical nonlinearities is explained by the fact that, in general, many other physical parameters, besides γ, of the liquid crystals are modified as heavier (larger n) liquid crystals are synthesized. In the present case the other parameters are ℓ and A. As seen in Table 8.1, ℓ and A together lead to a reverse trend on the optical nonlinearity as γ is increased. Other parameters, such as the viscosity, shape parameter, and molecular volume, are also greatly changed as we go to heavier liquid crystal molecules in the nCB series, and they could adversely affect the resulting nonlinear optical response.

8.3. MOLECULAR REORIENTATIONS IN THE NEMATIC PHASE

In the nematic phase field-induced reorientation of the director axis arises as a result of the tendency of the total system to assume a new configuration with the

Table 8.1. Orientational Nonlinearity of Liquid Crystals Near T_c

Liquid Crystal	$\ell\gamma^2/A$	Observed Nonlinearity $(T - T^*)\chi^{(3)}$(esu)
10CB	0.49	1.36×10^{-9}
5CB	0.84	2.71×10^{-9}
3CB	1.51	5.0×10^{-9}

minimum free energy.[1–5,19] The total free energy of the system consists of the distortion energy F_d and the optical dipolar interaction energy $\mathbf{F}_{op}$, which are given by, respectively,

$$F_d = \tfrac{1}{2} K_1 (\mathbf{V} \cdot \hat{n})^2 + \tfrac{1}{2} K_2 (\hat{n} \cdot \mathbf{V} \times \hat{n})^2 + \tfrac{1}{2} K_3 (\hat{n} \times \mathbf{V} \times \hat{n}) \qquad (8.34)$$

and

$$F_{op} = -\frac{1}{4\pi} \int \mathbf{D} \cdot d\mathbf{E} = -\frac{\varepsilon_\perp}{8\pi} E^2 - \frac{\varepsilon_a \langle (\hat{n} \cdot \mathbf{E})^2 \rangle}{8\pi}, \qquad (8.35)$$

where the angle brackets $\langle\ \rangle$ denote a time average. If the optical field E_{op} is a plane wave [i.e., $E_{op} = \hat{p}|E|\cos(\omega t - kz)$, where $\hat{p}$ is a unit vector along the polarization direction], then $\langle E_{op}^2 \rangle = |E|^2/2$. [Note that Equation (8.35) is written in cgs units.]

The first term on the right-hand side of Equation (8.35) is independent of the director axis orientation, and hence, it may be ignored when we consider the reorientation process. The second term indicates that the system (if $\varepsilon_a > 0$) favors a realignment of the director axis along the optical field polarization. In analogy to the elastic torque, an optical torque

$$\mathbf{m}_{op} = \frac{\Delta\varepsilon}{4\pi} \langle (\hat{n} \cdot \mathbf{E})(\hat{n} \times \mathbf{E}) \rangle \qquad (8.36)$$

is associated with this free-energy term. This is illustrated by the following example.

8.3.1. Simplified Treatment of Optical Field-Induced Director Axis Reorientation

Consider, for example, the interaction geometry depicted in Figure 8.7, where a linearly polarized laser is incident on a homeotropically aligned nematic liquid crystal. The propagation vector $\mathbf{K}$ of the laser makes an angle $(\beta + \theta)$ with the perturbed director axis: θ is the reorientation angle. For this case, if the reorientation angle θ is small, then only one elastic constant K_1 (for splay distortion) is involved. A minimization of the total free energy of the system yields a torque balance equation:

$$K_1 \frac{d^2\theta}{dz^2} + \frac{\Delta\varepsilon \langle E_{op}^2 \rangle}{8\pi} \sin 2(\beta + \theta) = 0. \qquad (8.37)$$

In the small θ approximation, this may be written as

$$2\xi^2 \frac{d^2\theta}{dz^2} + (2\cos 2\beta)\theta + \sin 2\beta = 0, \qquad (8.38)$$

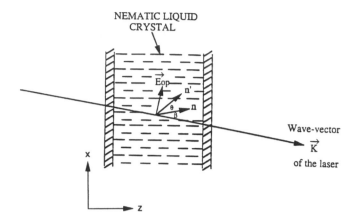

Figure 8.7. Interaction of a linearly polarized (extraordinary ray) laser with a homeotropically aligned nematic liquid crystal film.

where $\xi^2 = 4\pi K_1/[\Delta\varepsilon\langle E^2_{op}\rangle]$.

Because of the large birefringence of nematic liquid crystals, a small director axis reorientation will give rise to sufficiently large refractive index change to generate observable optical effects. Accordingly, we will continue our discussion based on the small θ limit. In this case, using the so-called hard-boundary condition [i.e., the director axis is not perturbed at the boundary ($\theta = 0$ at $z = 0$ and at $z = d$)], the solution of Equation (8.38) is

$$\theta = \frac{1}{4\xi^2}\sin 2\beta(dz - z^2),\qquad(8.39)$$

that is, the reorientation is maximum at the center and vanishingly small at the boundary, as shown in Figure 8.8.

As a result of this reorientation, the incident laser (an extraordinary wave) experiences a *z-dependent refractive index change* given by

$$\Delta n = n_e(\beta + \theta) - n_e(\beta),\qquad(8.40)$$

where $n_e(\beta + \theta)$ is the extraordinary ray index

$$n_e(\beta + \theta) = \frac{n_\parallel n_\perp}{\left[n_\parallel^2\cos^2(\beta + \theta) + n_\perp^2\sin^2(\beta + \theta)\right]^{1/2}}.\qquad(8.41)$$

For small θ, the change in the refractive index Δn is proportional to the square modulus of the optical electric field, that is,

$$\Delta n = n_2(z)\langle E^2_{op}\rangle\qquad(8.42)$$

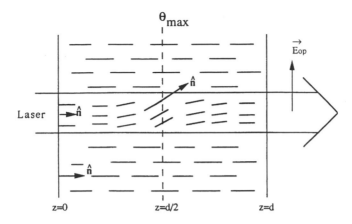

Figure 8.8. Director axis reorientation profile in a nematic film.

or

$$\Delta n - \alpha_2(z)I, \tag{8.43}$$

with $\alpha_2(z)$ given by

$$\alpha_2(z) = \frac{(\Delta\varepsilon)^2 \sin^2(2\beta)}{4Kc}(dz - z^2). \tag{8.44}$$

Note that, when averaged over the sample thickness, the factor $(dz - z^2)$ gives $d^2/6$.

What is truly unique about nematic liquid crystals is the enormity of α_2 compared to most other nonlinear optical materials. Let us denote the average of the value of α_2 over the film thickness by $\bar{\alpha}_2$. For a film thickness $d=100$ µm, $\Delta\varepsilon\sim0.6$, $K=10^{-6}$, and $\beta=45°$, we have

$$\bar{\alpha}_2 = 5 \times 10^{-3} \frac{\text{cm}^2}{\text{W}}. \tag{8.45}$$

In electrostatic units, this corresponds to a third-order susceptibility $\chi^{(3)}$ on the order of $9.54 \times 5 \times 10^{-3} \sim 5 \times 10^{-2}$ esu (see Chapter 11 on unit equivalence). This nonlinear coefficient is about eight orders of magnitude larger than that of CS_2 and about seven orders of magnitude larger than the isotropic phase liquid crystal reorientational nonlinearity discussed in Section 8.2.3.

In Chapter 5 where we discussed light scattering, such nonlinear response associated with director axis perturbation is traced back to the extreme sensitivity of the nematic phase to the optical electric field. In fact, the estimate made in Section 5.5.9 for the interaction geometry as shown in Figure 8.7 demonstrates that it is possible to get nonlinear index coefficient $n_2 \gg 1$ cm²/W.

8.3.2. More Exact Treatment of Optical Field-Induced Director Axis Reorientation

If the anisotropies of the elastic constants and optical field propagation characteristics in the birefringent nematic film are taken into account, the resulting equations for the optical field, elastic and optical torques, and so on become more complicated. The starting point of the analysis is the Euler–Lagrange equation associated with a small director axis reorientation angle $\theta(z)$:

$$\frac{\partial F}{\partial \theta} - \frac{d}{dz}\frac{\partial F}{\partial \theta'} = 0, \tag{8.46}$$

where $F = F_d + F_{op}$ from Equations (8.34) and (8.35). (Note that in some cases one needs to take into account surface anchoring energy as well.) In the present treatment we will assume the hard-boundary condition and ignore the surface interaction term. For reorientation in the x-z plane, only the elastic energies associated with the splay (K_1) *and* bend (K_3) are involved.

Writing $\hat{n} = \sin\theta''\,\hat{x} + \cos\theta''\,\hat{z}$, we have

$$(\mathbf{V}\cdot\hat{n})^2 = \sin^2\theta''\left(\frac{\partial\theta}{\partial z}\right)^2 \tag{8.47a}$$

and

$$(\hat{n}\times\mathbf{V}\times\hat{n}) = \cos^2\theta''\left(\frac{d\theta}{dz}\right)^2, \tag{8.47b}$$

where $\theta'' = \theta + \beta$. Also, the optical field may be expressed as

$$\mathbf{E}_{op} = (E_x\hat{x} + E_z\hat{z}), \tag{8.48}$$

where E_x and E_z remain to be calculated from the Maxwell equation:

$$\mathbf{V}(\mathbf{V}\cdot\mathbf{E}_{op}) - \mathbf{V}^2\mathbf{E}_{op} - \frac{\omega^2}{c^2}\bar{\bar{\varepsilon}}\,\mathbf{E}_{op} = 0. \tag{8.49}$$

This is because the dielectric tensor ε is dependent on the optical field $\mathbf{E}_{op}$ as a result of the optical field-induced director axis reorientation. In other words, Equations (8.46) and (8.49) have to be solved in a self-consistent manner to yield $\theta(z)$ and $\mathbf{E}_{op}(z)$.

Using Equations (8.47) and (8.48), Equation (8.46) can be explicitly written as

$$(K_1\sin^2\theta'' + K_3\cos^2\theta'')\frac{d^2\theta}{dz^2} - (K_3 - K_1)\sin\theta''\cos\theta''\left(\frac{d\theta}{dz}\right)^2$$
$$+ \frac{\Delta\varepsilon}{16\pi}\left[\sin 2\theta''(|E_x|^2 - |E_y|^2) + \cos 2\theta''(E_xE_z^* + E_z^*E_x)\right] = 0. \tag{8.50}$$

The self-consistent solutions of E_{op} and $\theta(z)$ from these equations are quite complex.[19] Nevertheless, it is interesting to note that in the limit of very small director axis reorientation, that is, $\theta \ll 1$, which is usually the case, the preceding equation reduces to the simplified one given in Section 6.4.1.

In the case where the optical field is incident perpendicularly to the sample, that is, $\beta = 0$, Equation (8.50) becomes greatly simplified and yields an interesting result. Setting $\beta = 0$, we get

$$K_3 \frac{d^2\theta}{dz^2} + \frac{\Delta\varepsilon}{16\pi} \left[2\theta \left(|E_x|^2 - |E_z|^2 \right) + |E_x E_z| + |E_x E_z| \right] = 0. \tag{8.51}$$

From $\nabla \cdot \vec{D} = 0$, we get

$$E_z \simeq -E_x \left(\frac{\Delta\varepsilon}{\varepsilon_\parallel} \right) \theta. \tag{8.52}$$

Equation (8.51) therefore becomes

$$K_3 \frac{d^2\theta}{dz^2} + \frac{\Delta\varepsilon}{8\pi} \frac{\varepsilon_\perp}{\varepsilon_\parallel} |E_x|^2 \theta = 0. \tag{8.53}$$

we find that Equation (8.53) is analogous to the equation for the dc field-induced Freedericksz transition, recalling that we have made the approximation $\sin\theta = \theta$. We can thus define a so-called optical Freedericksz field E_F given by

$$|E_F|^2 = \frac{8\pi^3 K_3 \varepsilon_\parallel}{\Delta\varepsilon \varepsilon_\perp} \left(\frac{1}{d^2} \right). \tag{8.54}$$

For $|E_x| < |E_F|$, $\theta = 0$. For $|E_x| > |E_F|$, director axis reorientation will take place.

8.3.3. Nonlocal Effect and Transverse Dependence

The preceding discussion is based on the assumption that the incident laser is a plane wave. If the laser is a focused Gaussian beam, with a beam size ω_0 comparable to or smaller than the film thickness, transverse correlation effects will arise. Molecules situated "outside" the laser beam will exert torques on molecules "inside" the beam; conversely, molecules inside the beam could also exert torques on those on the outside. The result is that the transverse dependence of the reorientation profile is not the same function as the transverse profile of the incident laser beam (e.g., Gaussian). Put in another way, one may recognize that Equation (8.53) is basically a diffusion equation, where the elastic term plays the role of the diffusive mechanism. As a result of this diffusive effect, as in many other physical processes, the spatial profile of the response is not the same as the excitation profile.

As shown in the detailed calculations given in the work of Khoo et al.:[20]

1. For $\beta = 0$, there is a threshold intensity for finite reorientation to occur. The threshold intensity depends on both the thickness of the film and the beam size ω_0. For ω_0. the threshold intensity increases dramatically (compared with the value for a plane wave). There is no threshold intensity for field-induced reorientation in the $\beta \neq 0$ case.

2. For a Gaussian laser beam input, the reorientation profile is not Gaussian, although it is still a bell-shaped function with a half-width ω_θ, which is different from ω_0. In general, for the $\beta = 0$ case, the half-width ω_θ is always larger than ω_0 for all values of ω_0, approaching ω_0 for large values of ω_0/d (see Fig. 8.9a). On the other hand, for the $\beta = 0$ case, ω_θ can be smaller than ω_0, depending on whether ω_0 is smaller or larger than d (see Fig. 8.9b).

The transverse dependence of the reorientation becomes important in the studies of self-phase modulation, self-focusing and self-guiding processes, and spatial solitons in liquid crystalline media (see Chapter 12).

8.4. NEMATIC PHASE REORIENTATION DYNAMICS

The dynamics of molecular reorientation are described by balancing all the prevailing torques acting on the director axis. For the interaction geometry given in Figure 8.4, the molecular reorientation involves the viscous, elastic, and optical torques and the resulting equation is given by

$$\gamma \frac{\partial \theta}{\partial t} = K \frac{\partial^2 \theta}{\partial z^2} + \frac{\Delta\varepsilon \langle E_{op}^2 \rangle}{8\pi} \sin(2\beta + 2\theta), \tag{8.55}$$

where we have introduced an effective viscosity coefficient γ.

For simplicity, we have again assumed that θ is small and that we can use the one-elastic-constant approximation. Writing $\sin \theta \approx \theta$ and $\cos \theta \approx 1$, Equation (8.55) becomes

$$\gamma \frac{\partial \theta}{\partial t} = K \frac{\partial^2 \theta}{\partial z^2} + \frac{\Delta\varepsilon \langle E_{op}^2 \rangle}{8\pi} \sin 2\beta + \theta \frac{\Delta\varepsilon \langle E_{op}^2 \rangle}{4\pi} \cos 2\beta. \tag{8.56}$$

8.4.1. Plane Wave Optical Field

Assuming that E_{op}^2 is a plane wave, we may write $\theta(t,z) = \theta(t)\sin(\pi z/d)$, and substituting it in Equation (8.56), we get

$$\gamma \dot{\theta} = -\frac{K\pi^2}{d^2}\theta + \frac{\Delta\varepsilon \langle E_{op}^2 \rangle \sin 2\beta}{8\pi} + \theta \frac{\Delta\varepsilon \langle E_{op}^2 \rangle}{4\pi} \cos 2\beta. \tag{8.57}$$

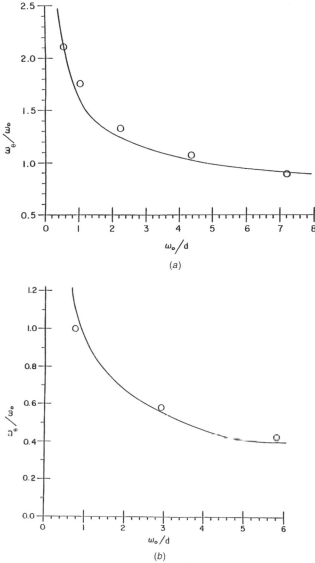

Figure 8.9. (a) Plot of ω_θ/ω_0 as a function of ω_0/d for $\beta \neq 0$. (b) Plot of ω_θ/ω_0 as a function of ω_0/d for $\beta = 0$.

A good understanding of the reorientation dynamics can be obtained if we separate it into two regimes: (1) optical torque $\gg$ elastic torque

$$\left| \frac{\Delta\varepsilon\langle E_{op}^2 \rangle}{8\pi} \sin(2\beta+2\theta) \right| \gg \left| K \frac{\partial^2\theta}{\partial z^2} \right|,$$

and (2) optical torque $\ll$ elastic torque

$$\left| \frac{\Delta\varepsilon\langle E_{op}^2 \rangle}{8\pi} \sin(2\beta + 2\theta) \right| \ll \left| K \frac{\partial^2 \theta}{\partial z^2} \right|.$$

For case (1) we may ignore the elastic term in Equation (8.57) and get an equation for $\dot{\theta}$ of the form

$$\dot{\theta} = a + b\theta, \tag{8.58}$$

where

$$a = \frac{\Delta\varepsilon\langle E_{op}^2 \rangle}{8\pi\gamma} \sin 2\beta, \tag{8.59}$$

$$b = \frac{\Delta\varepsilon\langle E_{op}^2 \rangle}{4\pi\gamma} \cos 2\beta. \tag{8.60}$$

If E_{op}^2 is associated with a square laser pulse (i.e., $E_{op}^2=0$ for $t < 0$; $E_{op}^2=E_0^2$ for $0 < t < \tau_p$), the solution for θ is therefore, for $0 < t < \tau_p$,

$$\theta(t) = \frac{a}{b}(e^{bt} - 1). \tag{8.61}$$

From Equation (8.61) we can see that $\theta(\tau_p)$ is appreciable only if $b\tau_p$ is appreciable. In other words, if the laser pulse duration is short (e.g., nanosecond), it has to be very intense in order to induce an appreciable reorientation effect. In this respect and because the surface elastic torque is not involved, the dynamical response of a nematic liquid crystal is quite similar to its isotropic phase counterpart. However, the dependence on the geometric factor $\sin 2\beta$ is a reminder that the nematic phase is, nevertheless, an (ordered) aligned phase, and its overall response is dependent on the direction of incidence and the polarization of the laser.

The effective optical nonlinearity in the transient case, compared to the steady-state value, can be estimated from Equation (8.61).

In the short time limit (i.e., $bt \ll 1$), $\theta(t) \sim at$. Therefore, we have

$$\frac{\theta(t)}{\langle \theta \rangle} = \frac{12}{\pi^2} \left(\frac{K_1 \pi^2}{\gamma d^2} \right) t, \tag{8.62}$$

where $\langle\theta\rangle$ is the averaged (over the sample thickness) value of θ given in Equation (8.39) for the steady-state case (e.g., cw laser). This may be expressed in another way:

$$\frac{\theta(t)}{\langle\theta\rangle} = \frac{12}{\pi^2}\frac{t}{\tau_r} \sim \frac{t}{\tau_r}, \tag{8.63}$$

where τ_r is the nematic axis reorientation time, $\tau_r = \gamma d^2/K\pi^2$. Using typical values in the estimate of $\bar{\alpha}_2$ in Equation (8.45): $\gamma = 0.1$ P, $d = 100$ μm, and $K = 10^{-6}$ dyne, we have $\tau_r \approx 1$ s. If we define a so-called effective nonlinear coefficient α_2^t for transient orientational nonlinearity, then

$$\alpha_2^t = \left(\frac{t}{\tau_r}\right)\bar{\alpha}_2. \tag{8.64}$$

For a 10 ns (10^{-8}) laser pulse, the effective nonlinearity is on the order of 5×10^{-11} esu.

For case (2), which naturally occurs when the *laser pulse is over*, we have

$$\gamma\dot{\theta} = -\frac{K\pi^2}{d^2}\theta, \tag{8.65}$$

that is,

$$\theta = \theta_{max}e^{-t/\tau_r}, \tag{8.66}$$

where

$$\alpha_{max} = \frac{a}{b}(e^{b\tau_p} - 1). \tag{8.67}$$

The preceding discussion and results apply to the case where an extraordinary wave laser is obliquely incident on the (homeotropic) sample (i.e., $\beta \neq 0$). For the case where a laser is perpendicularly incident on the sample (i.e., its optical electric field is normal to the director axis), there will be a critical optical field E_F, the so-called Freedericksz transition field [see Eq. (8.54)], below which molecular reorientation will not take place. Second, the turn-on time of the molecular reorientation depends on the field strength above E_F (i.e., on $E_{op}-E_F$). For small $E_{op}-E_F$, the turn-on time can approach many minutes! Studies with nanosecond and picosecond lasers[4,9] have shown that under this perpendicularly incident (i.e., $\beta=0$) geometry, it is very difficult to induce molecular reorientation through the mechanism discussed previously.

8.4.2. Sinusoidal Optical Intensity

In many nonlinear optical wave mixing processes involving the interference of two coherent beams, the resulting optical intensity imparted on the liquid crystal is a sinusoidal function. This naturally induces a spatially oscillatory director axis reorientational effect. Molecules situated at the intensity maxima will undergo reorientation, while those in the "dark" region (intensity minima) will stay relatively unperturbed. In analogy to the laser transverse intensity effect discussed at the end of the last section, molecules in these regions will exert torque on one another, and the resulting relaxation time constant will be governed by the characteristic length of these sinusoidal variations as well as the thickness d of the nematic film.

Consider, for example, an optical intensity function of the form $\langle E_{op}^2 \rangle \sim E^2$ $(1+\cos qy)$. The induced reorientational angle will have a corresponding modulated and a spatially uniform component. The spatially modulated reorientation angle is of the form

$$\theta \sim \theta(t)\sin\left(\frac{\pi z}{d}\right)\cos qy. \tag{8.68}$$

This gives rise to an elastic torque term given by

$$K\frac{\partial^2\theta}{\partial y^2} = -Kq^2\theta \tag{8.69}$$

on the right-hand side of Equation (8.56). Accordingly, the orientational relaxation dynamics when the optical field is turned off now becomes

$$\gamma\dot{\theta} = -K\left(\frac{\pi^2}{d^2}+q^2\right)\theta. \tag{8.70}$$

The corresponding orientational relaxation time constant is

$$\tau_r = \frac{\gamma}{K}\left(\frac{1}{(\pi^2/d^2)+q^2}\right). \tag{8.71}$$

Writing the wave vector q in terms of the grating constant $\Lambda=2\pi/q$, we have

$$\tau_r = \frac{\gamma}{K}\left(\frac{1}{(\pi^2/d^2)+(4\pi^2/\Lambda^2)}\right). \tag{8.72}$$

From Equation (8.72) we can see that if $\Lambda \ll d$, the orientational relaxation dynamics is dominated by Λ (i.e., the intermolecular torques); conversely, if $d \ll \Lambda$, the dynamics is decided by the boundary elastic torques.

These influences of the intermolecular and the elastic torques in cw-laser-induced nonlinear diffraction effects in nematic films are reported in the work by Khoo.[21] There it is also noted that the optical nonlinearity associated with nematic director axis reorientation is proportional to the factor $\Delta\varepsilon^2/K$ [see Eq. (8.44)] typical of orientational fluctuations induced by light scattering processes (see Chapter 5). Although both $\Delta\varepsilon$ and K are strongly dependent on the temperature, the combination $\Delta\varepsilon^2/K$ is not. This is because $\Delta\varepsilon$ is proportional to the order parameter S, whereas K is proportional to S^2.

This is indeed verified in the experimental study of the temperature dependence of nonlinear diffraction (see Fig. 8.10). The (orientational component) signal stays quite flat up to about $1°$ near T_c; near T_c, the nematic alignment begins to deteriorate and the signal diminishes. The factor $\Delta\varepsilon^2/K$ also appears in linear scattering associated with director axis fluctuations, as discussed in Chapter 5. On the other hand, one can see that the thermal component diverges as the temperature approaches T_c following the thermal index dependence (cf. Chapters 2 and 3).

8.5. LASER-INDUCED DOPANT-ASSISTED MOLECULAR REORIENTATION AND TRANS-CIS ISOMERISM

In nematic liquid crystals doped with some absorbing dye molecules, studies[22–27] have shown that the excited dye molecules could exert an intermolecular torque τ_{mol} on the liquid crystal molecules that could be stronger than the optical torque τ_{op}. In particular, Janossy et al.[22] have observed that some classes of anthraquinone dye molecules, when photoexcited, will exert a molecular torque $\tau_{mol} \sim \eta\tau_{op}$, with η that

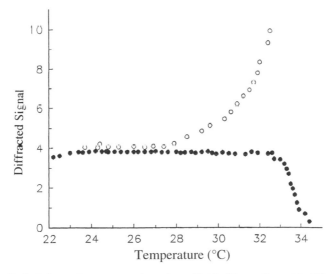

Figure 8.10. Typical observed temperature dependence (dots) of the nonlinear side diffraction owing to the nematic axis reorientation effect (from Khoo[21]). Circles are thermal index change effects discussed in Chapter 9.

can be as large as 100. On the other hand, Gibbons et al. and Chen and Brady[23] have observed that, under prolonged exposure, dye-doped liquid crystals (DDLCs) will align themselves in a direction orthogonal *to the optical electric field and the propagation wave vector* (i.e., in the $\hat{y}$ direction with reference to Fig. 8.11). Under suitable surface treatment conditions, such reorientational effects can be made permanent (but erasable).

Khoo et al.[24] report the observation of a negative reorientational effect that occurs in the transient regime under short laser pulse illumination. Using a variety of polarization configurations between the pump and probe beams in the dynamic grating diffraction study, the authors have established that the negative change in the refractive index (i.e., negative nonlinearity) is associated with the *liquid crystal director axis realigning toward the z direction*. The efficiency of this reorientation process is governed by the types of dye molecules used as dopants.

An interesting and important point is that, at low optical power, the orientational effect actually is the dominating one, even in such highly absorptive material. In other recent studies[25–29] of dye-doped nematic liquid crystals, observed nonlinearities have approached the so-called supranonlinear scale characterized by refractive index coefficients $n_2 \gg 1$ cm^2/W. The observed nonlinearity is orders of magnitude larger than the so-called giant optical nonlinearity of pure liquid crystals. In particular, methylred dye-doped nematic liquid crystals were first observed to have a nonlinear index

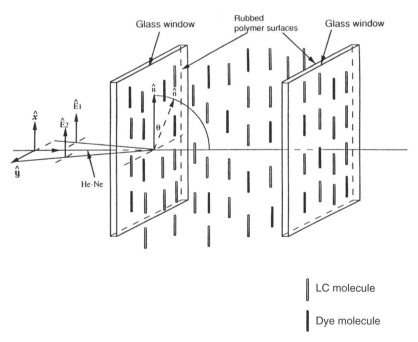

Figure 8.11. Schematic diagram of the optical fields and their propagation direction in a planar aligned dye-doped liquid crystal.[24] The two fields $\mathbf{E}_1$ and $\mathbf{E}_2$ are coherent beams derived from splitting a pump laser to induce dynamic grating in the liquid crystal sample. $\mathbf{E}_3$ is the probe beam.

n_2 as large as 6 cm^2/W. An important and useful feature of the nonlinear response in methyl-red-doped sample is that it can be enhanced with an applied *low-frequency* ac field. On the other hand, a *high-frequency* field will quench the reorientational effects. Studies of these on–off switching dynamics of the diffraction from the film as the applied ac field frequency is switched back and forth between 300 Hz and 30 kHz, at $V_{pp} = 20$ V, show that the on time is on the order of 12 ms, and the off time is about 17 ms. Faster response is possible at higher V_{pp}, or using nematics of lower viscosity and higher dielectric anisotropy. Such dependence on the ac field frequency may be useful for dual-frequency switching/modulation application.[28]

Dye molecules and organic materials such as azobenzene are also known[23,26,30] to exhibit the so-called *trans-cis* isomerism upon photoexcitation (see Fig. 8.12). In this process, a ground-state azomolecule in the *trans* form will assume the *cis* form in the excited state. If such azo compounds are dissolved in a nematic liquid crystal, for example, their *trans-cis* isomeric change will result in modifying the order parameter, and therefore changes the optical properties (e.g., refractive index) of the nematics. Studies of azobenzene liquid crystals and azobenzene liquid crystal doped nematic liquid crystals have shown that this is also an effective mechanism to create large optical nonlinearities.[26] Moreover, since the *trans-cis* isomerism can happen quite rapidly (possibly in nanoseconds), the resulting order parameter change and optical nonlinearities can be created just as rapidly.[30]

8.6. DC FIELD AIDED OPTICALLY INDUCED NONLINEAR OPTICAL EFFECTS IN LIQUID CRYSTALS: PHOTOREFRACTIVITY

As one can see from the preceding discussions on optical field induced director axis reorientation in liquid crystals, the torque exerted by the optical field on the director axis is basically quadratic in the field amplitude. Except for its dispersion influence on the optical dielectric constant $\varepsilon(\omega)$, the frequency of the electric field is basically not involved. Furthermore, if two or more fields are acting on the director axis, the resulting torque exerted on the director axis is simply proportional to the square amplitude of the total fields. Accordingly, it is possible to enhance the optical field induced effect by application of a low-frequency ac or dc electric field, much as the optically addressed liquid crystal spatial light modulator discussed in Chapter 6. In the latter, the responsible mechanism is the photoconduction generated by the incident optical field in the semiconductor layer adjacent to the liquid crystals.

In this section, we discuss a mechanism for enhanced optical nonlinearity in which the photoinduced charges and fields occur within the liquid crystal. It is well known that dc field induced current flow in nematic liquid crystals, which possess anisotropic conductivities, could lead to nematic flows and director axis reorientation, and to the creation of a space-charge field.[31] The charged carriers responsible for the electrical conduction come from impurities present in the otherwise purely dielectric nematic liquid crystal. If these impurities are photoionizable, an incident optical intensity [e.g., an intensity grating created by the interference of two coherent optical beams (see Fig. 8.13)], it is possible, therefore, to create a space-charge

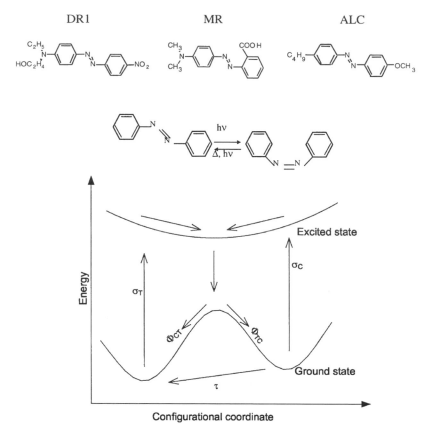

Figure 8.12. (Top) Molecular structures of a commercial dye DR1, mehtyl-red (MR) dye, and azobenzene liquid crystal (ALC). (Middle) Molecular structural changes associated with *trans-cis* isomerization of azomolecules upon optical illumination. Bottom diagram shows the excitation energy versus the molecular coordinate associated with these photoexcited processes and the various *trans-cis* cross sections.

density grating via the photoinduced spatial conductivity anisotropy. The space-charge field thus created, together with the applied dc field, will cause a refractive index change through any of the field-induced index change effects. In particular, these fields could give rise to director axis reorientation, and effectively, a new mechanism for optical nonlinearity.

This process of photoinduced charged-carrier generation, space-charge field, and refractive index change is analogous to photorefractive (PR) effect occurring in electro-optically active materials. There is, however, an important difference. In those so-called photorefractive materials, such as $BaTiO_3$, the induced index change Δn is linearly related to the total electric field E present:[32]

$$\Delta n = \gamma_{\text{eff}} E. \tag{8.73}$$

The sign and magnitude of the effective electro-optic coefficient γ_{eff} depends on the symmetry class of the crystal and the direction of the electric field.

Nematic liquid crystals, on the other hand, possess centrosymmetry. The field-induced refractive index change is quadratically related to the electric field, that is,

$$\Delta n = n_2 E^2. \tag{8.74}$$

A useful feature of such quadratic dependence is that it allows the mixing of the applied dc field with the space-charge field for enhanced director axis reorientation effect.

Since its discovery in 1994,[33,34] there have been numerous studies focusing on various aspects of the orientational photorefractivity in liquid crystalline systems.[28,35–40] One of the interesting aspects of the effect is that, just as in inorganic photorefractive crystals, the nonlinearity is nonlocal because of the phase shift between the space-charge fields and the optical-intensity grating functions. Such nonlocality gives rise to strong two-beam coupling effects at very modest applied dc field strength (≈ 100 V/cm) and optical power, in contrast to polymers and inorganic PR crystals, which require high field strength (several thousands of V/cm). This is mainly due to the large optical birefringence and dielectric anisotropy of nematics, and the easy susceptibility of the director axis orientation to external fields. The nonlinear index coefficient $n_2(I)$ associated with this process is on the order of 10^{-1}–10^{-3}cm^2/W, depending on the dopants used.[28]

8.6.1. Orientational Photorefractivity: Bulk Effects

In this section, we summarize the basic theories and mechanisms responsible for the orientational photorefractivity in liquid crystals. In particular, we consider the case where the incident optical intensity is sinusoidal as in a two-wave mixing configuration (see Fig. 8.13a), in which two equal power mutually coherent pump and probe beams with p-type polarization are overlapped on the homeotropically aligned liquid crystal cell. The two beams are obliquely incident on the sample at a small wave mixing angle. A small dc voltage is applied across the cell window, that is, parallel to the initial director axis direction. Taking into account the dc and optical fields, the total free energy of the system becomes

$$F = \frac{k}{2}\left\{\left[\vec{\nabla}\cdot n(\vec{r})\right]^2 + \left[\vec{\nabla}\times n(\vec{r})\right]^2\right\} - \frac{\Delta\varepsilon}{8\pi}\left[\vec{E}\cdot n(\vec{r})\right]^2 - \frac{\Delta\varepsilon_{\text{op}}}{8\pi}\left[\vec{E}_{\text{op}}\cdot n(\vec{r})\right]^2, \tag{8.75}$$

where k is the elastic constant (assuming the single constant approximation). $\Delta\varepsilon$ is the dc field anisotropy and $\Delta\varepsilon_{\text{op}}$ is the optical dielectric anisotropy. Denoting the reoriented director axis by $\hat{n} = (\sin\theta, 0, \cos\theta)$ and minimizing the free energy

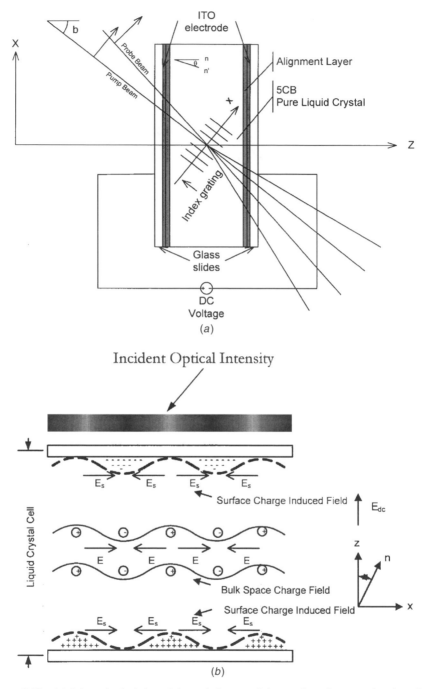

Figure 8.13. (a) Schematic depiction of the optical wave mixing configuration to study orientational photorefractivity of nematic liquid crystal under ac bias. (b) Schematic showing various space charge fields involved in the orientational photorefractive effect in nematic liquid crystals.

with respect to the reorientation angle θ yield the well-known Euler–Lagrange equation for θ:

$$
k\frac{d^2\theta}{dz^2} + k\frac{d^2\theta}{d\xi^2} + \frac{\Delta\varepsilon}{4\pi}\left[\sin\theta\cos\theta(E_x^2 - E_z^2) + \cos(2\theta)E_xE_z\right]
$$

$$
+ \frac{\Delta\varepsilon_{op}\cdot E_{op}^2}{8\pi}\sin 2(\beta + \theta) = 0.
$$

(8.76)

[Note: See Figure 8.7 for the definition of θ and β in Equation (8.76)].

Owing to the impurities and other photocharge producing agents present in the liquid crystal cells, three forms of space-charge fields are created within the bulk of the cell: a typical photorefractive component E_{ph} and two other components $E_{\Delta\varepsilon}$ and $E_{\Delta\sigma}$ caused by the dielectric and conductivity anisotropies[33,34] in conjunction with the reoriented director axis (see Fig. 8.13b).

Following the treatment by Rudenko and sukhov,[34] the photorefractive-like space-charge field induced by charge separation can be expressed as

$$
E_{ph} = E_{ph}^{(0)}\cos(q\xi) = \left(\frac{mk_BT}{2e}qv\frac{\sigma - \sigma_d}{\sigma}\right)\cos(q\xi),
$$

(8.77)

where m is the optical modulation factor, k_B is the Boltzmann constant, σ is the conductivity under illumination, σ_d is the dark state conductivity, $\upsilon=(D^+-D^-)/(D^++D^-)$, where D^+ and D^- are the diffusion constants for positive and negative ions, respectively, and $q=2\pi/\Lambda$ is the grating wave vector, with Λ the grating constant. Note that E_{ph} [$\sim\cos(q\zeta)$] is $\pi/2$ phase shifted from the imparted optical intensity grating $E_{op}^2 \sim \sin(q\xi)$.

Meanwhile, the applied field E_{dc} along the z direction, in conjunction with the conductivity and dielectric anisotropies and the reoriented (by an angle θ) director axis, creates a transverse (along the x direction) electric field component,[33,40] which in turn will further reorient the director axis. These space-charge fields are of the form[31,33,40]

$$
E_{\Delta\sigma} = -\frac{\left[(\sigma_{\|} - \sigma_{\perp})\sin\theta\cos\theta\right]}{\sigma_{\|}\sin^2\theta + \sigma_{\perp}\cos^2\theta}E_{dc}, \quad E_{\Delta\varepsilon} = -\frac{\left[(\varepsilon_{\|} - \varepsilon_{\perp})\sin\theta\cos\theta\right]}{\varepsilon_{\|}\sin^2\theta + \varepsilon_{\perp}\cos^2\theta}E_{dc}.
$$

(8.78)

For small angle approximation, the above equations can be linearized to yield

$$
E_{\Delta\sigma} \approx -\frac{\Delta\sigma}{\sigma_{\perp}}\theta E_{dc}, \quad E_{\Delta\varepsilon} \approx -\frac{\Delta\varepsilon}{\varepsilon_{\perp}}\theta E_{dc}.
$$

(8.79)

The total electric field then becomes

$$
E_{\text{total}} = \left[-\left(\frac{\Delta\sigma}{\sigma_\perp} + \frac{\Delta\varepsilon}{\varepsilon_\perp} \right) E_{\text{dc}} \theta \cos\beta - E_{\text{ph}} \cos\beta, \quad 0, \right.
$$

$$
\left. E_{\text{dc}} - \left(\frac{\Delta\sigma}{\sigma_\perp} + \frac{\Delta\varepsilon}{\varepsilon_\perp} \right) E_{\text{dc}} \theta \sin\beta - E_{\text{ph}} \sin\beta \right]
$$

$$
= \left[-\left(E_\Delta \cdot \theta + E_{\text{ph}} \right) \cos\beta, \quad 0, \quad E_{\text{dc}} - \left(E_\Delta \cdot \theta + E_{\text{ph}} \right) \sin\beta \right], \qquad (8.80)
$$

where

$$
E_\Delta = \left(\frac{\Delta\pi}{\pi_\perp} + \frac{\Delta\varepsilon}{\varepsilon_\perp} \right) E_{\text{dc}}.
$$

The torque balance equation thus becomes

$$
k \frac{d^2\theta}{dz^2} + k \frac{d^2\theta}{d\xi^2} + \frac{\Delta\varepsilon}{4\pi} [E_\Delta E_z \cos(\beta) \cdot \theta + E_z E_{\text{ph}} \cos(\beta)]
$$

$$
+ \frac{\Delta\varepsilon \cdot E_{\text{op}}^2}{8\pi} [\sin(2\beta) + 2\cos(2\beta) \cdot \theta] = 0. \qquad (8.81)
$$

Assuming that "hard" boundary condition exists, we find that a solution for θ is of the form

$$
\theta = \theta_0 \sin\left(\frac{\pi z}{d} \right) \cos(q\xi), \qquad (8.82)
$$

where d is the sample thickness and θ_0 is the NLC director maximum reorientation angle. A solution of Equation (8.81) is

$$
\theta_0 = \frac{\dfrac{\Delta\varepsilon_{\text{op}} \cdot E_{\text{op}}^2}{8\pi} \sin(2\beta) + \dfrac{\Delta\varepsilon \cdot E_{\text{dc}} E_{\text{ph}}^{(0)}}{4\pi} \cos(\beta)}{\left[\dfrac{\Delta\varepsilon}{4\pi} E_\Delta E_{\text{dc}} \cos(\beta) + \dfrac{\Delta\varepsilon_{\text{op}}}{4\pi} E_{\text{op}}^2 \cos(2\beta) \right] - \left[\left(\dfrac{\pi}{d} \right)^2 + q^2 \right] k}, \qquad (8.83)
$$

which is similar to those obtained previously,[33,40] if we neglect the term coming from the E_{op}^2.

In terms of the Freedericksz transition voltage, $V_F = \pi\sqrt{K/\varepsilon_0 \Delta\varepsilon}$ (in mks units), the solution for θ can be written as

$$\theta_0 = \frac{\dfrac{1}{2}\dfrac{\Delta\varepsilon_{op}}{\Delta\varepsilon} E_{op}^2 \sin(2\beta) + E_{ph}^o E_{dc}\cos(\beta)}{\left[E_\Delta E_{dc}\cos(\beta) + \dfrac{\Delta\varepsilon_{op}}{\Delta\varepsilon} E_{op}^2 \cos(2\beta)\right] - E_F^2\left[1 + \left(\dfrac{qd}{\pi}\right)^2\right]}. \tag{8.84}$$

To get a stable solution, we require that the denominator in Equation (8.84) be positive, i.e.,

$$E_{dc} \geq E_F\left[\frac{\left[1 + \left(\dfrac{qd}{\pi}\right)^2\right] - \left(\dfrac{\Delta\varepsilon_{op}}{\Delta\varepsilon}\right)\left(\dfrac{E_{op}}{E_F}\right)^2\cos(2\beta)}{\left(\dfrac{\Delta\sigma}{\sigma_\perp} + \dfrac{\Delta\varepsilon}{\varepsilon_\perp}\right)\cdot\cos\beta}\right]^{1/2}. \tag{8.85}$$

This allows us to identify a threshold field E_{th},

$$E_{th} = \alpha\cdot E_F = \left[\frac{\left[1 + \left(\dfrac{qd}{\pi}\right)^2\right] - \left(\dfrac{\Delta\varepsilon_{op}}{\Delta\varepsilon}\right)\left(\dfrac{E_{op}}{E_F}\right)^2\cos(2\beta)}{\left(\dfrac{\Delta\sigma}{\sigma_\perp} + \dfrac{\Delta\varepsilon}{\varepsilon_\perp}\right)\cdot\cos\beta}\right]^{1/2}\cdot E_F. \tag{8.86}$$

For 5CB, $k\sim10^{-11}$N, $\Delta\varepsilon\sim11$ ($\varepsilon_\parallel\sim16$, $\varepsilon_\perp\sim5$), $\varepsilon_0=8.85\times10^{-12}$ F/m, $\Delta\sigma/\sigma_\perp\sim0.5$, and a typical wave mixing geometry as depicted in Figure 8.13 ($qd\sim2\pi$, and the internal angle $\beta=22.5°$), Equation (8.86) gives $V_{th} = \alpha V_F$ with $\alpha \sim 1.5$ and $V_F \sim 1$ V (cf. the expression for V_F above Equation (8.84)).

We note here that the above estimate for V_F applies to the case where the applied field is an ac field, whereas in the studies of photorefractive effect, dc fields are used. The dc Freedericksz transition voltage V_F^{dc} has been shown from other studies[41,42] to be different (generally higher) from V_F, mostly due to the formation of electric double layers near the aligning surface arising from dc field induced separation of charge carriers. The reported V_F^{dc} value ranges from 1.5 V for a LC cell with nonphotoconductive alignment layers[41] to 4 V with photoconductive layers.[42] Accordingly, we expect that the photorefractive threshold voltage $V_{th} = \alpha V_F$ (from Eq. 8.86 above) to be more than (1.45×1.5)V ~ 2.3 V. This is indeed consistent with the experimental observations where the V_{th} values are about 3 V. Also, since the optical intensity used is generally in the mW/cm^2 regime, $E_{op}/E_F \ll 1$, and the factor α is essentially independent of the optical intensity.

8.6.2. Some Experimental Results and Surface Charge/Field Contribution

Recent studies reconfirm these results predicted from the bulk theories described above. In these studies, the laser used is the 514.5 nm line of an argon laser. The wave mixing angle ($\alpha \sim 1.5°$) corresponds to an optical intensity grating constant $\Lambda \approx 20~\mu m$ The writing beams are obliquely incident on the nematic cells, making an external angle of 45° (internal angle $\beta \approx 22.5°$) with the cell normal.

As shown in Figure 8.14a, with a dc voltage larger than a certain threshold (3.4 V), observable self-diffracted beams are generated. The orientational nature of the index grating was manifested by the strong anisotropy in the diffraction efficiency: self-diffraction is only observed for p-polarized beam mixing, while s-polarized beam mixing gives no diffraction. The observed threshold voltage of 3.4 V is in fair agreement with the theoretical estimate of > 2.3 V, in view of several approximations made in the calculation and uncertainties associated with the values for k, ε's, σ's, $\Delta\sigma$, $\Delta\varepsilon$, and so on, of the actual liquid crystal used. The observed quadratic

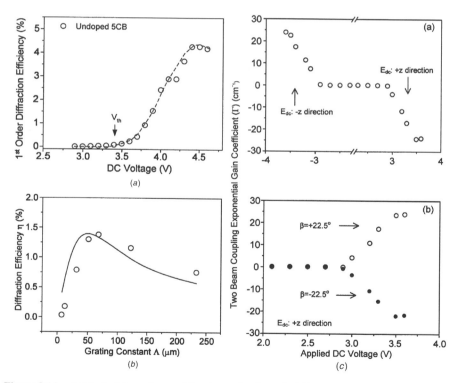

Figure 8.14.   (a) Typical dependence of the photorefractive self-diffraction efficiency as a function of applied dc voltage. (b) Dependence of photorefractive self-diffraction efficiency on the grating constant. Sample is 25 μm thick homeotropic C60-doped 5CB cell. Beam power is 10 mW (beam size: 2 mm). (c) Two beam coupling effect (unidirectional transfer of energy from one beam to the other) via photorefractive effect in nematic liquid crystals. Upper plot depicts dependence on direction of dc field; lower plot shows the dependence on the tilt of the sample w.r.t. the incidence beams. Sample is 25 μm thick homeotropic 5CB cell. Beam power is 1.5 mW (beam size: 2 mm).

dependence of the diffraction power on the applied dc voltage just above the threshold also corroborated the bulk theory. Other evidence supporting the role of the above bulk space-charge field model, such as the maximum diffraction efficiency at $qd \sim 2\pi$ (see Fig. 8.14b), two-beam coupling effect (see Fig. 8.14c) due to the phase-shifted induced grating, enhancement effect by photosensitive dopants, various dependence on the applied dc/ac fields, etc., have also been demonstrated.[28, 33, 39, 40] We further note that according to Equation 8.86 the bulk effect (Carr–Helfrich) shows that for a thinner sample, the numerator decreases. In the limit of larger grating constant, and smaller thickness so that $qd > 1$, the threshold field will be smaller, in agreement with the results.[40]

Nevertheless, recent studies by many groups have revealed that the bulk theories presented above need to be supplemented by considerations of surface effects, especially in cases involving photoactive surface alignment films[35–40] and the formation of "permanent" orientational gratings. A recent study,[39] for example, has shown that the photoinduced surface space-charge field contribution will result in an optical-intensity-dependent lowering of the threshold voltage required to initiate the photorefractive effect.

8.7. REORIENTATION AND NONELECTRONIC NONLINEAR OPTICAL EFFECTS IN SMECTIC AND CHOLESTERIC PHASES

8.7.1. Smectic Phase

The basic physics of laser-induced molecular reorientations, as well as the thermal, density, and flow phenomena in the smectic phase, is similar to that occurring in the nematic phase; the main difference lies in the magnitude of the various physical parameters that distinguish the smectic from the nematic phase. In general, smectic liquid crystals are highly viscous (i.e., they do not flow as easily as nematics). Their tendency to have layered structures (i.e., positional ordering among molecules in a plane) also imposes further restrictions on the reorientation of the molecules by an external field. In the smectic phase the order parameter dependence on the temperature is less drastic than in the nematic phase, and thus the refractive indices $n_{\parallel}$ and $n_{\perp}$ are not sensitively dependent on the temperature. Smectic liquid crystals do possess one important intrinsic advantage over nematic liquid crystals. As a result of the presence of a higher degree of order and less molecular orientation fluctuations in this phase, light-scattering loss in the smectic phase is considerably less than in the nematic phase. This will be important for optical processes that may require longer interaction lengths.

To date most studies on nonlinear optics are conducted in nematic liquid crystals, because of their special properties discussed before. Since the basic physics in smectics is similar, we refer the reader to the literature quoted in the following discussion for the details. Nevertheless, there are some interesting studies worth special mention here.

The possibility of laser-induced director axis reorientations in the smectic phase was first theoretically studied[5] in 1981. For smectic-A and smectic-B, one can see that a reorientation of the director axis will involve a change in the layer spacing. As shown in Chapter 4, such a distortion will involve a tremendous amount of energy and is therefore not observable under finite field strength. On the other hand, in smectic-C, it is possible to reorient the azimuthal component of the director axis of the molecule (see Fig. 8.15). This transition involves only a rotation of the director about the normal to the layer and does not involve a distortion of the layer spacing.

The optical nonlinearity associated with such a reorientational process in the presence of an external orienting dc magnetic field is estimated[5] to be comparable to that in the nematic phase. In a later publication, Ong and Young[43] presented a detailed theory of a purely optically induced reorientation effect. Some preliminary observations of such a reorientation process in a freely suspended smectic-C film were reported by Lippel and Young.[44]

Detailed theory and experimental observation of laser-induced director axis reorientation in a chiral smectic-C* (ferroelectric) liquid crystal were reported by Macdonald et al.[45] This study was conducted with a surface-stabilized ferroelectric liquid crystal in a planar oriented "bookshelf-like" configuration, where the director axis of the molecules is parallel, and the smectic layer perpendicular, to the cell walls (Fig. 8.16). Typically, the induced reorientation angle by a laser of intensity on the order of 3000 W/cm^2 is about 23°, with a switching time measured to be on the order of a millisecond or less. At such high optical intensity, the authors also noticed strong influence from laser heating of the sample.

8.7.2. Cholesteric Phase

Director axis reorientation in the cholesteric phase of liquid crystals was also first theoretically studied by Tabiryan and Zeldovich.[5] The cholesteric phase is unusual in that the director axis is spatially spirally distributed with a well-defined pitch, resulting in selective reflection of light. The basic physics of optically induced director axis distortion in the cholesteric phase is analogous to its nematic counterpart.

In the work by Lee et al.,[46] a theory and some qualitative experimental confirmation are presented on the optical retro-self-focusing effect associated with optically induced pitch dilation. The laser used has a Gaussian radial intensity distribution, which induces a radially varying pitch dilation effect, in analogy to the radially varying director axis reorientation induced by a Gaussian beam (cf. Section 8.3.3).

A quantitative study of laser-induced orientational effects in cholesterics was performed by Galstyan et al.[47] The director axis reorientation is induced by two

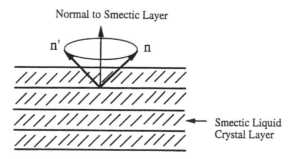

Figure 8.15.   Smectic-C liquid crystal axis azimuthal rotation by an external field.

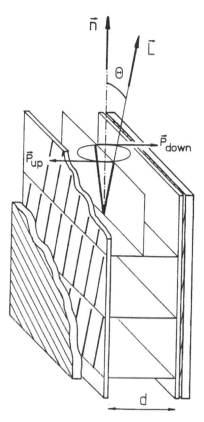

Figure 8.16. "Bookshelf-like" configuration of a ferroelectric liquid crystal film used for the laser-induced director axis rotation effect (from Macdonald et al.[45]) Here h is the normal to the smectic layer plane and L is the director axis direction. P is the spontaneous polarization.

counterpropagating waves of opposite circular polarizations (right- and left-handed). The signal beam originates as scattered noise from the pump beam and is amplified via stimulated scattering. In their experiment laser pulses on the order of 800 μs, with energy up to several hundred millijoules focused to a spot diameter of 35 μm, were required to generate observable stimulated scattering effects. This corresponds to a laser intensity on the order of a few MW/cm^2, which seems to be the usual intensity level needed to observe stimulated scatterings in nematics as well.[48]

REFERENCES

1. Khoo, I. C. 1981. Optically induced molecular reorientation and third order nonlinear optical processes in nematic liquid crystal. *Phys. Rev. A.* 23:2077.

2. Durbin, S. D., S. M. Arakelian, and Y. R. Shen. 1981. *Phys. Rev. Lett.* 47:1411.

3. Herman, R. M., and R. J. Serinko. 1979. Nonlinear-optical processes in nematic liquid crystals near Freedericksz transitions. *Phys. Rev. A.* 19:1757.

4. Khoo, I. C., R. G. Lindquist, R. R. Michael, R. J. Mansfield, and P. G. LoPresti. 1991. Dynamics of picosecond laser-induced density, temperature, and flow-reorientation effects in the mesophases of liquid crystals. *Appl. Phys.* 69:3853.

5. Tabiryan, N. V., and B. Ya. Zel'dovich. 1981. In a series of articles published in *Mol. Cryst. Liq. Cryst.* 62:637; 69:19,31, on nematic, smectic, and cholesteric liquid crystals.

6. Eichler, H. J., and R. Macdonald. 1991. Flow alignment and inertial effects in picosecond laser-induced reorientation phenomena of nematic liquid crystals. *Phys. Rev. Lett.* 67:2666.

7. Jackson, J. D., 1985. *Classical Electrodynamics.* New York: Wiley.

8. "Flow" effect, where the liquid crystal molecules are displaced by the intense optical field, has also been observed in the smectic phase.

9. Khoo I. C., and R. Normandin. 1984. Nanosecond laser-induced transient and erasable permanent grating diffractions and ultrasonic waves in a nematic film. *Appl. Phys.* 55:1416; see also Kiyano, K., and J. B. Ketterson. 1975. Ultrasonic study of liquid crystals. *Phys. Rev. A.* 12:615.

10. Wong, G. K. L., and Y. R. Shen. 1974. Study of pretransitional behavior of laser-field-induced molecular alignment in isotropic nematic substances. *Phys. Rev. A.* 10:1277; see, however, Deeg, F. W., and M. D. Fayer. 1989. Analysis of complex molecular dynamics in an organic liquid by polarization selective subpicosecond transient grating experiments. *Chem. Phys.* 91:2269, and references therein, where the observed individual molecular reorientation effect is associated with nuclear reorientation whose dynamics is more complex than the Debye relaxation process described by Wong and Shen.

11. Flytzanis, C., and Y. R. Shen. 1974. Molecular theory of orientational fluctuations and optical Kerr effect in the isotropic phase of a liquid crystal. *Phys. Rev. Lett.* 33:14.

12. Rao, D. V. G. L. N., and S. Jayaraman. 1974. Pretransitional behaviour of self-focusing in nematic liquid crystals. *Phys. Rev. A.* 10:2457.

13. Prost, J., and J. R. Lalanne. 1973. Laser-induced optical Kerr effect and the dynamics of orientational order in the isotropic phase of a nematogen. *Phys. Rev. A.* 8:2090; see also Lalanne, J. R., B. Martin, and B. Pouligny. 1977. Direct observation of picosecond reorientation of molecules in the isotropic phases of nematogens. *Mol. Cryst. Liq. Cryst.* 42:153.

14. Landau, L. D. 1965. *Collected Papers of L. D. Landau.* D. Ter Haar (ed.). New York: Gordon and Breach.

15. deGennes, P. G., 1974. *The Physics of Liquid Crystals.* Oxford: Clarendon Press.

16. Madden, P. A., F. C. Saunders, and A. M. Scott. 1986. Degenerate four-wave mixing in the isotropic phase of liquid crystals: The influence of molecular structure. *IEEE J. Quantum Electron.* QE-22:1287.

17. Berne, J., and R. Pecora. 1976. *Dynamic Light Scattering with Applications to Chemistry, Biology, and Physics.* New York: Wiley (Interscience).

18. Ibrahim, I. H., and W. Haase. 1981. On the molecular polarizability of nematic liquid crystals. *Mol. Cryst. Liq. Cryst.* 66:189; see also Luckhurst, G. R., and G. W. Gray. 1979. *The Molecular Physics of Liquid Crystals.* London: Academic Press.

19. Tabiryan, N. V., A. V. Sukhov, and B. Ya. Zel'dovich. 1986. The orientational optical nonlinearity of liquid crystals. *Mol. Cryst. Liq. Cryst.* 136:1. In this review article, various optical field-nematic axis configurations and the associated nonlinearities are discussed.

20. Khoo, I. C., P. Y. Yan, and T. H. Liu. 1987. Nonlinear transverse dependence of optically induced director axis reorientation of a nematic liquid crystal film: Theory and experiment. *Opt. Soc. Am. B.* 4:115, and references therein.

21. Khoo, I. C., 1983. Reexamination of the theory and experimental results of optically induced molecular reorientation and nonlinear diffraction in nematic liquid crystals: Spatial frequency and temperature dependence. *Phys. Rev. A.* 27:2747.

22. Janossy, I., L. Csillag, and A. D. Lloyd. 1991. Temperature dependence of the optical Freedericksz transition in dyed nematic liquid crystals. *Phys. Rev. A.* 44:8410–8413; Janossy, I., and T. Kosa. 1992. Influence of anthraquinone dyes on optical reorientation of nematic liquid crystals. *Opt. Lett.* 17:1183–1185.

23. Gibbons, W. M., P. J. Shannon, S.-T. Sun, and B. J. Swetlin. 1991. Surface-mediated alignment of nematic liquid crystals with polarized laser light. *Nature (London).* 351:49–50; see also Chen, A. G.-S., and D. J. Brady. 1992. Surface-stabilized holography in an azo-dye-doped liquid crystal. *Opt. Lett.* 17:1231–1233.

24. Khoo, I. C., H. Li, and Yu Liang. 1993. Optically induced extraordinarily large negative orientational nonlinearity in dye-doped liquid crystal. *IEEE J. Quantum Electron.* QE-29:1444.

25. Khoo, I. C., S. Slussarenko, B. D. Guenther and W.V. Wood. 1998. Optically induced space charge fields, dc voltage, and extraordinarily large nonlinearity in dye-doped nematic liquid crystals. *Opt. Lett.* 23: 253–255.

26. Khoo, I. C., P. H. Chen, M. Y. Shih, A. Shishido, and S. Slussarenko. 2001. Supra optical nonlinearities of methyl-red and azobenzene liquid crystal–doped nematic liquid crystals. *Mol. Cryst. Liq. Cryst.* 358:1–13; 364:141–149.

27. Khoo, I. C., M. Y. Shih, A. Shishido, P.H. Chen, and M. V. Wood. 2001. Liquid crystal photorefractivity: Towards supra-optical nonlinearity. *Opt. Mater.* 18:85–90; Khoo, I. C., M. Y. Shih, M. V. Wood, and P. H. Chen. 2000. Extremely nonlinear photosensitive nematic liquid crystal film. *Synth. Met.* 7413:1–6.

28. Khoo, I. C., M. Y. Shih, M. V. Wood, B. T. Guenther, P. H. Chen, F. Simoni, S. S. Slussarenko, O. Francescangeli, and L. Lucchetti. 1999. Dye-doped photorefractive liquid crystals for dynamic and storage holographic grating formation and spatial light modulation. *Proc. IEEE.* 87(11): 1987–1911; see also Khoo, I. C., Yana Zhang Williams, B. Lewis, and T. Mallouk. 2005. Photorefractive CdSe and gold nanowire-doped liquid crystals and polymer-dispersed liquid-crystal photonic crystals. *Mol. Cryst. Liq. Cryst.* 446:233–244.

29. Lucchetti, L., M. Di Fabrizio, O. Francescangeli, and F. Simoni. 2004. Colossal optical nolinearity in dye-doped liquid crystals. *Opt. Commun.* 233:417.

30. Shishido, A. T., O. Tsutsumi, A. Kanazawa, T. Shiono, T. Ikeda, and N. Tamai. 1997. Rapid optical switching by means of photoinduced change in refractive index of azobenzene liquid crystals detected by reflection-mode analysis. *J. Am. Chem. Soc.* 119: 7791–7796 and references therein.

31. Helfrich, W. 1969. Conduction-induced alignment of nematic liquid crystals: Basic model and stability consideration. *J. Chem. Phys.* 51:4092.

32. Gunter, P., and J. P. Huignard, eds., 1989. *Photorefractive Materials and Their Applications*, Vols. I and II. Berlin: Springer-Verlag.

33. Khoo, I. C., H. Li, and Y. Liang. 1994. Observation of orientational photorefractive effects in nematic liquid crystals. *Opt. Lett.* 19:1723.

34. Rudenko E. V., and A. V. Sukhov. 1994. Optically induced spatial charge separation in a nematic and the resultant orientational nonlinearity. *JETP.* 78(6): 875–882.

35. Kaczmarek, M., A. Dyadyusha, S. Slussarenko, and I. C. Khoo. 2004. The role of surface charge field in two-beam coupling in liquid crystal cells with photoconducting polymer layers. *J. Appl. Phys.* 96(5): 2616–2623.

36. Mun, J., C. S. Yoon, H.-W. Kim, S.-A, Choi, and J.-D Kim. 2001. Transport and trapping of photocharges in liquid crystals placed between photoconductive polymer layers. *Appl. Phy. Lett.* 79(13): 1933–1935.

37. Pagliusi, P., and G. Cipparrone. 2002. Surface-induced photorefractive-like effect in pure liquid crystals. *Appl. Phys. Lett.* 80(2): 168–170; Charge transport due to photoelectric interface activation in pure nematic liquid-crystal cells. *J. Appl. Phys.* 92(9): 4863–4869; 2003. Extremely sensitive light induced reorientation in nondoped nematic liquid crystal cells due to photoelectric activation of the interface. *ibid.* 93(11): 9116–9122.

38. Tabiryan, N. V., and C. Umeton. 1998. Surface-activated photorefractive and electro-optic phenomena in liquid crystals. *J. Opt. Soc. Am. B.* 15(7): 1912–1917.

39. Khoo, I. C., Kan Chen, and Y. Williams. 2006. Orientational photorefractive effect in undoped and CdSe nano-rods doped nematic liquid crystal: Bulk and interface contributions. *J. Sel. Top. Quantum Electron* 12(3): 443–450; see also Khoo, I. C., Yana Zhang Williams, B. Lewis, and T. Mallouk. 2005. Photorefractive CdSe and gold nanowire-doped liquid crystals and polymer-dispersed-liquid-crystal photonic crystals. *Mol. Cryst. Liq. Cryst.* 446:233–244.

40. Khoo, I. C. 1996. Orientational photorefractive effects in nematic liquid crystal films. *IEEE J. Quantum Electron.* 32: 525–534.

41. Sugimura, A., and Ou-Yang Zhong-can. 1992. Anomalous photocurrent transients in nematic liquid crystals: The nonlinear optical Pockel's effect induced by the Freedericksz transition. *Phys. Rev. A.* 45(4): 2439–2448.

42. Boichuk, V., S. Kucheev, J. Parka, V. Reshetnyak, Y. Reznikov, I. Shiyanovskaya, K. D. Singer, and S. Slussarenko. 2001. Surface-mediated light-controlled Freedericksz transition in a nematic liquid crystal cell. *J. Appl. Phys.* 90:5963–5967.

43. Ong, H. L., and C. Y. Young. 1984. Optically induced molecular reorientation in smectic-C liquid crystal. *Phys. Rev. A.* 29:297.

44. Lippel, P. H., and C. Y. Young. 1983. Observation of optically induced molecular reorientation in films of smectic-C liquid crystal. *Appl. Phys. Lett.* 43:909.

45. Macdonald, R., J. Schwartz, and H. J. Eichler. 1992. Laser-induced optical switching of a ferroelectric liquid crystal. *Int. J. Nonlinear Opt. Phys.* 1:119.

46. Lee, J.-C., S. D. Jacobs, and A. Schmid. 1987. Retro-self-focusing and pinholing effect in a cholesteric liquid crystal. *Mol. Cryst. Liq. Cryst.* 150B:617.

47. Galstyan, T. V., A. V. Sukhov, and R. V. Timashev. 1989. Energy exchange between optical waves counterpropagating in a cholesteric liquid crystal. *Sov. Phys. JETP.* 68(5):1001.

48. Khoo, I. C., R. R. Michael, and P. Y. Yan. 1987. Simultaneous occurrence of phase conjugation and pulse compression in simulated scatterings in liquid crystal mesophases. *IEEE J. Quantum Electron.* QE23:1344; see also Roy, D. N. Ghosh, and D. V. G. L. N. Rao. 1986. Optical pulse narrowing by backward, transient stimulated Brillouin scattering. *Appl. Phys.* 59:332.

9

Thermal, Density, and Other Nonelectronic Nonlinear Mechanisms

9.1. INTRODUCTION

In Chapter 5 we discussed how the electrostrictive effect gives rise to density fluctuations in liquid crystals, which are manifested in frequency shifts and broadening in the spectra of the scattered light. In absorbing media, density fluctuations are also created through the temperature rise following the absorption of the laser. The nature of optical absorption in liquid crystals, as in any other material, depends on the laser wavelength.

The (linear) transmission spectrum of a typical liquid crystal is shown in Figures 9.1 and 9.2, where dips in the curve correspond to strong single-photon absorption. The linear absorption constant α is quite high at wavelengths near or shorter than the ultraviolet ($\alpha \approx 10^2 - 10^3 \text{cm}^{-1}$), where the absorption band begins; in the visible and near-infrared regions, the absorption constant is typically small (with $\alpha \leq 1 \text{ cm}^{-1}$ and $\alpha \leq 10 \text{ cm}^{-1}$, respectively); in the midinfrared and longer wavelength regions, the absorption constant is higher ($\alpha \approx 10 - 10^2 \text{ cm}^{-1}$).

Under intense laser illumination, two- and multiphoton absorption processes will occur, as depicted in Figure 9.3. In this case a so-called nonabsorbing material in the single-photon picture outlined previously could actually be quite absorptive, if the two- or multiphoton process corresponds to a real transition to an excited state.

From the standpoint of understanding laser-induced temperature and density changes in liquid crystals, these photoabsorption processes may be simply represented as a means of transferring energy to the molecule. Figure 2.5 schematically depicts the scenario following the absorption of the incoming photons by the liquid crystalline molecules.

The equation describing the rate of change of energy density u is given by

$$\frac{\partial u}{\partial t} = \frac{nc\alpha E^2}{4\pi} + D\nabla^2 u - \frac{u}{\tau}. \tag{9.1}$$

The first term on the right-hand side denotes the rate of absorption of the light energy, the second term denotes the rate of energy diffusion and the third term describes the

Liquid Crystals, Second Edition   By Iam-Choon Khoo
Copyright © 2007 John Wiley & Sons, Inc.

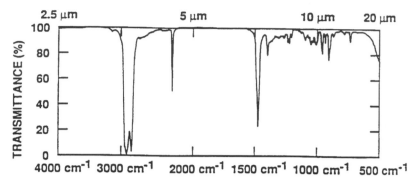

Figure 9.1.   Transmission spectrum of a nematic liquid crystal (3 CCH) in the 2.5 to 20 μm regime.

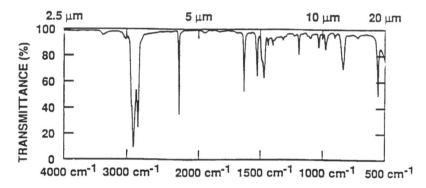

Figure 9.2.   Transmission spectrum of a nematic liquid crystal (7 PCH) in the 2.5 to 20 μm regime.

thermalization via inter- and intramolecular relaxation processes—usually collisions.

Liquid crystal molecules are large and complex. The inter- and intramolecular relaxation processes following photoabsorption are extremely complicated. However, it is well known that *individual* molecular motions in both crystalline and isotropic phases are all characterized by very fast picosecond relaxation times. One can therefore argue that, following photoabsorption, the thermalization time it takes to convert the absorbed energy into heat is quite short, typically on the order of picoseconds. Accordingly, if our attention is on the nonlinear optical responses of liquid crystals that are characterized by relaxation times in the nanosecond and longer time scale, we may adopt a formalism that ignores the thermalization process. Therefore, we may assume that the rate of energy density transfer reaches a steady state very quickly (i.e., $\partial u/\partial t \rightarrow 0$), and the diffusion process has no time to act. In this case we have $u/\tau = nc\alpha E^2/4\pi$.

The equations describing the laser-induced temperature T and density ρ changes are coupled. This is because both ρ and T are functions of the entropy S and the pressure P.[1]

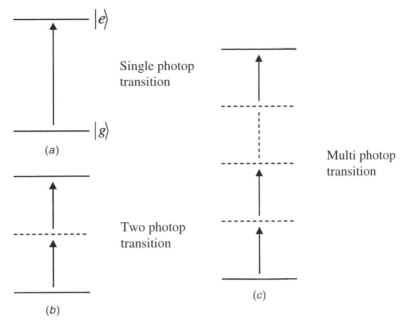

Figure 9.3. (a) Single-photon absorption process; (b) two-photon absorption process; (c) multiphoton absorption process.

Writing the temperature $T(\mathbf{r},t)$ and the density $\rho(\mathbf{r},t)$ as

$$T(\mathbf{r},t) = T_0 + \Delta T(\mathbf{r},t), \qquad (9.2)$$

$$\rho(\mathbf{r},t) = \rho_0 + \Delta\rho(\mathbf{r},t), \qquad (9.3)$$

the coupled hydrodynamical equations are given by [1,2]

$$-\frac{\partial^2}{\partial t^2}(\Delta\rho) + v^2\nabla^2(\Delta\rho) + v'\beta_T\rho_0\nabla^2(\Delta T) + \frac{\eta}{\rho_0}\frac{\partial}{\partial t}\nabla^2(\Delta\rho) = \frac{\gamma^e}{8\pi}\nabla^2(E^2) \quad (9.4)$$

and

$$\rho_0 C_v \frac{\partial}{\partial t}(\Delta T) - \lambda_T \nabla^2(\Delta T) - \frac{(C_p - C_v)}{\beta_T}\frac{\partial}{\partial t}(\Delta\rho) = \frac{u}{\tau} = \frac{\alpha n c}{4\pi}E^2, \quad (9.5)$$

where ρ_0 is the unperturbed density of the liquid crystal, C_p and C_v are the specific heats, λ_T is the thermal conductivity, η is the viscosity, v is the speed of sound, γ^e is the electrostrictive coefficient [$\gamma^e = \rho_0(\rho\varepsilon/\partial\rho)_T$], β_T is the coefficient of volume expansion.

Equation (9.4) describes the thermal expansion and electrostrictive effects on the density change, whereas Equation (9.5) describes the photoabsorption and the resulting temperature rise and heat diffusion process.

We must emphasize here that in these equations the effective values for most of these parameters are, of course, dependent on the particular phase (isotropic, nematic, smectic, etc.) and temperature,[3,4] as well as the laser beam polarization, propagation direction, and nematic axis alignment (i.e., the geometry of the interaction).

9.2. DENSITY AND TEMPERATURE CHANGES INDUCED BY SINUSOIDAL OPTICAL INTENSITY

As an example of how one may gain some insight into such a complicated problem, and for other practically useful reasons, we now consider the case where the optical intensity is a spatially periodic function (i.e., an intensity grating). Such an intensity function may be derived from the coherent superposition of two laser fields on the liquid crystal (see Fig. 9.4).

As a result of the spatially periodic intensity function, a spatially periodic refractive index change (i.e., an index grating) is induced. If this index grating is probed by a cw laser, side diffractions in the directions $\pm\theta$, $\pm2\theta$, and so on will be generated. This is shown in Figure 9.5 which depicts a typical so-called dynamic grating setup for studying laser-induced nonlinear diffractions.

The diffractions from the probe beam in the $\pm\theta$ directions are termed first-order diffractions. The efficiency of the diffraction η, defined by the ratio of the intensity of the diffraction to the zero-order (incident) laser intensity, is given by [2]

$$\eta \approx J_1^2\left(\frac{\pi\Delta n d}{\lambda}\right). \tag{9.6}$$

If Δn, the index grating amplitude, is small [i.e., $\pi\Delta n d/\lambda) \ll 1$],

$$\eta = \left(\frac{\pi\Delta n d}{\lambda}\right)^2. \tag{9.7}$$

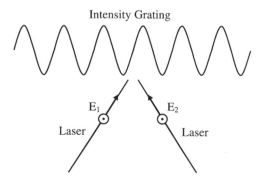

Figure 9.4. Sinusoidal optical intensity profile produced by interference of two co-polarized coherent lasers.

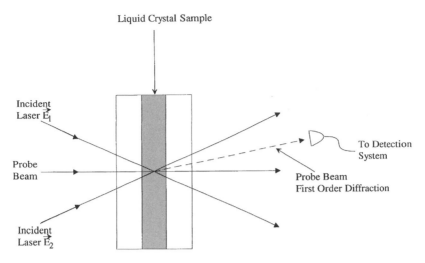

Figure 9.5. Experimental setup for probing the dynamics of laser-induced transient refractive index changes in liquid crystals.

The optical intensity inside the liquid crystal is of the form $E^2 = E_1^2 + E_2^2 + 2|E_1E_2|$ $\cos((\mathbf{k}_1 - \mathbf{k}_2)\cdot\mathbf{r})$ in the plane wave approximation. The dc part of E^2 gives rise to spatially uniform changes in ρ and T, and they do not contribute to the diffraction of the beam. We may therefore consider only the spatially periodic part and write $E^2 = 2|E_1E_2|\cos(\mathbf{q}\cdot\mathbf{y})$, where $\mathbf{q} = \mathbf{k}_1 - \mathbf{k}_2$. Furthermore, for simplicity as well as convenience, let $E_1 = E_2 = E_0$. This gives

$$E^2 = 2E_0^2(1 + \cos\mathbf{q}\cdot\mathbf{y}).\tag{9.8}$$

Correspondingly, $\Delta\rho$ and ΔT are of the form

$$\Delta\rho = \rho(t)\cos\mathbf{q}\cdot\mathbf{y},\tag{9.9}$$

$$\Delta T = T(t)\cos\mathbf{q}\cdot\mathbf{y},\tag{9.10}$$

where $\rho(t)$ and $T(t)$ are the density and temperature grating amplitudes. Substituting Equations (9.8) and (9.10) into Equations (9.4) and (9.5), one could solve for the grating amplitudes $\rho(t)$ and $T(t)$. Letting the initial conditions be $t = 0$ and $\rho(0) = T(0) = 0$ and ignoring the term proportional to $(C_p - C_v)$ in Equation (9.5), a straightforward but cumbersome calculation yields the following results.

For $0 < t < \tau_p$, where τ_p is the duration of the laser pulse (assumed to be a square pulse),

$$T(t) = \left(\frac{\alpha cnE_0^2}{4\pi\rho_0 C_v\Gamma_R}\right)[1 - \exp(-\Gamma_R t)],\tag{9.11}$$

$$\rho(t) = \left(\frac{\gamma^e E_0^2}{4\pi v^2}\right)[1 - \exp(-\Gamma_B t)\cos \Omega t] - \left(\frac{\beta_T \alpha cn E_0^2}{4\pi C_v \Gamma_R}\right)[1 - \exp(-\Gamma_R t)]. \quad (9.12)$$

In the preceding equations the thermal decay constant Γ_R is given by

$$\Gamma_R = \frac{\lambda_T q^2}{\rho C_v}, \quad (9.13)$$

the Brillouin decay constant Γ_B is given by

$$\Gamma_B = \frac{\eta q^2}{2\rho_0}, \quad (9.14)$$

and Ω, the sound frequency, is given by

$$\Omega = \sqrt{q^2 v^2 - \Gamma_B^2}. \quad (9.15)$$

The corresponding relaxation time constants are, respectively,

$$\tau_R = \Gamma_R^{-1} = \frac{\rho_0 C_v}{\lambda_T q^2} \quad (9.16)$$

and

$$\tau_B = \Gamma_B^{-1} = \frac{2\rho_0}{\eta q^2}. \quad (9.17)$$

In liquid crystals the typical values for the various parameters[3,4] in units are $n=1.5$, $\eta=7\times10^{-2}$ kg m^{-1} s^{-1}, $v=1540$ m s^{-1}, $\rho_0=10^3$ kg m^{-3}, $\lambda_T/\rho_0 C_v=0.79\times10^{-7}$ m^2/s. If a grating constant ($\Lambda=2\pi |K_1-K_2|^{-1}$) of 20 μm is used in the experiment, we have $\tau_R\approx100$ μs. For the same set of parameters, we have $\tau_B\approx200$ ns. These widely different time scales of the thermal and density effects provide a means to distinguish their relative contributions in the nonlinear dynamic grating diffraction experiment.

It is important to note here that the density change $\rho(t)$ given in Equation (9.12) is the sum of two distinct components:

$$\rho(t) = \rho^e(t) + \rho^T(t), \quad (9.18)$$

where

$$\rho^e(t) = \frac{\gamma^e E_0^2}{4\pi v^2}[1 - \exp(-\Gamma_B t)\cos \Omega t] \qquad (9.19)$$

and

$$\rho^T(t) = \frac{-\beta_T \alpha c n E_0^2}{4\pi C_v \Gamma_R}[1 - \exp(-\Gamma_R t)]. \qquad (9.20)$$

The component $\rho^e(t)$ is due to the electrostrictive effect (the movement of mole-cules under intense electric field); it is proportional to γ^e and is characterized by the Brillouin relaxation constant Γ_B and frequency Ω. From Equations (9.9) and (9.19) one can see that this ρ^e component gives rise to a propagating wave. On the other hand, the component $\rho^T(t)$ is due to the thermoelastic contribution (propor-tional to β^T) and is characterized by the thermal decay constant Γ_R; it is nonprop-agative.

More detailed solutions of the coupled density and temperature equations [Eqs. (9.4) and (9.5)] may be found in the work by Batra et al.[1] The preceding simple example, however, will suffice to illustrate the basic processes following photoab-sorption and their time evolution characteristics.

9.3. REFRACTIVE INDEX CHANGES: TEMPERATURE AND DENSITY EFFECTS

Because of these temperature and density changes, there are corresponding refractive index changes given, respectively, by

$$\Delta n_T = \frac{\partial n}{\partial T}T'(t) \qquad (9.21)$$

and

$$\Delta n_\rho = \frac{\partial n}{\partial \rho}\rho(t) = \frac{\partial n}{\partial \rho}(\rho^e + \rho^T). \qquad (9.22)$$

The thermal index component $\partial n/\partial T$ arises from two effects. One is the spec-tral shift as a result of the rise in the temperature of the molecule. This effect occurs within the thermalization time τ (i.e., in the picosecond time scale), and its contribution is usually quite small[5,6] for the ordered as well as the liquid phases.

In the nematic phase, another effect contributing to Δn_T arises from the temperature dependence of the order parameter,

$$\frac{\partial n}{\partial T} = \frac{\partial n}{\partial S}\frac{\partial S}{\partial T}. \tag{9.23}$$

As we have seen in earlier chapters, this effect is particularly dominant at temperatures in the vicinity of the nematic–isotropic transition. Since the order parameter S exhibits critical slowing down behavior near T_c, this component of the refractive index change also exhibits similar behavior. This was discussed in Chapter 2, and it will be discussed in more details in the next sections.

The density contributions, consisting of the electrostrictive component ρ^e and the thermoelastic component ρ^T, are effects typically experienced by solids and ordinary liquids; they are not strongly coupled to the order parameter.

The propagation of the electrostrictive component ρ^e will interfere with the non-propagating thermal component. Consider Figure 9.6 which depicts the presence of both a "static" refractive index grating due to Δn_T and Δn_p (ρ^T), and a "propagative" one from Δn_p (ρ^e). The latter is created by two counterpropagating waves right after the laser pulse, with wave vectors $\pm(\mathbf{k}_1 - \mathbf{k}_2)$ and a frequency Ω. Consequently, the maxima and minima of these two index gratings will interfere in time, leading to oscillations (at the frequency Ω of the sound wave) in the diffractions from the probe beam (see Fig. 9.7). Since the magnitude of the wave vector $|\mathbf{k}_1 - \mathbf{k}_2|$ is known and the oscillation frequency Ω can be directly measured, these observed oscillations in the diffraction will provide a means of determining sound velocities in a liquid crystal.[3,7] By varying the interaction geometry between the grating wave vectors $\mathbf{q}$ and the director or c axis of the liquid crystal sample, this dynamical diffraction effect could also be used to measure sound velocity and thermal diffusion anisotropies.

In the earlier experiments reported in the work by Khoo and Normandin,[3,7] nanosecond laser pulses from the second harmonic of a Q-switched Nd:YAG laser

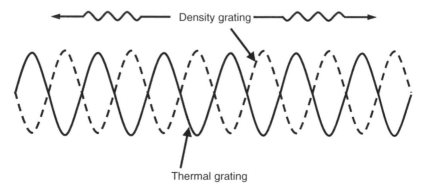

Figure 9.6. Interference of the propagative index gratings (arising from electrostrictive density changes) with the diffusive 9 (but not propagative) thermal grating.

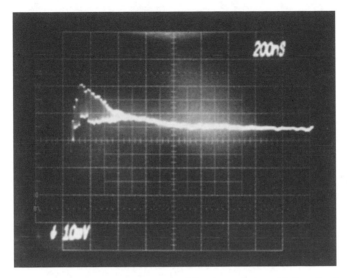

Figure 9.7. Oscilloscope trace of the observed probe diffraction evolution with time showing interference effect between the density and thermal components from a room-temperature nematic liquid crystal (E7).

are employed to excite these density and temperature (order parameter) interference effects in nematic and smectic-A liquid crystals.

For the nematic case,[3] the liquid crystal used is 5CB. The planar aligned sample is 40 μm thick. The sample is oriented such that the director axis is either parallel or orthogonal to the pump laser polarization direction. These interaction geometries allow the determination of the thermal diffusion constants $D_\parallel$ and $D_\perp$. Figures 9.8a and 9.8b are oscilloscope traces of the relaxation dynamics of the first-order probe beam diffraction from the thermal gratings for, respectively, thermal diffusions perpendicular and parallel to the director axis. In Figure 9.8a the thermal diffusion time constant $\tau_\perp$ is about 100 μs, while $\tau_\parallel$ from Figure 9.8b is about 50 μs. In the experimental set up, the crossing angle in air is $2°$, corresponding to a grating constant $\Lambda_\perp = 2\pi/q_\perp$ of 17 μm and $\Lambda_\parallel = 2\pi/q_\parallel$ of 15 μm.

Using Equation (9.16) for the thermal decay times and the values of the thermal diffusion constants $D_\parallel = 7.9 \times 10^{-4}\,\mathrm{cm^2\,s^{-1}}$ and $D_\perp = 1.25 \times 10^{-3}\,\mathrm{cm^2\,s^{-1}}$[4], the theoretical estimates of $\tau_\parallel$ and $\tau_\perp$ are 110 and 55 μs, respectively. These are in good agreement with the experimental results. For smaller/larger grating constants, the thermal diffusion times have been observed to be decreasing/increasing roughly in accordance with the q^{-2} dependence.

For the same sample the sound velocity anisotropy was too small to be detected with the precision of the instruments used. Nevertheless, the crossing-angle dependence of the period of oscillation (caused by acoustic–thermal grating interference, see Fig. 9.7) in the probe beam diffraction has been measured. The period $T = 2\pi/\Omega$ is related to the grating constant $\Lambda_{\parallel\perp} = \lambda/[2n_{\parallel,\perp}\sin(\theta/2)]$ by $T = \Lambda/v_s$. (For 5CB, $n_\parallel \approx 1.72$ and $n_\perp \approx 1.52$). Figure 9.9 plots the experimentally observed (unpublished data from Khoo and Normandin[3]) oscillation period as a function of the crossing angle θ.

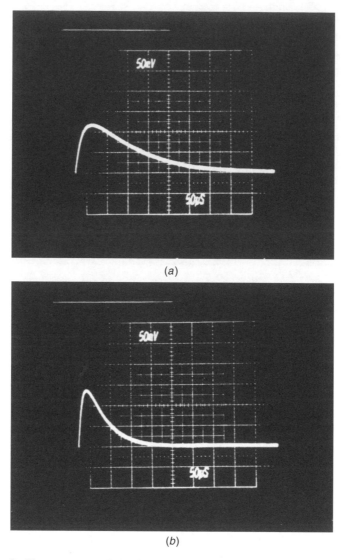

Figure 9.8. Oscilloscope traces of the thermal grating decay dynamics (time scale: 50 μs/div). (a) Grating wave vector is perpendicular to the director axis. (b) Grating wave vector is along the director axis, showing a faster decay.

In general, it follows the theoretical relationship (line) quite closely, with a sound velocity $v_s = 1.53 \times 10^3$ m s^{-1} determined using one of the experimental data points.

In the experiment with the smectic-A liquid crystal,[7] the dominant contribution to the grating diffraction seems to come from the density effect. The observed acoustic velocity and attenuation time are in accordance with theoretical expectation. An interesting effect reported in the work by Khoo and Normandin [7] is the formation of permanent gratings by intense excitation pulses. These gratings are erasable by

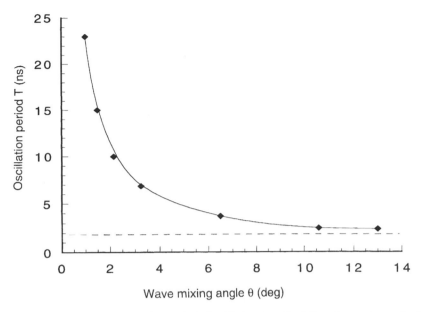

Figure 9.9.   Observed dependence of the period of oscillations (see Fig. 9.7) on the crossing angle of the excitation laser beams. Solid line is the theoretical dependence. Dotted lines shows the detection system dynamic response limit.

warming the sample through the isotropic phase and cooling it back to the aligned smectic-A state.

Since the refractive indices of nematic liquid crystals are very sensitively dependent on the temperature, through its dependence on the order parameter, we expect to see very dramatic dominance of the diffraction by the temperature component when the sample temperature is raised toward T_c. Figure 2.3b shows an oscilloscope trace of the probe beam diffraction for an E7 sample at about 20°C above room temperature (i.e., at 42°C). Clearly, the temperature component is greatly enhanced compared to the density component (the small "spike" detected at the beginning of the trace is the density component). As noted in Chapter 2, another interesting feature is that the time for the signal to reach its peak has also lengthened to about 20 μs, compared to about 100 ns at room temperature (cf. Fig. 9.7). Such lengthening of the thermal component is attributed to a critical slowing down of the order parameter as a phase transition is approached. In both cases the recorded thermal decay time is on the order of 100 μs.

These studies of dynamic grating diffraction in nematic liquid crystal (E7) film have also been conducted using short microsecond infrared (CO_2) laser pulses.[8] Since the wavelength of a CO_2 laser is around 10 μm, the absorption of the laser is via the molecular rovibrational modes of the electronic ground state. Figure 2.3a shows the observed diffracted signal. There is an initial spike which is attributed to the "fast" decaying density contributions. On the other hand, the "slower" thermal component associated with the order parameter exhibits a rather long buildup time of about 100 μs. This implies that, in nematic liquid crystals, the molecular correlation

effects leading to ordering among the molecules in the ground (electronic) rovibrational manifold are different from those in the excited (electronic) rovibrational
manifold. Such hitherto unexplored dependences of the order parameter dynamics on
the electronic and rovibrational excitational states clearly deserve further studies.

9.4. THERMAL AND DENSITY OPTICAL NONLINEARITIES OF NEMATIC LIQUID CRYSTALS IN THE VISIBLE–INFRARED SPECTRUM

Since both laser-induced thermal and density changes give rise to intensity-dependent
refractive index changes, they may be viewed as optical nonlinearities. From the experimental observations (see Fig. 9.7), we can say that, in general, the density and thermal
contributions to the probe diffraction are comparable for a sample maintained at a temperature far from T_c. In the vicinity of T_c the thermal component is much larger. The
absolute magnitude of the thermal nonlinearity, of course, depends on several factors.
From the preceding discussion, some of the obvious factors are the index gradient
dn/dT, which depends critically on the order parameter dependence on the temperature;
the absorption constant α, which varies over several orders of magnitude depending on
the laser wavelength (i.e., the spectral regime); the thermal decay constant Γ_R, which is
a function of the laser–nematic interaction geometry (e.g., laser spot size, sample thickness, and diffusion constant); and a collection of liquid crystalline parameters. Some of
the parameters for a few typical liquid crystals are listed in Table 9.1.

One of the most striking and important optical properties of liquid crystals is their
large birefringence throughout the whole optical spectrum (from near UV to the
infrared). The large thermal index gradients noted in Table 9.1 for the visible–infrared
spectrum are a consequence of that. In Figures 9.10 and 9.11 the measured refractive

Table 9.1   Typical Values for Some Nematic Liquid Crystal

Parameter	Value	Nematic
Absorption constant (α in cm^{-1})	<1, visible	E7
	23, near infrared	E7
	40 ~ 100, infrared	E7
	69, infrared	5CB
	44, infrared	E46 (EM Chemicals)
Diffusion constant $D = \lambda_T/\rho_0 C_v$	$D_\parallel = 1.95 \times 10^{-3}\,\mathrm{cm^2/s}$	E7
	$D_\perp = 1.2 \times 10^{-3}\,\mathrm{cm^2/s}$	E7
Thermal index gradients $dn_\perp/dT$	$10^{-3}\,\mathrm{K}$	5CB at 25 °C
	$10^{-2}\,\mathrm{K}$ (visible-infrared)	5CB near T_C
$Dn_\parallel/d_T$	$-2 \times 10^{-3}\,\mathrm{K}$	5CB at 25 °C
	$-10^{-2}\,\mathrm{K}$ (visible-infrared)	5CB near T_C

Visible → 0.5 μm;   near infrared → CO wavelength ≈ 5 μm;   infrared → 10 μm.

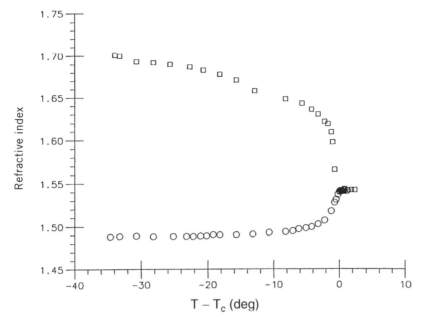

Figure 9.10.   Temperature dependence of the refractive indices of E7 at 10.6 μm.

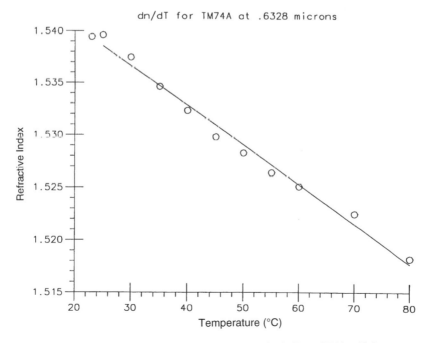

Figure 9.11.   Temperature dependence of the refractive indices of E46 at 10.6 μm.

indices of E7 and E46 (from EM Chemicals) in the infrared (10.6 μm CO_2 laser line) regime are shown.[9]

Besides these material parameters, perhaps the single most important factor governing the magnitude of the effective optical nonlinearities is the laser pulse duration relative to the material response time. If the laser pulse is too short, the material response time will be minimal, and the corresponding nonlinearity as "seen" by the laser will be small. This could be cast in more quantitative terms as follows.

We shall consider the following two limiting cases of thermal nonlinearities: (1) steady-state regime, where $\tau_p \gg \tau_R$ and τ_s (τ_s is the order parameter response time), and (2) transient regime, corresponding to $\tau_p \ll \tau_R$ and τ_s.

9.4.1. Steady-State Thermal Nonlinearity of Nematic Liquid Crystals

From these liquid crystalline parameters, one can estimate the steady-state thermal nonlinear coefficient $\alpha_2(T)$ defined by

$$\Delta n_T = \alpha_2^{SS}(T)I_{op}, \tag{9.24}$$

where the optical intensity I_{op} is measured in W/cm^2.

From Equations (9.11) and (9.21) we have, in the steady state when $t = \Gamma_R^{-1}$,

$$\Delta n_T = \frac{\alpha c n E_0^2}{4\pi \rho_0 C_v \Gamma_R} \frac{\partial n}{\partial T} = \alpha_2^{SS}(T)I_{op}, \tag{9.25}$$

that is,

$$\alpha_2^{SS}(T) = \frac{\alpha}{\rho_0 C_v \Gamma_R}\left(\frac{dn}{dT}\right). \tag{9.26}$$

Recall that $\Gamma_R = Dq^2$. Equation (9.26) thus gives

$$\alpha_2^{SS}(T) = \frac{\alpha}{\rho_0 C_v Dq^2}\left(\frac{dn}{dT}\right). \tag{9.27}$$

Since $q = 2\pi/\Lambda$, where Λ is the grating space of the temperature modulation, we may express $\alpha_2^{SS}(T)$ as

$$\alpha_2^{SS}(T) \approx \frac{\alpha \Lambda^2}{4\pi^2 \rho_0 C_v D}\left(\frac{dn}{dT}\right). \tag{9.28}$$

This expression is valid for the case where the two interference laser beam sizes are large compared to the liquid crystal cell thickness. If focused laser beams are used, we have to replace the grating constant Λ in Equation (9.28) by the characteristic diffusion length. This can be the laser-focused spot size or the thickness of the cell, whichever is associated with the dominant (i.e., shortest time) diffusion process. Therefore, we may write

$$\alpha_2^{SS}(T) = \frac{\alpha\Lambda_{th}^2}{4\pi^2\rho_0 C_v D}\left(\frac{dn}{dT}\right), \tag{9.29}$$

where Λ_{th} is the characteristic thermal diffusion length. Needless to say, this approach has grossly overlooked many of the details involved in laser-induced temperature and refractive index changes in an actual system, where the heat diffusion process is more likely a three-dimensional problem and the laser intensity distribution has a Gaussian envelope, and so on. Nevertheless, Equation (9.28) or (9.29) should provide us with a reasonable means of estimating the thermal nonlinearity. For example, if the characteristic diffusion length Λ_{th} is 20 μm, then, using typical liquid crystalline parameters, $\rho \approx 1$ g/c.c., $C_p\ [\approx C_v] \approx 2$ J/g/K, $D \approx 2\times10^{-3}$ cm^2/s, $\alpha = 100$ cm^{-1} (cf. Table 7.1 for CO_2 laser wavelength), and $dn/dT = 10^{-3}$ K^{-1}, we get

$$\begin{aligned}
\alpha_2^{SS}(T) &\approx \frac{100}{4\pi}\cdot\frac{(20\times10^{-4})^2}{1\cdot2\cdot2\cdot10^{-3}}\cdot10^{-3} \\
&= 2.5\times10^{-6}\ \text{cm}^2/\text{W}.
\end{aligned} \tag{9.30}$$

In esu units (see Chapter 10), the corresponding nonlinear third-order susceptibility is given by

$$\chi_{esu}^{(3)}(T) = 9.5\times n_2(I) = 2.4\times10^{-5}\ \text{esu}. \tag{9.31}$$

We remind the reader, again, that these expressions for thermal nonlinearity are rough estimates only. More detailed calculations are clearly needed if we desire more accurate quantitative information; they can be obtained by solving Equations (9.4) and (9.5) for the appropriate interaction geometries and boundary conditions.

9.4.2. Short Laser Pulse Induced Thermal Index Change in Nematics and Near-T_c Effect

As in the case of laser-induced molecular reorientation discussed in the preceding chapter, if the duration of the laser pulse is short compared to the thermal decay time τ_R^{-1}, the effective induced optical nonlinearity is diminished.

From Equations (9.11) and (9.12), it is straightforward to show that, for time $t \ll \tau_R^{-1}$, we may write the induced index change $\Delta n(t)$ by

$$\Delta n(t) = \alpha_2(t,T) I_{op},\tag{9.32}$$

where

$$\alpha_2(t,T) = \alpha_2^{ss}(T) \left(\frac{t}{\tau_R} \right).\tag{9.33}$$

For the same set of parameters used in estimating $\alpha_2^{ss}(T)$ in Equation (9.30), we note that $\tau_R = 0.5 \times 10^{-4}$ s. Therefore, for $t = 1$ μs (10^{-6} s), we have

$$\alpha_2(t,T) \approx (2 \times 10^{-2}) \alpha_2^{ss}(T) = 5 \times 10^{-8} \text{ cm}^2/\text{W}.\tag{9.34}$$

For $t = 1$ ns (10^{-9} s), we have

$$\alpha_2(t,T) \approx 5 \times 10^{-11} \text{ cm}^2/\text{W}.\tag{9.35}$$

From Equations (9.26) to (9.29), one can see that the diminished thermal nonlinearity for short laser pulses may be improved by the optimized choice of molecular and geometrical parameters such as the absorption constant α, the grating parameter q, and the thermal index. Since the thermal index dn/dT of a nematic liquid crystal is considerably enhanced near T_c, one would expect that $\alpha_2^{ss}(T)$ will be proportionately increased. This expectation, however, is not borne out in practice because of the critical slowing down[1, 3] in the order parameter S near T_c (see Chapter 2 and the discussion in the preceding section). As a result of the long buildup time of the thermal index change, owing to the slower response of the order parameter S, the nonlinear coefficient $\alpha_2(t,T)$ for *nanosecond* laser pulses cannot be significantly increased by maintaining the liquid crystal near T_c.

To put this in more quantitative terms, let us ascribe the laser-induced order parameter change a time dependence of the form

$$dS = dS_{ss} \left[1 - \exp\left(-\frac{t}{\tau_S} \right) \right],\tag{9.36}$$

where $dS_{ss} = (dS/dT)_{ss} dT$ is the steady-state order parameter change associated with a temperature change dT and τ_s is the order parameter response time. Near T_c, τ_s exhibits a critical slowing down behavior and is therefore of the form

$$\tau_S \sim (T - T_c)^{-\alpha},\tag{9.37}$$

where α is some positive exponent near unity.

For a short laser pulse (i.e., $\tau_p = \tau_s$), the induced index change due to the order parameter is therefore given by

$$dn \sim \left(\frac{\tau_p}{\tau_s}\right)\left(\frac{dS}{dT}\right)_{SS}. \tag{9.38}$$

From (2.20) we have

$$\left(\frac{dS}{dT}\right)_{SS} \sim (T - T_c)^{-0.78}. \tag{9.39}$$

Since

$$\left(\frac{\tau_p}{\tau_s}\right) \sim (T - T_c)^{\alpha}, \tag{9.40}$$

we therefore have

$$dn \sim (T - T_c)^{\alpha - 0.78}. \tag{9.41}$$

In other words, near T_c, the short ($\tau_p \ll \tau_s$) laser pulse induced index change will not be enhanced, in spite of the large thermal gradient, owing to the diminishing effect from the critical slowing down of the order parameter response. Since the response times of the order parameter can be as long as microseconds near T_c, we could expect enhanced optical nonlinearities to manifest themselves only for relatively long laser pulses (on the order of microseconds or longer). Indeed, experiments with microsecond infrared laser pulses in optical limiting studies[8] and millisecond laser pulses in limiting and wave mixing studies[4,10] have shown that the efficiency of the processes increases tremendously near T_c.

9.5. THERMAL AND DENSITY OPTICAL NONLINEARITIES OF ISOTROPIC LIQUID CRYSTALS

A laser-induced change in the temperature of an isotropic liquid crystal can modify its refractive index in two ways, very much as in the nematic phase. One is the change in density $d\rho$ due to thermal expansion. This is the thermal absorptive component discussed before [Eq. (9.18) for ρ^T]; this term may be written as $(\partial n/\partial \rho)\,\rho^T$. The other is the so-called internal temperature change dT which modifies the spectral dependence of the molecular absorption–emission process; we may express this contribution as $(\partial n/\partial T)_\rho\,dT$. A pure density change effect arises from the electrostrictive component ρ^e, which contributes a change in the refractive index by $(\partial n/\partial \rho)\,\rho^e$.

Therefore, the total change in the refractive index Δn is given by

$$\Delta n = \left(\frac{\partial n}{\partial T}\right)_\rho dT + \left(\frac{\partial n}{\partial \rho}\right) \rho^T + \left(\frac{\partial n}{\partial \rho}\right)_T \rho^e. \tag{9.42}$$

In most liquids the first term is important in the picosecond regime, that is, if the excitation laser pulse duration is in the picosecond regime.[5,6] For longer pulses (e.g., in the nanosecond regime), the second and third terms provide the principal contributions. A good discussion of how the first and the next two terms in Equation (9.42) affect nonlinear light-scattering processes may be found in the works by Mack[5] and Herman and Gray,[6] respectively.

The change in the refractive index dn caused by the laser-induced temperature rise dT, described by the term $(\partial n/\partial \rho) \rho^T$ in Equation (9.42), is often written in terms of the index gradient dn/dT. For most organic liquids, including isotropic liquid crystals, dn/dT is on the order of 10^{-4} K. Figure 9.12 shows the temperature dependence of the liquid crystal TM74A (from EM Chemicals) as measured in our laboratory. The liquid crystal is a mixture of four chiral nematic materials in the isotropic phase. The measured dn/dT of the material is about 5×10^{-4} K^{-1}.

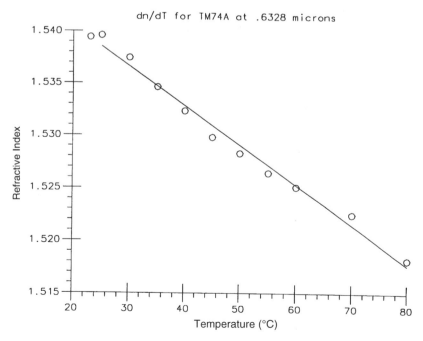

Figure 9.12. Experimentally measured steady-state refractive index dependence on temperature of an isotropic liquid crystal.

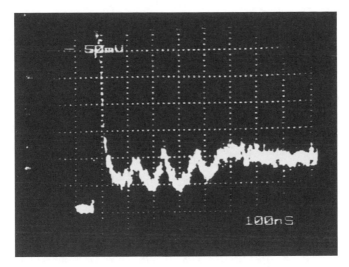

Figure 9.13.   Observed He-Ne probe diffraction from the isotropic liquid crystal TM74A under nanosecond Nd:Yag second harmonic laser pulse excitation.

The thermal optical nonlinearity of TM74A under short laser pulse excitation has also been studied by a dynamic grating technique.[11] For visible laser pulses (the second harmonic of a Nd:YAG laser at 0.53 μm) and an absorption constant $\alpha \sim 2 \text{ cm}^{-1}$ (determined by the absorbing dye dopant concentration), the measured nonlinear index coefficient n_2 is on the order of about -2×10^{-11} cm^2/W.

In these studies, as in the case of nematic liquid crystals, the dynamic grating diffraction also contains a fast decaying component due to the density contribution, $(\partial n/\partial \rho) \, \rho^e$ (see Fig. 9.13). This component decays in about 100 ns. Its peak magnitude is about twice that of the thermal component, which decays in a measured time of about 30 μs for the grating constants of 11 μm used in the experiment.

9.6.   COUPLED NONLINEAR OPTICAL EFFECTS IN NEMATIC LIQUID CRYSTALS

So far, we have singled out and discussed the various nonresonant physical mechanisms that contribute to optical nonlinearities. This approach allows us to understand their individual unique or special properties. However, in reality, these physical parameters are closely coupled to one another; perturbation of one parameter, when it reaches a sufficient magnitude, will inevitably lead to perturbations of the other parameters. These coupled responses could give rise to optical nonlinearities of differing signs and dynamical dependences and, consequently, complex behaviors in the nonlinear optical processes under study. In this section we discuss two examples of coupled liquid crystal responses to laser excitation: thermal-orientational coupling and flow-orientational coupling.

9.6.1.   Thermal-Orientational Coupling in Nematic Liquid Crystals

Consider the interaction of a linearly polarized extraordinary wave laser beam with a homeotropically aligned nematic liquid crystal as shown in Figure 9.14. The extraordinary refractive index as seen by a low-intensity laser is given by

$$n_e(\beta, T) = \frac{n_\perp(T)n_\parallel(T)}{\sqrt{n_\parallel^2(T)\cos^2\beta + n_\perp^2(T)\sin^2\beta}} \tag{9.43}$$

If the laser induces both molecular reorientation $\Delta\theta$ and a change in temperature, the resulting refractive index becomes

$$n_e(\beta + \Delta\theta, T + \Delta T) = \frac{n_\perp(T + \Delta T)n_\parallel(T + \Delta T)}{\sqrt{n_\parallel^2(T + \Delta T)\cos^2(\beta + \Delta\theta) + n_\perp^2(T + \Delta T)\sin^2(\beta + \Delta\theta)}}. \tag{9.44}$$

The resulting change in the extraordinary wave refractive index Δn_e is thus given by

$$\Delta n_e = n_e(\beta + \Delta\theta, T + \Delta T) - n_e(\beta, T). \tag{9.45}$$

Its magnitude and sign are determined by the initial value of β and T.

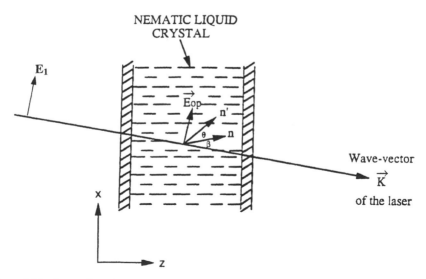

Figure 9.14.   Interaction of an extraordinary wave laser with a homeotropically aligned nematic liquid crystal at oblique incidence.

In general, if both reorientational and thermal effects are present, these processes will act in opposition to each other in terms of their contributions to the net refractive index seen by the optical beam.

An example of these competing nonlinear effects may be seen in the CO-laser-induced grating diffraction experiment performed by Khoo[9] and the nanosecond visible laser experiment of Hsiung et al.[12]

Under nanosecond laser pulse excitation, as we have shown in the preceding sections, the density contribution to the induced index change can be very large. When this occurs, a quantitative analysis of all the contributing effects, their signs and magnitudes, becomes rather complicated.[13]

9.6.2. FIow-Orientational Effect

Since liquid crystal molecules are highly anisotropic (requiring several viscosity coefficients to describe the viscous forces accompanying various flow processes), it will take a treatise here to formulate a quantitative dynamical theory to describe these flow phenomena which are also coupled to the complex process of density, temperature, and orientational changes. Nevertheless, one could design "simple" experiments to gain some insight into these processes, as exemplified in the work by two groups.[14,15] In these studies using dynamical grating diffraction techniques, picosecond laser pulses are used to induce density, temperature, and orientational-flow effects in nematic liquid crystals.

In the work by Khoo et al.,[14] the pump beams are copolarized and propagate in a plane parallel to the director axis; that is, the optical electric fields are perpendicular to the director axis (see Fig. 9.14, $\beta=0$ case). In this case there is no molecular reorientation effect. The principal nonlinear mechanisms are the thermal and density effects and flow. On the other hand, in the work by Eichler and Macdonald,[15] the two pump beams are cross-polarized and propagate at an angle $\beta \approx 22°$ with the director axis (see Fig. 9.14). Instead of having a sinusoidal intensity distribution (as in the work by Khoo et al.[14] with copolarized beams), one has a sinusoidal distribution in terms of the polarization state: the polarization evolves from circular, to elliptical, to linear, as shown in Figure 9.15, while the total intensity of the interfering pump beam is uniform across the distribution. In this case the principal nonlinear mechanism is due to the creation of a reorientation grating.

The flow-orientational coupling can be described by including an extra torque in the equation describing the director axis reorientational angle θ (cf. Chapter 3; also Eichler and Macdonald[15]),

$$\frac{\mu \partial^2 \theta}{\partial t^2} + \gamma_1 \frac{\partial \theta}{\partial t} + M_{op} + M_{el} + M_{fo} = 0, \qquad (9.46)$$

where M_{op} is the optical torque, M_{el} is the elastic torque, and M_{fo} is the torque due to the flow-orientational coupling. μ is the moment of inertia of the collection of

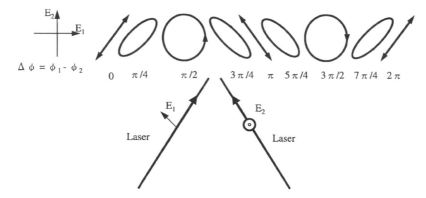

Figure 9.15.   Polarization grating generated by interfering two coherent cross-polarized laser beams.

molecules (as opposed to a single molecule) undergoing the flow-reorientational process, and γ_1 is a viscosity coefficient.

These torques obviously depend on the interaction geometry. For the geometry used in Figure 9.16, and assuming that the flow and reorientational direction are principally along the x direction, we have

$$M_{op} = \frac{\varepsilon_\perp}{\varepsilon_\parallel} \frac{\Delta \varepsilon}{16\pi} |E_{op}|^2 \sin 2(\beta + \theta), \tag{9.47}$$

$$M_{el} = -K \frac{\partial^2 \theta}{\partial z^2}, \tag{9.48}$$

$$M_{fo} = -\frac{1}{2}(\gamma_1 - \gamma_2 \cos^2 \theta) \frac{\partial V_x}{\partial z}. \tag{9.49}$$

Under these approximations (i.e., the flow is along the x direction), the force creating the flow F_x is obtainable from the Maxwell stress tensor:

$$\mathbf{F} = (\mathbf{D} \cdot \nabla)\mathbf{E}^* - \tfrac{1}{2} \nabla (\mathbf{E} \cdot \mathbf{D}^*). \tag{9.50}$$

Under the preceding assumption, this was shown[15] to be

$$F_{x'} = \tfrac{1}{2} \varepsilon_0 \varepsilon_\perp q E^2 \cos^2 qy, \tag{9.51}$$

where q is the magnitude of the grating wave vector ($q = |\mathbf{k}_1 - \mathbf{k}_2|$, $\mathbf{k}_1$ and $\mathbf{k}_2$ are the wave vectors of the two pump beams).

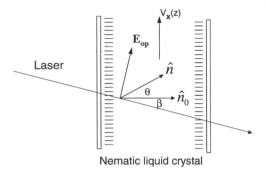

Figure 9.16. Geometry of interaction for pulsed laser-induced flow-orientational coupling in a nematic liquid crystal film.

The time constant characterizing the flow process may be estimated from the dynamical equation for the flow velocity field $V(r,t)$ discussed in Chapter 3. Assuming that the viscosity coefficient involved is η, the velocity field $V(r,t)$ obeys a greatly simplified equation:

$$\rho V(r,t) - \eta \nabla^2 V(r,t) = F_{stress}. \tag{9.52}$$

To estimate the flow *relaxation* time scale, we simply set $F_{stress} = 0$ in the preceding equation. This is valid because the picosecond laser pulses, as well as the time scales associated with all the physical processes such as thermal expansion and density change that *initiate* the flow process, are all much shorter than the flow response time.

In the work by Khoo et al.,[14] the flow process is assumed to be radial, with a characteristic flow length defined by the width ω_0 of the laser beam; that is, $\nabla^2 V_\beta$ in Equation (9.52) is on the order of $\eta/\omega_0^2 V_\beta$. This gives a flow damping constant $\Gamma_{flow} \approx \eta \omega_0^{-2}/\rho_0$. Using (mks units) $\rho_0 = 1$ kg/l, $\eta = 7 \times 10^{-2}$, and $\omega_0 = 0.75$ mm $- 750$ μm, we have a flow relaxation time $\Gamma_{flow}^{-1} \approx 8$ ms. This is in agreement with the experimental observation (see Fig. 9.17). The flow is initiated by the large temperature and density changes and is manifested in the initial rise portion of the observed slowly varying component in Figure 9.17. The flow process is terminated in a time on the order of 8 ms, when the accompanying reorientation process also ceases. Then the molecules relax to the initial alignment in a time scale characterized by the usual reorientation relaxation time constant $\tau_\theta = Kq^2/\eta \approx 50$ ms (for $K = 10^{-11}$ kg ms^{-1}; experimental values of $q = 2\pi/17$ μm and $\eta = 7 \times 10^{-2}$, in mks units).

In the work by Eichler and Macdonald,[15] the flow is along the x direction and is characterized by two wave vectors: q_t ($=0.20$ μm^{-1}) along the y direction and q_1 ($=0.25$ μm^{-1}) along the z direction. Accordingly, the flow damping constant $\Gamma_{flow} = \eta(q_1^2 + q_r^2)/\rho$. Note that this damping constant is analogous to the density damping constant [cf. Eq. (9.14)]; this is because they both originate from a common origin, namely, the translational movement of the molecules. The flow relaxation

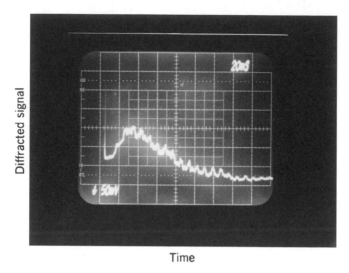

Figure 9.17. Oscilloscope trace of the probe diffraction showing the observed flow-reorientational effect (time scale: 20 ms/div). The initial spike is the density and thermal components (see Figs. 9.7 and 9.8). The slowly varying component in the milliseconds time scale is the flow-reorientational effect (after Ref. 14).

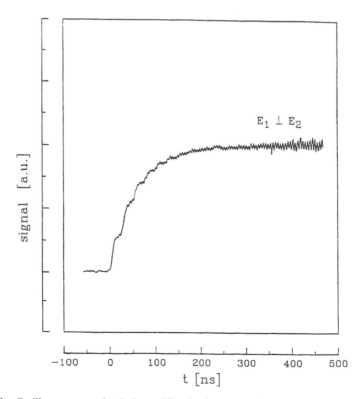

Figure 9.18. Oscilloscope trace of probe beam diffraction from cross-ploarized pump beam (after Ref.15).

time Γ_{flow}^{-1} constant may again be estimated as before to give $\Gamma_{flow}^{-1} = 139$ ns, using the values $\eta = 7 \times 10^{-2}$ and $\rho = 1$ quoted previously. This is on the order of the typical observed time constant characterizing the initial "rise" portion (see Fig. 9.18) of the flow-reorientational process.[15]

As these preceding discussions have demonstrated, laser-induced flow processes and their couplings to other nonlinear mechanisms are very complex problems. These studies have, at best, reviewed only some qualitative aspects of the problems. They do, however, serve as good starting points for more quantitative investigations.

REFERENCES

1. Batra, I. P., R. H. Enns, and D. Pohl. 1971. Stimulated thermal scattering of light. *Phys. Status Solidi.* 48:11.

2. Eichler, H. J., P. Gunter, and D. N. Pohl. 1986. *Laser Induced Dynamic Grating*. Berlin: Springer-Verlag.

3. Khoo, I. C., and R. Normandin. 1985. The mechanism and dynamics of transient thermal grating diffraction in nematic liquid crystal film. *IEEE J. Quantum Electron.* QE21:329.

4. Khoo, I. C., and S. T. Wu. 1993. *Optics and Nonlinear Optics of Liquid Crystals*. Singapore: World Scientific. Khoo, I. C. 1988. In *Progress in Optics*. E. Wolf (ed.). Vol. 26. Amsterdam: North-Holland Publ. Garland, C. W., and M. E. Huster. 1987. Nematic–smectic-C heat capacity near the nematic-A–smectic-C point. *Phys. Rev. A.* 35:2365.

5. Mack, M. E. 1969. Stimulated thermal light scattering in the picosecond regime. *Phys. Rev. Lett.* 22:13.

6. Herman, R. M., and M. A. Gray. 1967. Theoretical prediction of the stimulated thermal Rayleigh scattering in liquids. *Phys. Rev. Lett.* 19:824.

7. Khoo, I. C., and R. Normandin. 1984. Nanosecond laser induced transient and permanent gratings and ultrasonic waves in smectic film. *J. Appl. Phys.* 55:1416; see also Mullen, M. E., B. Uthi, and M. J. Stephen. 1972. Sound velocity in nematic liquid crystal. *Phys. Rev. Lett.* 28:799; Lord, Jr., A. E. 1972. Anisotropic ultrasonic properties of smectic liquid crystal. *ibid.* 29:1366.

8. Lindquist, R. G., P. G. LoPresti, and I. C. Khoo. 1992. Infrared and visible laser induced thermal and density optical nonlinearities in nematic and isotropic liquid crystals. *Proc. SPIE Int. Soc. Opt. Eng.* 1692:148.

9. Khoo, I. C. 1990. The infrared nonlinearities of liquid crystals and novel two-wave mixing processes. *J. Mod. Opt.* 37:1801.

10. See, for example, Khoo, I. C., P. Y. Yan, G. M. Finn, T. H. Liu, and R. R. Michael. 1988. Low power (10.7 μm) laser beam amplification via thermal grating mediated degenerate four wave mixings in a nematic liquid crystal film. *J. Opt. Soc. Am. B.* 5:202.

11. Khoo, I. C., Sukho Lee, P. G. LoPresti, R. G. Lindquist, and H. Li. 1993. Isotropic liquid crystalline film and fiber structures for optical limiting application. *Int. J. Nonlinear Opt. Phys.* 2(4).

12. Hsiung, H., L. P. Shi, and Y. R. Shen. 1984. Transient laser-induced molecular reorientation and laser heating in a nematic liquid crystal. *Phys. Rev. A.* 30:1453.

13. Khoo, I. C., J. Y. Hou, G. L. Din, Y. L. He, and D. F. Shi. 1990. Laser induced thermal, orientational and density nonlinear optical effects in nematic liquid crystals. *Phys. Rev. A.* 42:1001.

14. Khoo, I. C., R. G. Lindquist, R. R. Michael, R. J. Mansfield, and P. G. LoPresti. 1991. Dynamics of picosecond laser-induced density, temperature, and flow-reorientation effects in the mesophases of liquid crystals. *J. Appl. Phys.* 69:3853.

15. Eichler, H. I., and R. Macdonald. 1991. Flow alignment and inertial effects in picosecond laser-induced reorientation phenomena of nematic liquid crystals. *Phys. Rev. Lett.* 67:2666.

10

Electronic Optical Nonlinearities

10.1. INTRODUCTION

In this chapter we treat those nonlinear optical processes in which the electronic wave functions of the liquid crystal molecules are significantly perturbed by the optical field. Unlike the nonelectronic processes discussed in the previous chapters, these electronic processes are very fast; the active electrons of the molecules respond almost instantaneously to the optical field in the form of an induced electronic polarization. Transitions from the initial level to some final excited state could also occur.

Such processes are obviously dependent on the optical frequency and the resonant frequencies of the liquid crystal constituent molecules. They are also understandably extremely complicated, owing to the complex electronic and energy level structure of liquid crystal molecules. Even calculating such basic quantities as the Hamiltonian, the starting point for quantum mechanical calculations[1] of the electronic wave function and energy levels and linear optical properties, requires very powerful numerical computational techniques.

We shall adopt a greatly simplified approach where the liquid crystal molecule is represented as a general multilevel system. Only dipole transitions among the levels are considered. Using this model, we quantitatively illustrate some important basic aspects of the various electronic nonlinear optical processes and their accompanying nonlinearities. Special features pertaining to liquid crystalline materials are then discussed.

10.2. DENSITY MATRIX FORMALISM FOR OPTICALLY INDUCED MOLECULAR ELECTRONIC POLARIZABILITIES

Consider the multilevel system depicted in Figure 10.1; the optical transition between any pair of levels (i and j) is mediated by an electric dipole. In the density matrix formalism,[2] the expectation value of the dipole moment $\langle \mathbf{d}(t) \rangle$ at any time t

Liquid Crystals, Second Edition   By Iam-Choon Khoo
Copyright © 2007 John Wiley & Sons, Inc.

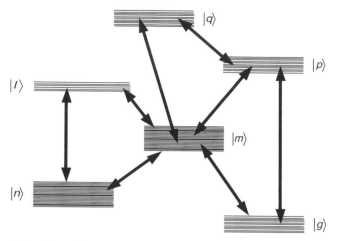

Figure 10.1. Schematic depiction of the energy level structure of a multilevel molecule.

following the turn-on of the interaction between the molecule and an optical field is given by

$$\langle \mathbf{d}(t) \rangle = \sum_{i=1}^{N} \sum_{j=1}^{N} d_{ij} \rho_{ij}(t). \tag{10.1}$$

The quantum mechanical equation of motion for the ij component of the density matrix $\rho_{ij}(t)$ is obtainable from the Schrödinger equation:

$$i\hbar \frac{\partial}{\partial t} \rho_{ij} = [H, \rho]_{ij}, \tag{10.2}$$

where H is the total Hamiltonian describing the molecule and its interaction with the optical field, that is,

$$H = H_0 - \mathbf{d} \cdot \mathbf{E}, \tag{10.3}$$

where H_0 is the Hamiltonian describing the unperturbed molecule and $-\mathbf{d} \cdot \mathbf{E}$ is the dipolar interaction between the optical field $\mathbf{E}$ and the molecule.

In general, because of the complexity of the Hamiltonian H_0 and the large number of energy levels involved, an analytical solution of Equation (10.3) is neither possible nor instructive. On the other hand, we can solve Equation (10.4) in a perturbative manner and obtain a solution for the induced dipole moment, and therefore the electric polarization, in a power series of the interaction $-\mathbf{d} \cdot \mathbf{E}$. This is valid if the dipolar interaction is of a small perturbation magnitude compared to H_0. Accordingly, we attach a perturbation parameter λ ($\lambda = 1$) to $-\mathbf{d} \cdot \mathbf{E}$ and rewrite Equation (10.3) as

$$H = H_0 - \lambda \mathbf{d} \cdot \mathbf{E}. \tag{10.4}$$

Also, $\rho_{ij}(t)$ is expanded in powers of λ as follows:

$$\rho_{ij}(t) = \rho_{ij}^{(0)} + \lambda\rho_{ij}^{(1)} + \lambda^2\rho_{ij}^{(2)} + \lambda^3\rho_{ij}^{(3)}. \tag{10.5}$$

The reader can easily verify that, upon substituting ρ_{ij} in Equation (10.5) and H in Equation (10.4) in the equation of motion and equating terms containing equal powers of λ, we have

$$\lambda^0: \quad i\hbar\frac{\partial}{\partial t}\rho_{ij}^{(0)} = \left[H_0, \rho^{(0)}\right]_{ij}, \tag{10.6}$$

$$\lambda^1: \quad i\hbar\frac{\partial}{\partial t}\rho_{ij}^{(1)} = \left[H_0, \rho^{(1)}\right]_{ij} - \left[\mathbf{d}\cdot E, \rho^{(0)}\right]_{ij}, \tag{10.7}$$

$$\lambda^2: \quad i\hbar\frac{\partial}{\partial t}\rho_{ij}^{(2)} = \left[H_0, \rho^{(2)}\right]_{ij} - \left[\mathbf{d}\cdot E, \rho^{(1)}\right]_{ij}, \tag{10.8}$$

$$\lambda^3: \quad i\hbar\frac{\partial}{\partial t}\rho_{ij}^{(3)} = \left[H_0, \rho^{(3)}\right]_{ij} - \left[\mathbf{d}\cdot E, \rho^{(2)}\right]_{ij}, \tag{10.9}$$

The first term on the right-hand side of Equations (10.6)–(10.9) may be explicitly written in terms of the energy difference $E_i - E_j$:

$$\begin{aligned}\left[H_0, \rho^{(n)}\right]_{ij} &= \langle i\,|\,H_0\rho^{(n)} - \rho^{(n)}H_0\,|\,j\rangle\\ &= E_i\langle i\,|\,\rho^{(n)}\,|\,j\rangle - \langle i\,|\,\rho^{(n)}\,|\,j\rangle E_j\\ &= (E_i - E_j)\rho_{ij}^{(n)},\end{aligned} \tag{10.10}$$

where we have made use of the fact that $|i\rangle$ and $|j\rangle$ are the eigenfunctions of the Hamiltonian H_0, and so $H_0\,|i\rangle = E_i|i\rangle$ and $H_0\,|j\rangle = E_j|i\rangle$.

Equations (10.7)–(10.9) become

$$\rho_{ij}^{(n)} = -i\omega_{ij}\rho_{ij}^{(n)} - \frac{1}{i\hbar}\left[\mathbf{d}\cdot E, \rho^{(n-1)}\right], \quad n = 1, 2, 3,\dots. \tag{10.11}$$

The zeroth-order equation (10.6) gives

$$\rho_{ij}^{(0)} = -i\omega_{ij}\rho_{ij}^{(0)}. \tag{10.12}$$

If the Hamiltonian H_0 of the molecule and therefore its wave functions and eigenvalues are known, as in the case of simple atomic systems, the preceding equations can be solved to any desired order. In dealing with the complex molecules comprising liquid crystals, in fact for any molecular system, the actual determination of the Hamiltonian H_0 is itself quite a feat. In the following discussion we will assume that both H_0 and the relevant energy levels involved in the interaction of a molecule with the incident optical field are known (through some quantum mechanical calculations or experiments).

10.2.1. Induced Polarizations

An important point regarding the response of a multilevel molecule to the optical field can be immediately deduced from the preceding equation. The driven part of the solution to $\rho_{ij}^{(n)}$ is proportional to the nth power of the optical field E. From the definition for the dipole moment, Equation (10.1), one can see that the induced dipole moment, obtained from the driven part of the solution to $\rho_{ij}^{(n)}$, is of the form of a power series in E:

$$\mathbf{d}(t) \sim \alpha : E + \beta : EE + \gamma : EEE, \tag{10.13}$$

where the double dots signify tensorial operation between α, β, and γ with the vector fields E.

Since E is a vector described by three Cartesian components (E_i, E_j, E_k), α, β, and γ are tensors of second, third, and fourth ranks, respectively; $\alpha = \{\alpha_{ij}\}$, $\beta = \{\beta_{ijk}\}$, and $\gamma = \{\gamma_{ijkh}\}$ are, respectively, linear, second-order, and third-order molecular polarizabilities. The induced electric polarization P, defined as the dipole moment per unit volume, is therefore of the form

$$P = \varepsilon_0 \chi^{(1)} : E + \chi^{(2)} : EE + \chi^{(3)} : EEE. \tag{10.14}$$

The first, second, and third terms on the right-hand side of Equation (10.14) are, respectively, the linear, second-order, and third-order nonlinear polarizations, with $\chi^{(1)}$, $\chi^{(2)}$, and $\chi^{(3)}$ as the respective susceptibility tensors. The connection between the macroscopic susceptibility tensor $\chi^{(n)}$ and the microscopic polarizabilities α, β, and γ is the molecular number density weighted by the local field correction factor. We will discuss this in more detail in the following sections.

10.2.2. Multiphoton Absorptions

The density matrix formalism described previously also shows that, starting from a particular molecular state, the solutions for the population density of the state are generally of the following form:

$$\left\langle \rho_{jj}^{(2n)}(t) \right\rangle \neq 0, \quad n = 1,2,3,\dots, \tag{10.15}$$

while the expectation values of all odd-power diagonal elements are vanishing. Since the nth-order driven part of the density matrix element is proportional to the nth-order power of E, we thus have

$$\left\langle \rho_{jj}^{(2)} \right\rangle \sim E^2 \sim I, \tag{10.16a}$$

$$\left\langle \rho_{jj}^{(4)} \right\rangle \sim E^4 \sim I^2, \tag{10.16b}$$

$$\left\langle \rho_{jj}^{(4)} \right\rangle \sim E^6 \sim I^3. \tag{10.16c}$$

Equation (10.16a) shows that the probability of a molecule being excited to the jth state, starting from the initial state $|i\rangle$, for example, is linearly dependent on the intensity I of the optical field. This is the familiar linear absorption process in which the states $|i\rangle$ and $|j\rangle$ are connected by a single-photon absorption (see Fig. 10.2). As one may deduce from the detailed calculations given in the following section, $\rho_{ij}^{(2)}$ contains a resonant denominator $(\omega - \omega_{ij})$, the so-called single-photon resonance encountered in traditional absorption spectra.

The next-order nonvanishing diagonal term $\rho_{jj}^{(4)}$ given in Equation (10.16b) is proportional to the square of the optical intensity, I^2, and contains resonant denominators of the form $(\omega_{ij} - 2\omega)$ or $(\omega_{in} - \omega)(\omega_{nj} - \omega)$. This is the so-called two-photon transition (or absorption) process where the initial state $|i\rangle$ is connected to the final state $|j\rangle$ by the simultaneous absorption of two photons (see Fig. 10.3) such that their sum energy 2ω equals ω_{ij}. Note that by the dipole selection rule, if the $|i\rangle$ and $|j\rangle$ states are connected by a single-photon transition, then the two-photon process is vanishing

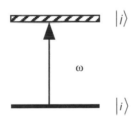

Figure 10.2. Single-photon transitions.

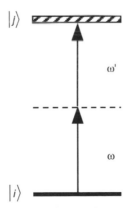

Figure 10.3. Two-photon absorption process involving simultaneous absorption of two photons at the same frequency.

and vice versa. In the same manner, the $\rho_{ij}^{(6)}$ term given in Equation (10.16c) is proportional to the third-power intensity of the optical field, I^3, and is known as a three-photon transition (or absorption) process. The initial state $|i\rangle$ is connected to the final state $|j\rangle$ by a simultaneous absorption of three photons and is characterized by a three photon resonant denominator of the form $(\omega_{ij} - 3\omega)$ (see Fig. 10.4) or $(\omega_{in} - \omega)(\omega_{nm} - \omega)(\omega_{mj} - \omega)$. By the dipole selection rule, if $|i\rangle$ and $|j\rangle$ are connected by a dipole transition, then three-, five-, and higher-order odd-power absorption processes are allowed, whereas all even-power terms are vanishing and vice versa.

Figure 10.5 depicts these photonic absorption processes that could occur in liquid crystal molecules, as well as various inter- and intramolecular nonradiative processes discussed in the preceding chapters. Two- and three-photon absorption processes in a liquid crystal (5CB) have been studied previously by Deeg and Fayer[3] and Eichler et al.[4] In the work by Deeg and Fayer,[3] two-photon absorptions of the isotropic phase of 5CB were studied with subpicosecond laser pulses. It was determined that the first singlet state reached by the two-photon transition relaxes mainly through a radiative mechanism. On the other hand, the study conducted by Eichler et al.[4] with a picosecond laser in the same liquid crystal (in the isotropic as well as the nematic phase) probes the three-photon absorption processes. The excited state reached by such a three-photon transition was found to relax quite substantially through a nonradiative mechanism, resulting in the generation of a large thermal effect. These observations may explain the generation of thermal gratings in the otherwise nonabsorbing (in single-photon language) 5CB material.

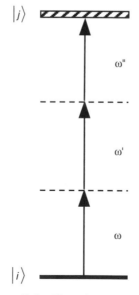

Figure 10.4.   Three-photon transitions.

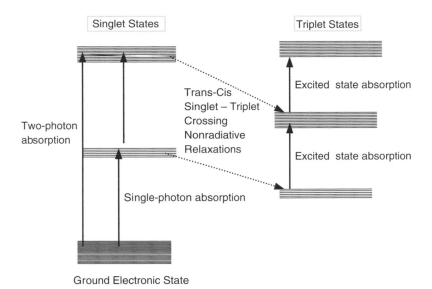

Figure 10.5. A schematic depiction of various electronic, inter- and intra molecular process in liquid crystal.

More recently, these nonlinear absorption processes have been actively investigated in the contexts of nonlinear transmission and optical limiting of ultrashort laser pulses. A detailed discussion will be given in Chapter 12.

10.3. ELECTRONIC SUSCEPTIBILITIES OF LIQUID CRYSTALS

10.3.1. Linear Optical Polarizabilities of a Molecule with No Permanent Dipole

To obtain explicit expressions for the linear and nonlinear polarizabilities α_{ij}, β_{ijk}, and γ_{ijkh}, consider the solution of $\rho_{ij}(t)$ to the corresponding order. To simplify the calculations, we consider a single molecule illuminated by an optical field $E(t)$ of the form

$$E(t) = \sum_{\text{all } \omega} E(\omega) e^{-i\omega t}. \tag{10.17}$$

For example, if there are three frequency components ω_1, ω_2, and ω_3 in $E(t)$, Equation (10.17) gives

$$E(t) = \left[E(\omega_1) e^{-i\omega_1 t} + E(\omega_2) e^{-i\omega_2 t} + E(\omega_3) e^{-i\omega_3 t} \right] + \text{c.c.} \tag{10.18}$$

The zeroth-order solution of $\rho_{nm}^{(o)}(t)$ is $\rho_{nm}^{(o)} e^{-\omega_{nm} t}$. From Equation (10.11), the solution of $\rho_{ij}^{(1)}(t)$ is given by

$$\rho_{nm}^{(1)}(t) = \frac{i}{h} \int_{-\infty}^{t} e^{-i\omega_{nm}(t-t')} \left[\mathbf{d} \cdot E(t'), \rho^{(0)} \right]_{nm} dt'. \tag{10.19}$$

The commutator in the preceding equation can be evaluated as

$$[\mathbf{d} \cdot \mathbf{E}(t'), \rho^{(0)}(t')]_{nm} = \sum_l [d_{nl}\rho_{lm}^{(0)}(t') - \rho_{nl}^{(0)}(t')d_{lm}] \cdot \mathbf{E}(t'). \tag{10.20}$$

We make a simplifying assumption that there is initially no coherence among the molecular states; that is, all the off-diagonal elements $\rho_{ij}^{(0)} = 0$ for $i \neq j$. [Note that the diagonal elements $\rho_{jj}^{(0)}$ give the population of the ith state and some of them are non-vanishing.] In this case the commutator in Equation (10.20) gives

$$[\mathbf{d} \cdot \mathbf{E}(t'), \rho^{(0)}]_{nm} = [\rho_{mm}^{(0)} - \rho_{nn}^{(0)}]\mathbf{d}_{nm} \cdot \mathbf{E}(t'). \tag{10.21}$$

Substituting Equation (10.21) in Equation (10.19) for $\rho_{nm}^{(1)}(t)$ gives

$$\rho_{nm}^{(1)}(t) = e^{-i\omega_{nm}t} \frac{i}{\hbar} [\rho_{mm}^{(0)} - \rho_{nn}^{(0)}]\mathbf{d}_{nm} \cdot \int_{-\infty}^{t} \mathbf{E}(t')e^{i\omega_{nm}t'} dt', \tag{10.22}$$

$$\rho_{nm}^{(1)}(t) = \frac{i}{\hbar} [\rho_{mm}^{(0)} - \rho_{nn}^{(0)}] \sum_{\omega} \frac{\mathbf{d}_{nm} \cdot \mathbf{E}(\omega)e^{-i\omega t}}{\omega_{nm} - \omega}. \tag{10.23}$$

The expectation value of the induced first-order dipole moment $\langle \mathbf{d}^{(1)}(t) \rangle$ is thus given by

$$\langle \mathbf{d}^{(1)}(t) \rangle = tr[\rho^{(1)}d] = \sum_{nm} \rho_{nm}^{(1)}\mathbf{d}_{mn}$$

$$= \sum_{\omega} e^{-i\omega t} \sum_{m,n} \left[\frac{\rho_{mm}^{(0)} - \rho_{nn}^{(0)}}{\hbar} \right] \frac{\mathbf{d}_{mn}[\mathbf{d}_{nm} \cdot \mathbf{E}(\omega)]}{\omega_{nm} - \omega}. \tag{10.24}$$

The induced dipole moment that oscillates at a particular frequency component ω is thus given by

$$\mathbf{d}(\omega) = \frac{1}{\hbar} \sum_{m,n} [\rho_{mm}^{(0)} - \rho_{nn}^{(0)}] \frac{\mathbf{d}_{mn}[\mathbf{d}_{nm} \cdot \mathbf{E}(\omega)]}{\omega_{nm} - \omega}. \tag{10.25}$$

If we write $\mathbf{d}$ and $\mathbf{E}$ in terms of their Cartesian components [i.e., as (d^i, d^j, d^k) and (E_i, E_j, E_k), respectively], Equation (10.25) gives

$$d^i = \sum_j \alpha_{ij} E_j, \tag{10.26}$$

where the (linear) polarizability tensor component α_{ij} is given by

$$\alpha_{ij} = \frac{1}{\hbar} \sum_{m,n} [\rho_{mm}^{(0)} - \rho_{nn}^{(0)}] \frac{d_{mn}^i d_{nm}^j}{\omega_{nm} - \omega}. \tag{10.27}$$

If there are N independent molecules per unit volume, the induced polarization $P = N\mathbf{d}$. From Equation (10.26), the linear susceptibility $\chi_{ij}^{(1)}$ is thus given by $\varepsilon_0\chi_{ij}^{(1)}=N\alpha_{ij}$, using the definition $P_i=\varepsilon_0\chi_{ij}^{(1)}E_j$.

Equation (10.27) shows that the polarizability is dependent on the level populations $\rho_{mm}^{(0)}$ and $\rho_{nn}^{(0)}$. This dependence is more clearly demonstrated if we note that the summation over all values of m and n in Equation (10.27) allows us to rewrite it as [by interchanging the "dummy" indices m and n in the second term of Eq. (10.27)]

$$\alpha_{ij} = \frac{1}{\hbar}\sum_{m,n}\rho_{mm}^{(0)}\left(\frac{d_{mn}^i d_{nm}^j}{\omega_{nm}-\omega}+\frac{d_{nm}^i d_{mn}^j}{\omega_{nm}+\omega}\right). \tag{10.28}$$

Note that $\omega_{nm}=-\omega_{mn}$. If the molecule is in the ground state $|g\rangle$ [i.e., $\rho_{gg}^{(0)}\neq0$, $\rho_{mm}=0$ for $m \neq g$] and if we are interested in the contribution from a particular excited state $|n\rangle$ connected to the ground state $|g\rangle$ by the dipole transition, we have

$$\alpha_{ij}(\omega_n) = \frac{1}{\hbar}\left(\frac{d_{gn}^i d_{ng}^j}{\omega_{ng}-\omega}+\frac{d_{ng}^i d_{gn}^j}{\omega_{ng}+\omega}\right). \tag{10.29}$$

In actuality, the ground state $|g\rangle$ may consist of many degenerate or nearly degenerate levels, or a band. In this case it is more appropriate to invoke the so-called oscillator strength (rather than the dipole amount) connecting n and g.

The oscillator strength f_{ng} is related to d_{ng} by

$$f_{ng} = \frac{2m\omega_{ng}\left|\mathbf{d}_{ng}\right|^2}{3e^2\hbar}. \tag{10.30}$$

If we note that the average value of $d_{ng}^i d_{ng}^j$ in Equation (10.29) is $|\mathbf{d}_{ng}|^2/3$, we may rewrite it as

$$\alpha_{ij}(\omega_n) = \frac{f_{ng}e^2}{2m\omega_{ng}}\left(\frac{1}{\omega_{ng}-\omega}+\frac{1}{\omega_{ng}-\omega}\right). \tag{10.31}$$

The general multilevel linear electric polarizability [Eq. (10.27)] can thus be written as

$$\alpha_{ij} = \sum_n\alpha_{ij}^{(n)}(\omega_n). \tag{10.32}$$

The preceding results are obtained assuming that all the molecular states are sharp. Singularities will occur therefore if the optical interactions with the molecule are at resonances, that is, when the denominators involving frequency differences approach vanishing values (e.g., $\omega_{ng}\approx\omega$). In actuality, the molecular levels are broadened by a variety of homogeneous or inhomogeneous relaxation mechanisms (collisions, natural lifetime broadening, Doppler effects, etc.). These relaxation processes could be phenomenologically accounted for by ascribing a negative imaginary part $-i\gamma_{nm}$ to the frequencies ω_{nm} (i.e., replacing ω_{nm} by $\omega_{nm} - i\gamma_{nm}$), where the γ_{nm}'s are the relaxation rates associated with the n and m states. Note that although $\omega_{nm} = \omega_{mn}$, $\gamma_{nm} = \gamma_{mn}$. As a result of the inclusion of $-i\gamma_{nm}$, α_j becomes a complex quantity. Consequently, the macroscopic parameters such as the susceptibility $\chi^{(1)}$ and therefore the refractive indices are also complex.

Consider now Equation (10.31) for $\alpha_{ij}(\omega_{ng})$ (i.e., a particular transition involving the ground state). If we replace ω_{ng} by $\Omega_{ng} = \omega_{ng} - i\gamma_{ng}$ and write $\omega_{ng} = \omega_0$ and $\gamma_{ng} = \gamma_0$, $f_{ng} \equiv f_{05}$, and $\Delta\omega = \omega_0 - \omega$, we obtain

$$\alpha_{ij}(\omega_0) = \overline{\alpha}\left[\frac{\Delta\omega}{(\Delta\omega)^2 + \gamma_0^2} + \frac{i\gamma_0}{(\Delta\omega)^2 + \gamma_0^2}\right], \tag{10.33}$$

where $\overline{\alpha} \equiv f_0 e^2 / 2m\omega_0$. If we have N independent molecules per unit volume in the medium, the macroscopic susceptibility χ_{ij} is given by $\varepsilon_0\chi_{ij} = N\alpha_{ij}$, that is, $\chi_{ij} = (N/\varepsilon_0)\alpha_{ij}(\omega_0)$. The refractive index n_{ii}, for example, is thus given by $n_{ii} = \sqrt{1 + \chi_{ij}} = n_0 + i\tilde{n}$. The real part n_0 and the imaginary part $\tilde{n}$ of n_{ii} are given by

$$n_0(\omega) = 1 + \frac{N}{\varepsilon_0}\overline{\alpha}\left[\frac{\Delta\omega}{(\Delta\omega)^2 + \gamma_0^2}\right]. \tag{10.34}$$

and

$$\tilde{n}(\omega) = \frac{N}{\varepsilon_0}\overline{\alpha}\left[\frac{\gamma_0}{(\Delta\omega)^2 + \gamma_0^2}\right]. \tag{10.35}$$

The imaginary part $\tilde{n}$ gives a (linear) attenuation factor in the propagation of a plane optical wave $\exp(-\omega/c)\tilde{n}(\omega)l$, where l is the propagation length into the medium. The attenuation constant $(\omega/c)\tilde{n}$ is usually referred to as the linear absorption constant in units of (length)$^{-1}$. $n_0(\omega)$ is the dispersion of the medium (see Fig. 10.6a). $\tilde{n}(\omega)$ gives the linear absorption spectral line shape which is a Lorentzian (see Fig. 10.6b).

It is important to note that, in the preceding derivation of the refractive index $n = n_0 + i\tilde{n}$, we have accounted for only one particular transition $n \to g$; the transition frequency $\omega_{ng} = \omega_0$ is the frequency of interest in the context of resonance with the laser frequency ω. There is, of course, a whole collection (in fact, an infinite number) of transitions that are off resonance with respect to the laser frequency. Collectively, these transitions give rise to a nonresonant background refractive index n_b, which is obtainable by performing the sums in Equations (10.32) and (10.31). n_b is, of course, also complex and may be written as $n_b = n_{b0} + i\tilde{n}_b$.

10.3.2. Second-Order Electronic Polarizabilities

Second order molecular electronic polarizabilities are obtained by solving for $\rho^{(2)}(t)$. Substituting Equation (10.22) for $\rho^{(1)}(t)$ in the right-hand side of Equation

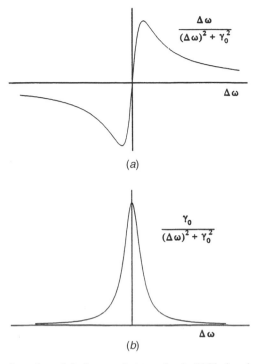

Figure 10.6.   (a) Index dispersion $n_0(\omega)$ of a ground-state molecule. (b) The imaginary part of the refractive index $\tilde{n}_0(\omega)$.

(10.8), evaluating the commutator $[\mathbf{d} \cdot \mathbf{E}(t), \rho^{(1)}(t)]$ as before [now with $\mathbf{E}(t)$ expressed as $\mathbf{E}(t) = \sum_\omega \mathbf{E}(\omega')e^{-i\omega't}$], and performing a simple integration, we obtain, for an exemplary frequency component $(\omega' + \omega)$ of $\rho_{nm}^{(2)}(t)$,

$$\rho_{nm}^{(2)}(t) = e^{-i(\omega' + \omega)t} \left\{ \sum_l \frac{\rho_{mm}^{(0)} - \rho_{ll}^{(0)}}{\hbar^2} \frac{\left[\mathbf{d}_{nl} \cdot \mathbf{E}(\omega')\right]\left[\mathbf{d}_{lm} \cdot \mathbf{E}(\omega)\right]}{\left[\omega_{nm} - (\omega' + \omega)\right](\omega_{lm} - \omega)} \right.$$
$$\left. + \sum_l \frac{\rho_{nn}^{(0)} - \rho_{ll}^{(0)}}{\hbar^2} \frac{\left[\mathbf{d}_{lm} \cdot \mathbf{E}(\omega')\right]\left[\mathbf{d}_{nl} \cdot \mathbf{E}(\omega)\right]}{\left[\omega_{nm} - (\omega' + \omega)\right](\omega_{nl} - \omega)} \right\}. \qquad (10.36)$$

There are, of course, as many frequency components in $\rho_{nm}^{(2)}(t)$ as there are possibilities of combining ω and ω' from the optical field. Here we focus our attention on a particular example in order to explicitly illustrate the physics. By definition, the second-order induced dipole moment is

$$\langle \mathbf{d}^{(2)} \rangle \equiv \sum_{nm} \rho_{nm}^{(2)} \mathbf{d}_{mn}. \qquad (10.37)$$

The ith Cartesian component of $\langle \mathbf{d}^{(2)} \rangle$, which we shall henceforth write as $d_i^{(2)}$, is of the form

$$d_i^{(2)} = \beta_{ijk}\big[-(\omega'+\omega);\omega',\omega\big]E_j(\omega')E_k(\omega), \tag{10.38}$$

with the second-order polarizability tensor β_{ijk} given by

$$\beta_{ijk}\big[-(\omega'+\omega);\omega',\omega\big] = \sum_{n,m,l}\left[\frac{\rho_{mm}^{(0)} - \rho_{ll}^{(0)}}{\hbar^2}\right]\frac{d_{mn}^i d_{nl}^j d_{lm}^k}{\big[\Omega_{nm} - (\omega'+\omega)\big]\big(\Omega_{lm}-\omega\big)}$$
$$-\sum_{n,m,l}\left[\frac{\rho_{ll}^{(0)} - \rho_{nn}^{(0)}}{\hbar^2}\right]\frac{d_{mn}^i d_{lm}^j d_{nl}^k}{\big[\Omega_{nm} - (\omega'+\omega)\big]\big(\Omega_{nl}-\omega\big)}, \tag{10.39}$$

where we have used $\Omega_{nm} = \omega_{nm} - i\gamma_{nm}$ instead of ω_{nm} in order to account for the n–mth-level relaxation mechanism.

The expression for β_{ijk} may be rewritten in many equivalent ways by appropriate relabeling of the indices n, m, and l. In particular, one can show in a manner analogous to the preceding section that, for a molecule in a ground state, the second-order polarizability tensor component β_{ijk} becomes

$$\beta_{ijk}^{\text{ground}}\big[-(\omega'+\omega);\omega',\omega\big] = \sum_{nl}\left[\frac{d_{gn}^i d_{nl}^j d_{lg}^k \hbar^{-2}}{\big[\Omega_{ng} - (\omega'+\omega)\big]\big(\Omega_{lg}-\omega\big)}\right.$$
$$+ \frac{d_{nl}^i d_{gn}^j d_{lg}^k \hbar^{-2}}{\big[\Omega_{ln} - (\omega'+\omega)\big]\big(\Omega_{lg}-\omega\big)}\bigg]$$
$$-\sum_{nm}\left[\frac{d_{mn}^i d_{ng}^j d_{gm}^k \hbar^{-2}}{\big[\Omega_{nm} - (\omega'+\omega)\big]\big(\Omega_{gm}-\omega\big)}\right.$$
$$+ \frac{d_{ng}^i d_{mn}^j d_{gm}^k \hbar^{-2}}{\big[\Omega_{gn} - (\omega'+\omega)\big]\big(\Omega_{gm}-\omega\big)}\bigg]. \tag{10.40}$$

Notice that, in general, both single-photon resonances (e.g., $\Omega_{gm}-\omega$) and two-photon resonances [e.g., $\Omega_{ln}-(\omega'+\omega)$] are involved.

10.3.3. Third-Order Electronic Polarizabilities

Using the same straightforward though cumbersome algebra, the third-order density matrix element $\rho_{nm}^{(3)}(t)$ and the corresponding nonlinear polarizability tensor component γ_{ijkh} can be obtained. It suffices to note that $\rho_{nm}^{(3)}(t)$ will involve a triple product of $\mathbf{d} \cdot \mathbf{E}$ and oscillate as $e^{-i(\omega+\omega'+\omega'')t}$, where ω, ω', and ω'' are the frequency components contained in the optical field $\mathbf{E}(t)$. The interested reader, with the practice gained from the preceding section, can easily show that if $\mathbf{E}(t)$ is now expressed as

$E(t)=\sum_{\omega}E(\omega'')e^{-i\omega''t}$ in the commutator $[\mathbf{d}\cdot E(t), \rho^{(2)}(t)]$, the component of $\rho_{nm}^{(3)}$ that oscillates at a particular frequency $(\omega+\omega'+\omega'')$ is given by

$$\rho_{nm}^{(3)}(t) = e^{-i(\omega''+\omega'+\omega)t} \sum_{pq} \left\{ \frac{1}{\hbar} \frac{\left[\mathbf{d}_{np} \cdot E(\omega'')\right]}{\left[\omega_{nm} - (\omega''+\omega'+\omega)\right]} \right.$$

$$\times \left[\frac{\left(\rho_{mm}^{(0)} - \rho_{qq}^{(0)}\right)}{\hbar^2} \frac{\left[\mathbf{d}_{pq} \cdot E(\omega')\right]\left[\mathbf{d}_{qm} \cdot E(\omega)\right]}{\left[\omega_{pm} - (\omega'+\omega)\right]\left(\omega_{qm} - \omega\right)} \right.$$

$$\left. -\frac{\left(\rho_{qq}^{(0)} - \rho_{pp}^{(0)}\right)}{\hbar^2} \frac{\left[\mathbf{d}_{qm} \cdot E(\omega')\right]\left[\mathbf{d}_{pq} \cdot E(\omega)\right]}{\left[\omega_{pm} - (\omega'+\omega)\right]\left(\omega_{pq} - \omega\right)} \right]$$

$$-\frac{1}{\hbar} \frac{\left[\mathbf{d}_{pm} \cdot E(\omega'')\right]}{\left[\omega_{nm} - (\omega''+\omega'+\omega)\right]}$$

$$\times \left[\frac{\left(\rho_{pp}^{(0)} - \rho_{qq}^{(0)}\right)}{\hbar^2} \frac{\left[\mathbf{d}_{nq} \cdot E(\omega')\right]\left[\mathbf{d}_{qp} \cdot E(\omega)\right]}{\left[\omega_{np} - (\omega'+\omega)\right]\left(\omega_{qp} - \omega\right)} \right.$$

$$\left. \left. -\frac{\left(\rho_{qq}^{(0)} - \rho_{nn}^{(0)}\right)}{\hbar^2} \frac{\left[\mathbf{d}_{qp} \cdot E(\omega')\right]\left[\mathbf{d}_{nq} \cdot E(\omega)\right]}{\left[\omega_{np} - (\omega'+\omega)\right]\left(\omega_{nq} - \omega\right)} \right] \right\}. \tag{10.41}$$

From this, the third-order polarizability tensor component $\gamma_{ijkh} [-(\omega''+\omega'+\omega); \omega'',$ $\omega', \omega]$ can be identified from

$$d_i(\omega'' + \omega' + \omega) = \gamma_{ijkh} E_j(\omega'')E_k(\omega')E_h(\omega), \tag{10.42}$$

where d_i is the ith Cartesian component of the expectation value dipole moment

$$d_i = \hat{i} \cdot \langle \mathbf{d} \rangle = \sum_{n,m} \rho_{nm}^{(3)} d_{mn}^i. \tag{10.43}$$

This gives

$$\gamma_{ijkh}\left[-(\omega'' + \omega' + \omega); \omega'', \omega', \omega\right] = \sum_{nmpq} \frac{\hbar^{-3} d_{mn}^i}{\left[\omega_{nm} - (\omega'' + \omega' + \omega)\right]}$$

$$\left\{ \frac{[\rho_{mm}^{(0)} - \rho_{qq}^{(0)}]d_{np}^j d_{pq}^k d_{qm}^h}{[\omega_{pm} - (\omega'+\omega)](\omega_{qm} - \omega)} - \frac{[\rho_{qq}^{(0)} - \rho_{pp}^{(0)}]d_{np}^j d_{qm}^k d_{pq}^h}{[\omega_{pm} - (\omega'+\omega)](\omega_{pq} - \omega)} \right.$$

$$\left. -\frac{[\rho_{pp}^{(0)} - \rho_{qq}^{(0)}]d_{pm}^j d_{nq}^k d_{qp}^h}{[\omega_{np} - (\omega'+\omega)](\omega_{qp} - \omega)} + \frac{[\rho_{qq}^{(0)} - \rho_{nn}^{(0)}]d_{pm}^j d_{qp}^k d_{nq}^h}{[\omega_{np} - (\omega'+\omega)](\omega_{nq} - \omega)} \right\}. \tag{10.44}$$

If we account for relaxation mechanisms, again, all the ω's will be replaced by the corresponding $\Omega = \omega - i\gamma$. From Equation (10.44) we can see that besides one and two-photon resonances, γ_{ijkh} also contain three-photon resonances, characterized by denominators of the form $\Omega_{nm}-(\omega''+\omega'+\omega)$.

From these expressions for the second- and third-order nonlinear polarizabilities, several observations can be made:

- The magnitudes and signs of the nonlinear responses of a molecule are highly dependent on the state [e.g., $\rho_{nm}^{(0)}$] it is in; an excited-state molecule obviously has a nonlinear response very different from a ground-state molecule.
- The magnitudes of the nonlinear polarizabilities depend on the dipole matrix moments involved, which are basically governed by the electronic structure of the molecules.
- The contribution from any pair of levels (e.g., p and q) to the nonlinear response will be greatly enhanced as the resonance condition $\omega_{pq}-\omega=0$ is approached.
- Away from resonances, the nonlinear response depends mainly on the oscillator strength, which is governed by the electronic structure of the molecule.

10.4. ELECTRONIC NONLINEAR POLARIZATIONS OF LIQUID CRYSTALS

As we can see from this simple but general consideration of a multilevel molecule, nonlinear electronic polarizations occur naturally in all materials illuminated by an optical field. The differences among the nonlinear responses of different materials are due to differences in their electronic properties (wave functions, dipole moments, energy levels, etc.) which are determined by their basic Hamiltonian H_0. For liquid crystals in their ordered phases, an extra factor we need to take into account are molecular correlations.

The expressions for the molecular polarizabilities derived in the last two sections can be applied to a liquid crystal molecule if the Hamiltonian (and wave functions) of the liquid crystal molecule is known. To obtain the right Hamiltonian function and solve for the wave functions and the dipole matrix elements required for evaluating α_{ij}, β_{ijk}, and γ_{ijkh} of the actual liquid crystal molecule is a quantum chemistry problem that requires a great deal of numerical computation.

Several methods[5,6] have been developed for calculating these molecular nonlinear polarizabilities. These methods have proven to be fairly reliable for predicting the second- and third-order molecular polarizabilities of organic molecules and polymers in general and liquid crystalline molecules in particular. For treating the ordered phases of liquid crystals, one has to account for molecular correlations and ordering by introducing an orientational distribution function.[7,8]

Other important factors that should be accounted for are the local field corrections and the anisotropy of the molecules. In general, liquid crystalline molecules are complex and possess permanent dipole moments. Furthermore, quadrupole moments could also contribute significantly in second-order nonlinear polarizations,[9] and thus the treatment presented in the preceding sections, which ignores these points, has to be appropriately modified.

In the rest of this chapter, we address these and other macroscopic symmetry properties of liquid crystalline electronic optical nonlinearities.

10.4.1. Local Field Effects and Symmetry

The molecular polarizabilities obtained in the last two sections are microscopic parameters that characterize an *individual* molecule. To relate them to the bulk or macroscopic parameters, namely, the susceptibilities $\chi^{(1)}, \chi^{(2)}, \chi^{(3)}$, and so forth, one needs to account for the intermolecular fields as well as the symmetry properties of the bulk materials. These intermolecular fields will result in a collective response that is different from the one obtained by treating the bulk materials as being composed of noninteracting molecules. In other words, one has to account for the so-called local field correction factors.

The local field correction factors for linear susceptibility were discussed in Chapter 3. In the usual studies[2] of local field correction factors for nonlinear susceptibilities, molecular correlations that characterize liquid crystals are not taken into account. With this gross approximation, the second and third-order nonlinear susceptibilities of an isotropic medium or a medium with cubic symmetry containing N molecules per unit volume are given by[2]

$$\chi_{ijk}^{(2)}\big[-(\omega'+\omega);\omega',\omega\big]$$
$$= NL^{(2)}(\omega'+\omega,\omega',\omega)\beta_{ijk}\big[-(\omega+\omega');\omega',\omega\big], \tag{10.45a}$$

where

$$L^{(2)}(\omega'+\omega,\omega',\omega) = \left[\frac{\varepsilon^{(1)}(\omega'+\omega)+2}{3}\right]$$
$$\left[\frac{\varepsilon^{(1)}(\omega')+2}{3}\right]\left[\frac{\varepsilon^{(1)}(\omega)+2}{3}\right]. \tag{10.45b}$$

Similarly,

$$\chi_{ijkh}^{(3)}\big[-(\omega''+\omega'+\omega);\omega'',\omega',\omega\big]$$
$$= NL^{(3)}(\omega''+\omega'+\omega,\omega'',\omega',\omega)$$
$$\gamma_{ijkh}\big[-(\omega''+\omega'+\omega);\omega'',\omega',\omega\big], \tag{10.46a}$$

where

$$L^{(3)} = \left[\frac{\varepsilon^{(1)}(\omega''+\omega'+\omega)+2}{3}\right]\left[\frac{\varepsilon^{(1)}(\omega'')+2}{3}\right]\left[\frac{\varepsilon^{(1)}(\omega')+2}{3}\right]\left[\frac{\varepsilon^{(1)}(\omega)+2}{3}\right]. \tag{10.46b}$$

10.4.2. Symmetry Considerations

Real liquid crystal molecules are much more complex than the multilevel molecule studied in the preceding section. In general, they are anisotropic and possess permanent dipoles (i.e., the molecules themselves are polar). Nevertheless, in bulk form, these liquid crystalline molecules tend to align themselves such that their collective dipole moment is of vanishing value.

Put another way, most phases of liquid crystals are characterized by centrosymmetry, due to the equivalence of the $-\mathbf{n}$ and $\mathbf{n}$ directions. This holds in the nematic phase (which belongs to the $D_{\infty h}$ symmetry group), the smectic-A phase (D_{∞} symmetry), and the smectic-C phase (C_{2h} symmetry). As a result of such centrosymmetry, the macroscopic second-order polarizability $\chi_{ijk} \equiv 0$.

The most well-known and extensively studied noncentrosymmetric liquid crystalline phase is the smectic~C* phase. The helically modulated smectic~C* system, with a spatially varying spontaneous polarization, is locally characterized by a C_2 symmetry; the average polarization of the bulk is still vanishing. By unwinding such a helical structure, the system behaves as a crystal with C_2 point symmetry. Such unwound SmC* phases possess sizable second-order nonlinear polarizabilities, with $\chi^{(2)}$ ranging from[10] 8×10^{-16} m/V for DOBAMBC to about 0.2×10^{-12} m/V for o-nitroalkoxyphenyl-biphenyl-carboxylate.[11]

10.4.3. Permanent Dipole and Molecular Ordering

The fact that liquid crystal molecules themselves could possess a permanent dipole moment (even though the bulk dipole moment has vanished) has been taken into account in the treatment of Saha and Wong[7] of the molecular correlation effects in third-order nonlinear polarizabilities. Their calculations show that, in general, the experimentally measurable third-order nonlinear polarizabilities $\bar{\gamma}_{xxxx}$, $\bar{\gamma}_{yyyy}$, and $\bar{\gamma}_{zzzz}$, for example, are related to their microscopic counterparts β_{ijk} and γ_{ijkl} by the following relationships:

$$\bar{\gamma}_{zzzz} = \gamma_{\text{iso}} + \tfrac{2}{7}\alpha\langle P_2 \rangle + \tfrac{8}{7}\beta\langle P_4 \rangle, \tag{10.47}$$

$$\bar{\gamma}_{xxxx} = \bar{\gamma}_{yyyy} = \gamma_{\text{iso}} - \tfrac{1}{7}\alpha\langle P_2 \rangle + \tfrac{3}{7}\beta\langle P_4 \rangle, \tag{10.48}$$

$$\begin{aligned}
\gamma_{\text{iso}} = &\tfrac{1}{5}\left(\gamma_{1111} + \gamma_{2222} + \gamma_{3333} + 2\gamma_{1122} + 2\gamma_{3322} + 2\gamma_{3311}\right) \\
&+ (5kT)^{-1}\left(\mu_3\beta_{333} + \mu_3\beta_{322} + \mu_3\beta_{311} + \mu_2\beta_{233} + \mu_2\beta_{222}\right. \\
&\left. + \mu_2\beta_{211} + \mu_1\beta_{133} + \mu_1\beta_{122} + \mu_1\beta_{111}\right),
\end{aligned} \tag{10.49}$$

$$\begin{aligned}
\alpha = &\left(2\gamma_{3333} - \gamma_{1111} - \gamma_{2222} - 2\gamma_{1122} + 2\gamma_{3322} + \gamma_{3311}\right) \\
&+ (kT)^{-1}\left(2\mu_3\beta_{333} + \tfrac{1}{2}\mu_3\beta_{322} + \tfrac{1}{2}\mu_3\beta_{311} + \tfrac{1}{2}\mu_2\beta_{233} - \mu_2\beta_{222}\right. \\
&\left. - \mu_2\beta_{211} + \tfrac{1}{2}\mu_1\beta_{133} - \mu_1\beta_{122} - \mu_1\beta_{111}\right),
\end{aligned} \tag{10.50}$$

where μ_i ($i = 1, 2, 3$) are the components of the permanent dipole moment in the major axes frame of the molecule, $\langle P_l \rangle$ are the moments of the molecular orientational distribution function $f(\theta)$ defined by

$$\langle P_l \rangle = \int_{-1}^{1} f(\theta) P_l(\cos\theta) \, d\cos\theta, \tag{10.51}$$

and P_l is the lth-order Legendre polynomial.

This follows from the definition for the orientational distribution[7,8] function $f(\theta)$:

$$f(\theta) = \sum_{l=0}^{\infty} \left(\frac{2l+1}{2} \right) a_l P_l(\cos\theta) \tag{10.52}$$

for uniaxial molecules (most liquid crystals are in this category).

10.4.4. Quadrupole Contribution and Field-Induced Symmetry Breaking

It is important to note at this juncture that so far our discussion of the nonlinear polarization is based on the dipole interaction [cf. Eq. (10.3)]. We have neglected quadrupole and higher-pole contributions as they are generally small in comparison. However, in situations where the dipolar contribution is vanishing, either because the molecules are symmetric[12] or the bulk is centrosymmetric, the quadrupole contribution is not necessarily vanishing and may give rise to sizable second-order nonlinear polarization.[13] Nematic liquid crystal second-order nonlinear susceptibilities associated with quadrupole moment contributions were studied in detail by Ou-Yang and Xie.[9] The starting point of their analysis is the general expression for second-order nonlinear polarization:

$$P_i^{\mathrm{NL}} = \chi_{ijk} E_j E_k + \gamma_{ijkl} E_j \frac{\partial E_l}{\partial n_k} + ..., \quad i,j,k,l = 1,2,3. \tag{10.53}$$

The first term on the right-hand side of Equation (10.53) vanishes in media with centrosymmetry, such as nematic liquid crystals. It is, however, nonvanishing if the nematic director $\mathbf{n}$ is distorted (e.g., curvature deformation that produces flexoelectric effects[14]). The second term is the quadrupole moment contribution, which is, in general, nonzero.

By considering the symmetry and coordinate transformation properties of a nematic liquid crystal under the action of an applied electric field, these authors were able to obtain explicit expressions for the susceptibility tensors χ_{ijk} and χ_{ijkl} and explain the experimental observations of second-harmonic generations by several groups.[7,15]

10.4.5. Molecular Structural Dependence of Nonlinear Susceptibilities

The magnitudes of the various tensorial components of the nonlinear susceptibilities of liquid crystals are determined fundamentally by their electronic structure. Since liquid crystal molecules are large and complex, quantitative numerical ab initio computations of these susceptibilities are rather involved.

Most studies of liquid crystal electronic nonlinear susceptibilities, as in other organic or polymeric molecular systems, are therefore phenomenological in nature. Early studies included the systematic examination of about 100 different organic compounds by Davydov et al.[16] They found the conjugated π electrons in the benzene ring to be responsible for large second-harmonic signal generation (see Fig. 10.7). This observation was also confirmed by other studies,[17] which showed that the delocalized electronics in the conjugated molecules could produce large second- and third-order nonlinearities.

In the past two decades, an extensive volume of research has been performed in the field of molecular nonlinear optics (for details see Refs. 1, 5, and 18). In spite of the tremendous variations in the molecular structures of the organic materials, the following conclusions seem to be well documented.

- The nonlinear optical response of thermotropic liquid crystals and similar organic materials is due mainly to the delocalization of the π-electron wave functions of the so-called polarizable core of a liquid crystal. The polarizable core usually consists of more than one benzene ring connected by a variety of linkages or bonds.
- The nonlinear polarizations of a molecule, as well as its linear optical properties (e.g., absorption spectrum, etc.), can be drastically modified by substituents. In general, the interactions of substituents with the σ bonds are rather mild, whereas their effects on the π electron are much more pronounced and extend over the entire delocalization range.
- There are two types of substituents: acceptors and donors, in analogy to semiconductor physics. If a substituent group has vacant low-lying π orbitals, it will attract electrons from the host conjugated molecule; such a substituent is called an acceptor. Examples include NO_2, $C\equiv N$, and the $CONH_2$ group. On the other hand, substituents that have occupied high-lying π orbitals will tend to share their electronic charges with the conjugated molecule; they are thus classified as donors. Examples include $N(CH_3)_2$, NH_2, and OH.
- In the engineering of new molecules, it has been observed that multiple substituents by weakly interacting substituents tend to produce an additive effect. On the other hand, when strong interacting substituents are used, for example, a strong

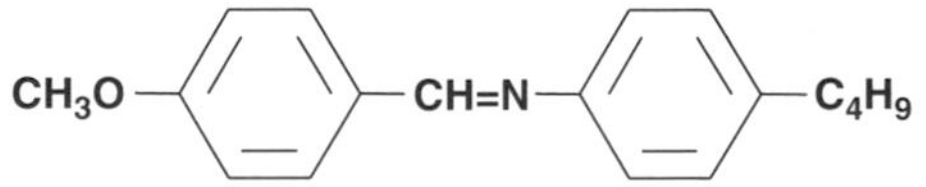

Figure 10.7. Molecular structure of MBBA showing the presence of two benzene rings. The six π electrons in each ring may be considered as free electrons confined to the conjugated bond.

donor and a strong acceptor linked to the same conjugated molecule as follows:

$$NO_2 —C_6H_4— NH_2,$$

acceptor   donor

the result could be a tremendous enhancement in the oscillator strength and susceptibilities to external (optical) fields.

- It is important to note that although the nonlinear polarizabilities (and susceptibilities) of a particular class of molecules may be improved by molecular engineering, other physical properties (e.g., viscosity, size, elastic constant, absorption, etc.) are also likely to be affected. These physical properties play an equally (in some cases even more) important role in actual nonlinear optical processes and applications. A simple example of this necessity to consider the overall picture, rather than an optimization procedure based solely on a particular molecular parameter, is discussed in Chapter 8.

- In some absorbing (at the 5145 Å line of an Ar^+ laser) liquid crystals [4-4'-bis(heptyloxy)azoxybenzene], a study[20] has shown peculiar polarization dependences on the optical fields. These effects are undoubtedly related to the electronic structures of these liquid crystals and their changes following photoabsorption, but the exact mechanisms remain to be ascertained.

- In some nematic liquid crystals (e.g., MBBA or azobenzene liquid crystals), theories and experiments[20] have shown that, when photoexcited, the molecules will undergo some structural or conformational changes (from *trans* state to *cis* state) and will exhibit unusually large orientational optical nonlinearities. Such electronic-orientational nonlinearities are discussed earlier in Chapter 8.

REFERENCES

1. See, for example, chapters dealing with Hamiltonian and quantum chemistry in Chemla, D. S., and J. Zyss. 1987. *Nonlinear Optical Properties of Organic Molecules and Crystals*. Orlando, FL: Academic Press.

2. Boyd, R. W. 1993. *Nonlinear Optics*. New York: Wiley (Interscience).

3. Deeg, F. W., and M. D. Fayer. 1989. *J. Chem. Phys.* 91:2269.

4. Eichler, H. I., R. Macdonald, and B. Trösken. 1993. *Mol. Cryst. Liq. Cryst.* 231:1–10.

5. See, for example, Chemla and Zyss[1] also Prasad, P. N., and D. J. Williams. 1990. *Introduction to Nonlinear Optical Effects in Molecules and Polymers*. New York: Wiley (Interscience).

6. LeGrange, J. D., M. G. Kuzyk, and K. D. Singer. 1987. *Mol. Cryst. Liq. Cryst.* 150B:567.

7. Saha, S. K., and G. K. Wong. 1979. *Appl. Phys. Lett.* 34:423.

8. Jen, S., N. A. Clark, P. S. Pershan, and E. B. Priestley. 1977. *J. Chem. Phys.* 66:4635.

9. Ou-Yang, Z.-C., and Y.-Z. Xie. 1985. *Phys. Rev. A.* 32:1189.

10. Shtykov, N. M., M. I. Barnik, L. A. Beresnev, and L. M. Blinov. 1985. *Mol. Cryst. Liq. Cryst.* 124:379.

11. Lin, J. Y., M. G. Robinson, K. M. Johnson, D. M. Wabba, M. B. Ros, N. A. Clark, R. Shao, and D. Doroski. 1991. *J. Appl. Phys.* 70:3426.

12. Prost, J., and J. P. Marcerou. 1977. *J. Phys. (Paris).* 38:315.

13. Bjorkholm, J. E., and A. E. Siegman. 1967. *Phys. Rev.* 154:851.

14. Meyer, R. B. 1969. *Phys. Rev. Lett.* 22:918; Gu, S.-J., S. K. Saha, and G. K. Wong. 1981. *Mol. Cryst. Liq. Cryst.* 69:287.

15. Arakelyan, S. M., G. L. Grigoryan, S. Ts. Nersisyan, M. A. Nshanyan, and Yu. S. Chulingaryan. 1978. *Pisma Zh. Eksp. Teor. Fiz.* 28:202; 1978. *JETP Lett.* 28:186.

16. Davydov, B. L., L. D. Derkacheva, V. V. Duna, M. E. Zhabotinskii, V. F. Zolin, L. 6. Kereneva, and M. A. Samokhina. 1970. *Pisma Zh. Eksp. Teor. Fiz.* 12:24; *JETP Lett.* 12:16.

17. Levine, B. F., and C. G. Bethen. 1975. *J. Chem. Phys.* 63:2666; Oudar, J. L. 1977. *ibid.* 67:446; see also Chemla and Zyss.[1]

18. A detailed review of recent work is that of Ledoux, I., and J. Zyss. 1994. *Int. J. Nonlinear Opt. Phys.* 3(2):

19. Khoo, I. C. 1991. *Mol. Cryst. Liq. Cryst.* 207:317.

20. See, for example, Shishido, A. T., O. Tsutsumi, A. Kanazawa, T. Shiono, T. Ikeda, and N. Tamai. 1997. Rapid optical switching by means of photoinduced change in refractive index of azobenzene liquid crystals detected by reflection-mode analysis. *J. Am. Chem. Soc.* 119: 7791–7796, and references therein.

11

Introduction to Nonlinear Optics

11.1. NONLINEAR SUSCEPTIBILITY AND INTENSITY-DEPENDENT REFRACTIVE INDEX

In Chapter 2, we discuss the refractive indices of liquid crystals in terms of the induced polarization P and the optical electric field E, where P is linearly related to E. Generally speaking, a material is said to be optically nonlinear when the induced polarization P is not linearly dependent on E. This could happen if the optical field is very intense. It could also happen if the physical properties of the material are easily perturbed by the optical field. In this chapter we describe the general theoretical framework for studying these processes. Specific nonlinear optical phenomena observed in liquid crystalline systems will be presented in Chapter 12.

There are two basic approaches. In many systems (e.g., atoms, molecules, and semiconductors), the primary processes responsible for nonlinear polarizations are associated with electronic transitions. To describe such processes and obtain the correct polarization, it is necessary to employ quantum mechanical theories. On the other hand, many processes are essentially classical in nature. In liquid crystals, for example, processes such as thermal and density effects, molecular reorientations, flows, and electrostrictive effects require only classical mechanics and electromagnetic theories. In this chapter the fundamentals of nonlinear optics are described within the framework of classical electromagnetic theories. Some of the quantum mechanical aspects of electronic nonlinearities were given in Chapter 10.

11.1.1. Nonlinear Polarization and Refractive Index

All optical phenomena occurring in a material arise from the optical field-induced polarization P. In general, the total polarization P may be written in the form,

$$\vec{P} = \vec{P}_L + \vec{P}_{NL}, \tag{11.1}$$

Liquid Crystals, Second Edition   By Iam-Choon Khoo
Copyright © 2007 John Wiley & Sons, Inc.

where the subscripts L and NL denote linear and nonlinear responses, respectively.

We discuss the linear part of the polarization P_L in Chapter 3. Following the same procedure to treat the linear polarization, the wave equation for the case when the nonlinear part of the polarization P, P_{NL} in Equation (11.1), is also included can be simply derived to yield

$$\nabla^2 E_i - \mu_i \varepsilon_i \frac{\partial^2 E_i}{\partial t^2} = \mu_0 \frac{\partial^2 P_{NL}^{(i)}}{\partial t}. \tag{11.2}$$

Since P_{NL} is not linear in E_i, its effects on the propagation go far beyond the simple velocity change caused by the linear polarization term P_L discussed before. Nevertheless, let us first consider here the simple but very important case where the effect of the nonlinear polarization is manifested in the form of a change in the refractive index. This is in analogy to the case of linear optics, where the refractive index (unity) of the vacuum is modified to a value (n_i) owing to the linear polarization. To see this more clearly, consider, for example, a commonly occurring P_{NL} that is proportional to the third power of E_i in the form

$$\vec{P}_{NL} = \chi_{NL} \langle E_i^2 \rangle \vec{E}_i. \tag{11.3}$$

Substituting Equation (11.3) in Equation (11.2) and assuming a plane wave form for $E_i(\mathbf{r},t)$ and P_{NL}, that is,

$$\begin{aligned} E_i(r,t) &= E_i \exp i(\vec{k} \cdot \vec{r} - \omega t), \\ P_{NL}(r,t) &= P_{NL} \exp i(\vec{k} \cdot \vec{r} - \omega t), \end{aligned} \tag{11.4}$$

we have

$$\nabla^2 E_i - \mu_i' \varepsilon_i' \frac{\partial^2 E_i}{\partial t^2} = 0, \tag{11.5}$$

where

$$\mu_i' \varepsilon_i' = \mu_i \varepsilon_i + \mu_0 \chi_{NL} \langle E_i^2 \rangle. \tag{11.6}$$

Equation (11.6) allows us to define an effective optical dielectric constant ε' given by

$$\varepsilon' = \varepsilon_i + \chi_{NL} \langle E_i^2 \rangle, \tag{11.7}$$

where we have let $\mu_i' = \mu_i = \mu_0$ (for nonmagnetic materials). The effective refractive index n is therefore given by

$$n^2 = \frac{\mu_i' \varepsilon_i'}{\mu_0 \varepsilon_0} = n_0^2 + \frac{\chi_{\mathrm{NL}} \langle E_i^2 \rangle}{\varepsilon_0}. \tag{11.8}$$

For the typical case where $\chi_{\mathrm{NL}} \langle E^2 \rangle / \varepsilon_0 = n_0^2$, we can approximate n in Equation (11.8) by

$$n \approx n_0 + n_2(E)\langle E_i^2 \rangle, \tag{11.9}$$

where

$$n_2(E) = \frac{\chi_{\mathrm{NL}}}{2\varepsilon_0 n_0}. \tag{11.10}$$

In the current literature $n_2(E)$ is often referred to as the nonlinear coefficient. Since $|E|^2$ is related to the optical intensity I_{op}:

$$I_{\mathrm{op}} = \frac{\varepsilon_0 n c}{2} \langle E^2 \rangle, \tag{11.11}$$

Equation (11.9) may also be written in the form

$$n = n_0 + n_2(I) I_{\mathrm{op}}, \tag{11.12}$$

with

$$n_2(E) = n_2(I)\left(\frac{n_0 \varepsilon_0 c}{2} \right). \tag{11.13}$$

In SI units the factor $\varepsilon_0 c / 2$ amounts to $1/753$. It is important to note here that the special form of the nonlinear polarization [see Eq. (11.3)], as well as the associated nonlinear coefficient $n_2(I)$ or $n_2(E)$, is appropriate for describing only a specific class of nonlinear optical phenomena. In the next section we will describe other more general forms of nonlinear polarization. For the present, however, we will continue with this form of polarization in discussing some conventional usage and systems of units.

11.1.2. Nonlinear Coefficient and Units

In electrostatic units, the nonlinear polarization is often written as

$$P_i = \chi_{esu}^{(3)} E_i E_i^* E_i, \tag{11.14}$$

where the superscript (3) on χ_{esu} signifies that this is a third-order nonlinear polarization; that is, the resulting polarization involves a triple product of the field E.

The corresponding polarization in SI units involves a third-order susceptibility $\chi_{SI}^{(3)}$, that is,

$$P_i = \chi_{SI}^{(3)} E_i E_i^* E_i. \tag{11.15}$$

Since $P_i(esu)/P_i(SI) = 3 \times 10^5$ and $E_i(esu) = (3 \times 10^4)^{-1} E_i(SI)$, we have

$$\chi_{esu}^{(3)} = 3^4 \times 10^{17} \chi_{SI}^{(3)}. \tag{11.16}$$

Also, from Equation (11.10) and noting that $\langle E^2 \rangle = |E_i|^2/2$ (using $\langle \cos^2 \omega t \rangle = \frac{1}{2}$), we have the following relationship:

$$\chi_{SI}^{(3)} = \varepsilon_0 n_0 n_2(E). \tag{11.17}$$

The difference between $\chi_{SI}^{(3)}$ here and χ_{NL} in Equation (11.10) is due to the different form in which the nonlinear polarization P_{NL} is written [cf. Eqs. (11.15) and (11.3)] and the fact that $\langle E_i^2 \rangle = (E_i E_i^*)/2$.

Another useful relationship to know is the one between the nonlinear coefficient $n_2(I)$ (usually expressed in standard units) and the nonlinear susceptibility $\chi^{(3)}$ (usually quoted in the literature in electrostatic units). Using Equations (11.16), (11.17), and (11.13), we have

$$\chi_{esu}^{(3)} = 3^4 \times 10^{17} \chi_{SI}^{(3)} = 3^4 \times 10^{17} \times \varepsilon_0 n_0^2 n_2(I) \cdot \frac{1}{753}$$
$$= (9.54 \times 10^4 n_0^2) n_2(I), \tag{11.18}$$

where $n_2(I)$ is in m^2/W.

If one uses the usual "mixed" unit of cm^2/w for $n_2(I)$ (owing to the prevalent usage of the mixed unit for the optical intensity I in W/cm^2), then $\chi_{esu}^{(3)}$ is given by, from Equation (11.18),

$$\chi_{esu}^{(3)} = 9.54 n_0^2 n_2(I)_{mixed}, \tag{11.19}$$

where the unit for $n_2(I)_{mixed}$ is cm^2/W.

Mixed systems of units have been employed by various researchers as some of the numerical examples in the preceding chapters have demonstrated. A mastery of these unit conversion techniques is almost a prerequisite in the study of liquid crystals and/or nonlinear optics.

11.2.   GENERAL NONLINEAR POLARIZATION AND SUSCEPTIBILITY

As we mentioned in the beginning of the chapter, in the electromagnetic approach to nonlinear optics, the nonlinear response of a medium to an applied electromagnetic field is described by a functional dependence of the form

$$P(\vec{r};t) = f\left[E(\vec{r},t), E(\vec{r}',t') \right]. \tag{11.20}$$

The nonlinear response at the space-time point (r,t) is due not only to the action of the applied fields at the same space-time point, but also of the fields at a different space-time point (r',t'). In other words, the response is a nonlocal function of time and space.

A more explicit form for $P(r,t)$ can be obtained if we make some simplifying assumptions. ,

An approximation that is often made is the expression of the resulting polarization as a power series in the fields. We have

$$P(\vec{r},t) = P^{(1)}(\vec{r},t) + P^{(2)}(\vec{r},t) + P^{(3)}(\vec{r},t) + \cdots. \tag{11.21}$$

The first term is the linear polarization discussed before and given by

$$P^{(1)}(\vec{r},t) = \varepsilon_0 \int_{-\infty}^{\infty} \overline{\overline{\chi}}^{(1)}(\vec{r} - \vec{r}', t - t') : E(\vec{r}',t')d^3\vec{r}dt', \tag{11.22}$$

where the double dots between $\overline{\overline{\chi}}^{(1)}$ and E signify a tensorial operation. Note that χ is a second-rank tensor, and E, a first-rank tensor.

Upon Fourier transformation, this gives

$$P(\vec{k};\omega) = \varepsilon_0 \chi^{(1)}(\vec{k};\omega) \cdot E(\vec{k};\omega), \tag{11.23}$$

which is the generalized nonlocal version of Equation (3.30a) [see also Eq. (10.14)].

In the same token, the second-order polarization term $P^{(2)}(\vec{r}, t)$ can be written as

$$P^{(2)}(\vec{r},t) = \int_{-\infty}^{\infty} \chi^{(2)}(\vec{r} - \vec{r}',t - t';\vec{r} - \vec{r}'',t - t'') : E(\vec{r}',t')E(\vec{r}'',t'')d\vec{r}'d\vec{r}''dt'dt'',$$

(11.24)

where $\chi^{(2)}$ is a third-rank tensor. Upon Fourier transformation, this gives

$$P^{(2)}(\vec{k},\omega) = \chi^{(2)}(k,\vec{k}',\vec{k}'';\omega,\omega',\omega'') : E(\vec{k}'.\omega')E(\vec{k}'',\omega''). \tag{11.25}$$

Similarly, we have

$$P^{(3)}(\vec{r},t) = \int_{-\infty}^{\infty} \chi^{(3)}(\vec{r} - \vec{r}',t - t',\vec{r} - \vec{r}'',t - t'',\vec{r} - \vec{r}''',t - t''')$$
$$E(\vec{r}',t')E(\vec{r}'',t'')E(\vec{r}''',t''')d\vec{r}'d\vec{r}''d\vec{r}'''dt'dt''dt''',$$

(11.26)

where $\chi^{(3)}$ is a fourth-rank tensor, and

$$P^{(3)}(\vec{k},\omega) = \chi^{(3)}(\vec{k},\vec{k}',\vec{k}'',\vec{k}''',\omega,\omega',\omega'',\omega''') : E(\vec{k}',\omega')E(\vec{k}'',\omega'')E(\vec{k}''',\omega'''), \qquad 11.27)$$

and so on for $P^{(n)}$, $\chi^{(n)}$.

If the nth-order nonlinear susceptibility $\chi^{(n)}(\vec{r}, t)$ is independent of $\vec{r}$, the Fourier transform $\chi^{(n)}(\vec{k}, \omega)$ will be independent of k. In this sense the nonlinear response is termed local. Also, if the response of the medium is instantaneous, the expressions for the nonlinear polarizations $P^{(2)}$, $P^{(3)}$, and so on become simple tensorial products of E:

$$P^{(2)}(t) = \chi^{(2)} : E(t)E(t) \tag{11.28}$$

and

$$P^{(3)}(t) = \chi^{(3)} : E(t)E(t)E(t). \tag{11.29}$$

11.3. CONVENTION AND SYMMETRY

Since the nonlinear polarization P_{NL} involves various powers of the optical electric field (in the forms EE, EEE, etc.) and if the total electric field E comprises many frequency components [e.g., $E = E(\omega_1)+E(\omega_2)+ \cdots +E(\omega_n)$], there are several possible combinations of the frequency components that will contribute to a particular frequency component for P_{NL}.

As an example, consider the second-order polarization $P^{(2)}$. Also, let us ignore all tensorial relationships and focus our attention on copolarized fields for simplicity. One can see that the frequency component $\omega_3 = \omega_1+\omega_2$ in P_{NL} can arise from two possible combinations: $E(\omega_1)E(\omega_2)$ and $E(\omega_2)E(\omega_1)$.

In actual calculation, it is easier to adopt the convention similar to that mentioned in the works by Shen[1] and Boyd,[2] where these various combinations are correctly distinguished by distinctive labeling of the nonlinear susceptibility χ involved. For example, the combinations $E(\omega_1)E(\omega_2)$ and $E(\omega_2)E(\omega_1)$ are associated with χ $(-\omega_3,\omega_1,\omega_2)$ and $\chi(-\omega_3,\omega_2,\omega_1)$, respectively. In this case, we need not invoke the degeneracy factor g in the definition of the nonlinear polarization. Also, it will account for possible differences between $\chi(-\omega_3,\omega_1,\omega_2)$ and $\chi(-\omega_3,\omega_2,\omega_1)$.

To express the relationships between P_{NL} and E in a more detailed form, we now write the electric field and nonlinear polarization in terms of their spatial coordinate components

$$E = \hat{i} E_i + \hat{j} E_j + \hat{k} E_k,$$ (11.30a)

$$P_{NL} = \hat{i} P_i + \hat{j} P_j + \hat{k} P_k.$$ (11.30b)

Also, the E_l's and P_l's ($l = i, j, k$) are expressed in the plane wave form

$$E_l = \sum_{\omega_n} \frac{1}{2}\left[A_l^{\omega_n}(r,t)e^{i(k\cdot r - \omega_n t)} + c.c.\right],$$ (11.31a)

$$P_l = \sum_{\omega_m} \frac{1}{2}\left[P_l^{\omega_m}(r,t)e^{i(k\cdot r - \omega_m t)} + c.c.\right].$$ (11.31b)

The nonlinear polarization $P_l^{\omega_3}$, responsible for the mixing of two frequency components ω_1 and ω_2 to produce a third wave of frequency ω_3 (see Fig. 11.1a), for example, will thus be given by

$$\tfrac{1}{2}P_i^{\omega_3} = 2^{-2}\left[\chi_{ijk}\left(-\omega_3,\omega_1,\omega_2\right)A_j^{\omega_1}A_k^{\omega_2} + \chi_{ijk}\left(-\omega_3,\omega_2,\omega_1\right)A_j^{\omega_2}A_k^{\omega_1}\right].$$ (11.32)

On the right-hand side of Equation (11.23), the two terms in the square brackets come from the two possible ways of ordering ω_1 and ω_2 (i.e., we can have $A_j^{\omega_1}A_k^{\omega_2}$ and $A_j^{\omega_2}A_k^{\omega_1}$). The denominator 2^2 comes from the factor $\frac{1}{2}$ involved in the definition of A [Eqs. (11.31a) and (11.31b)]. The subscripts i, j, k, and so forth signify the corresponding Cartesian components of the field. The frequencies ω_1, ω_2, and ω_3 appearing in the parentheses following χ_{ijk} represent a conventional way of expressing the fact that $\omega_3 = \omega_1 + \omega_2$ (i.e., $-\omega_3 + \omega_1 + \omega_2 = 0$). On the left-hand side, the factor $\frac{1}{2}$ comes from the definition of the amplitude $P_l^{\omega_m}$ in Equation (11.31b). With these factors properly accounted for, Equation (11.32) reads

$$P_i^{\omega_3} = \tfrac{1}{2}\left[\chi_{ijk}^{(2)}\left(-\omega_3,\omega_1,\omega_2\right)A_j^{\omega_1}A_k^{\omega_2} + \chi_{ijk}^{(2)}\left(-\omega_3,\omega_2,\omega_1\right)A_j^{\omega_2}A_k^{\omega_1}\right],$$ (11.33)

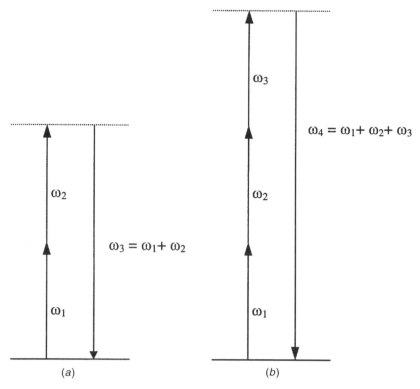

Figure 11.1. (a) Three-wave mixing process involving the mixing of two frequencies ω_1 and ω_2 to produce a new frequency $\omega_3 = \omega_1 + \omega_2$. (b) Four-wave mixing process involving $\omega_1, \omega_2,$ and ω_3 to produce $\omega_4 = \omega_1 + \omega_2 + \omega_3$.

where $\chi_{ijk}^{(2)}$ is a third-rank tensor.

If $\omega_1 = \omega_2 = \omega$ (i.e., $\omega_3 = \omega_1 + \omega_2 = 2\omega$, second-harmonic generation), then the number of distinct permutations of ω_1 and ω_2 is reduced to 1. We therefore have

$$P_i^{2\omega} = \tfrac{1}{2}\chi_{ijk}^{(2)}(-2\omega,\omega,\omega)A_j^\omega A_k^\omega. \tag{11.34}$$

It is important to note here that a totally different process is involved if we have $\omega_1 = \omega$ and $\omega_2 = -\omega$. This process is called optical rectification, since it results in a dc response with $\omega_3 = \omega_1 + \omega_2 = 0$. In this case the number of distinct permutations of ω_1 and ω_2 is 2. It is also important here to remind ourselves that $i, j,$ and k refer to the coordinate axes, and that in Equation (11.34) and other equations involving tensorial relationships, a sum over nonrepeated indices is implied [i.e., on the right-hand side of Eq. (11.34), for example, one needs to sum over $j = 1, 2, 3$ and $k = 1, 2, 3$].

In a similar manner, the nonlinear polarization component ω_4 generated by the mixing of three waves ω_1, ω_2, and ω_3 (i.e., $\omega_4 = \omega_1 + \omega_2 + \omega_3$) is given by

$$P_i^{\omega_4} = \frac{1}{2^2} \left[\chi_{ijkl}^{(3)}(-\omega_4, \omega_1, \omega_2, \omega_3) A_j^{\omega_1} A_k^{\omega_2} A_l^{\omega_3} + \cdots \right], \qquad (11.35)$$

where $\chi_{ijkl}^{(3)}$ is a fourth-rank tensor (see Fig. 11.1b). With a little practice, the reader can readily write down the corresponding nonlinear polarization components for n-wave mixing processes. It is important to note here that $\chi_{ijkl}^{(3)}$ is related to $\chi_{SI}^{(3)}$ [cf. Eq. (11.15)] by

$$\frac{1}{2^2} \chi_{ijkl}^{(3)} = \chi_{SI}^{(3)}. \qquad (11.36)$$

This relationship, generated by the different plane wave forms used for expressing the field polarization [cf. Eqs. (11.4) and (11.31)], *should be borne in mind* as one compares values of the nonlinear coefficients quoted or reported in the literature.

In accordance with current convention for n-wave mixing processes, n stands for the total number of interacting waves (counting the incident and the generated waves) within the nonlinear medium (Fig. 11.2). Therefore, interactions involving nonlinear polarization of the form given in Equations (11.33) and (11.34), where two incident waves combine to give a generated wave, are called three-wave mixing processes. Processes associated with the combination of three waves to yield a fourth one [cf. Eq. (11.35)] are called four-wave mixing processes. If all the frequencies involved are the same, the processes are called degenerate. We have, for example, degenerate four-wave mixing; otherwise they are called nondegenerate wave mixings.

In Cartesian coordinates obviously there are altogether 3^4 elements in the third-order susceptibility $\chi_{ijkl}^{(3)}$, a fourth-rank tensor, since (i, j, k, l) each has three components 1, 2, 3. In an isotropic medium with inversion symmetry, however, it can be shown[3] that there are only four different components, three of which are independent:

$$\begin{aligned}
&\chi_{1111} = \chi_{2222} = \chi_{3333}, \\
&\chi_{1122} = \chi_{1133} = \chi_{2211} = \chi_{2233} = \chi_{3311} = \chi_{3322}, \\
&\chi_{1212} = \chi_{1313} = \chi_{2121} = \chi_{2323} = \chi_{3232} = \chi_{3131}, \\
&\chi_{1221} = \chi_{1331} = \chi_{2112} = \chi_{2312} = \chi_{3113} = \chi_{3223}, \\
&\chi_{1111} = \chi_{1122} + \chi_{1212} + \chi_{1221}.
\end{aligned} \qquad (11.37)$$

In an anisotropic medium such as a liquid crystal, the symmetry considerations can be very complex.[2]

These nonlinear (microscopic) susceptibilities are derived from explicit calculations of the dipolar interaction between a particular atom or molecule with incident optical fields (see Chapter 10). In the presence of other atoms or molecules, and their polarizations, an atom or molecule will experience a net depolarization field in addition to

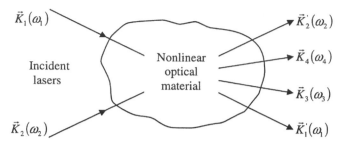

Figure 11.2. Temporal and spatial frequency wave mixing in a nonlinear optical material.

the incident optical field.[4] In condensed matter or inhomogeneous media, these depolarization fields are quite complex and, in general, very difficult to estimate. On the other hand, in simple liquids or gases (i.e., isotropic, homogeneous media), the depolarization field may be accounted for by multiplying each of the field components by a so-called Lorentz correction factor:[2,5]

$$L = \frac{\varepsilon(\omega_i) + 2}{3}. \tag{11.38}$$

The macroscopic susceptibilities $\chi^{(3)}$, $\chi^{(2)}$, and $\chi^{(1)}$ discussed previously are therefore related to the microscopic susceptibilities by

$$\chi^{(1)} = \left(\frac{\varepsilon(\omega) + 2}{3}\right)^2 \chi_{\text{mic}}^{(1)}(\omega),$$

$$\chi_{\text{mac}}^{(3)} = \left[\frac{\varepsilon(\omega_4) + 2}{3}\right]\left[\frac{\varepsilon(\omega_1) + 2}{3}\right]\left[\frac{\varepsilon(\omega_2) + 2}{3}\right]\left[\frac{\varepsilon(\omega_3) + 2}{3}\right] \times \chi_{\text{mic}}^{(3)}(-\omega_4, \omega_1, \omega_2, \omega_3).$$

$$\tag{11.39}$$

In liquid crystalline media, owing to large molecular correlation effects, these local field effects are considerably more complex.[6] Several formalisms have been developed to treat the various phases; some of these studies are pointed out in Chapter 10.

11.4. COUPLED MAXWELL WAVE EQUATIONS

The interaction of various optical fields inside a nonlinear medium are described by Maxwell equations. For n-wave mixing processes, there are n coupled Maxwell wave equations.

The solutions for the optical electric fields at any arbitrary penetration depth into the nonlinear medium depend on the initial boundary conditions on the incident field intensity and their phases and boundary conditions. Usually, in order to solve the equations in a meaningful way, some approximations are made.

An approximation that is generally valid is the so-called slowly varying envelope approximation. In effect, a particular frequency component E_j of the total electric field can be represented by a product of a fast (temporally and spatially) oscillatory term $\exp[i(\vec{k}_j \cdot \vec{r} - \omega_j t)]$ and an amplitude function $A_j(\vec{r}, t)$, whose temporal and spatial variations are on a much slower scale. Therefore, we have

$$E_j = \tfrac{1}{2}\Big[A_j(\vec{r};t)\exp\big[i(\vec{k}_j \cdot \vec{r} - \omega_j t)\big] + \text{c.c.}\Big], \quad j = 1, \ldots, n, \qquad (11.40)$$

where n is the number of frequency components present. Although these n interacting fields, in general, propagate in n different directions, one can define a reference direction, usually denoted as the z direction, to describe their interactions and propagation. Accordingly, the fast oscillatory term $\exp[i(\vec{k}_j \cdot \vec{r} - \omega t)]$ may be written as $\exp[i(k_j z - \omega t)]$ where k_j is now understood as the z component of the wave vector $\vec{k}_j$.

In more quantitative terms, the slowly varying envelope approximation (SVEA) translates into the following inequalities:

$$\left|\frac{\partial^2 A_j}{\partial z^2}\right| \ll \left|k_j \frac{\partial A_j}{\partial z}\right|, \qquad (11.41a)$$

$$\left|\frac{\partial^2 A_j}{\partial t^2}\right| \ll \left|\omega_j \frac{\partial A_j}{\partial t}\right|. \qquad (11.41b)$$

In the slowly varying approximation, the corresponding nonlinear polarization vector components are expressed as

$$P_{\text{NL}}^{(j)} = \tfrac{1}{2}\Big|P_j(\vec{r};t)e^{i(k_j^p z - \omega t)} + \text{c.c.}\Big|, \qquad (11.42)$$

that is, the total nonlinear approximation, $P_{\text{NL}} = \sum_j P_{\text{NL}}^{(j)}$, contains n frequency components. The superscript p on k_j^p in Equation (11.42) denotes that it is the wave vector associated with the jth component of the *polarization*; in general, it is *different* from the wave vector of the jth component of the electric field, as we will see presently.

Substituting Equations (11.40)–(11.42) in the wave Equation (11.2), which we rewrite here

$$\nabla^2 E - \mu\varepsilon\frac{\partial^2 E}{\partial t^2} = \mu_0 \frac{\partial^2 P}{\partial t^2}, \qquad (11.43)$$

we obtain

$$\nabla_{\perp}^2 A_j + 2ik_j \frac{\partial A_j}{\partial z} + 2i\mu\varepsilon\omega \frac{\partial A_j}{\partial t} = -\mu_0\omega^2 P_j^{\text{NL}} \exp i\Delta k_j z, \qquad (11.44)$$

where $\Delta k_j = k_j^p - k_j$ is the wave vector mismatch for the jth wave. Also, we have separated the three-dimensional ∇^2 into a transverse term ($\nabla_{\perp}^2$) and a longitudinal term ($\partial^2/\partial z^2$). Since $k_j = \omega_j/v_j$, Equation (11.44) becomes

$$\nabla_{\perp}^2 A_j + 2ik_j \left(\frac{\partial A_j}{\partial z} + \frac{1}{v_j} \frac{\partial A_j}{\partial t} \right) = -\mu_0\omega^2 P_j^{\text{NL}} \exp i\Delta k_j z . \qquad (11.45)$$

If we transform this (z,t). coordinate system to one that moves at a velocity v_j along z, that is, (z',t') given by

$$z' = z, \qquad t' = t - \frac{z}{v_j}, \qquad (11.46)$$

Equation (11.45) becomes

$$\nabla_{\perp}^2 A_j + 2ik_j \frac{\partial A_j}{\partial z} = -\mu_0\omega^2 P_j^{\text{NL}} \exp i\Delta k_j z \qquad (11.47)$$

for $j = 1, 2, \ldots$ (the number of interacting waves), where we have written z' as z since they are the same.

In deriving Equation (11.47) we have also assumed that the process under study is in a steady state; the time dependence of the process is "removed" by the substitution of $t' = t - z/v_j$ [cf. Eqs. (11.45) and (11.46)]. In many actual experiments, especially in those involving pulsed lasers, this is definitely not a correct assumption, and one has to revert to using the full Maxwell equation.

11.5. NONLINEAR OPTICAL PHENOMENA

Numerous distinctly different nonlinear optical phenomena[1,2] can be created via the nonlinear polarizations discussed previously. Here we discuss some widely studied exemplary ones.

11.5.1. Stationary Degenerate Four-Wave Mixing

Figure 11.3 shows a typical interaction geometry. Two equal-frequency laser beams $\vec{E}_1$ and $\vec{E}_2$ are incident on a nonlinear optical material of thickness d; they intersect

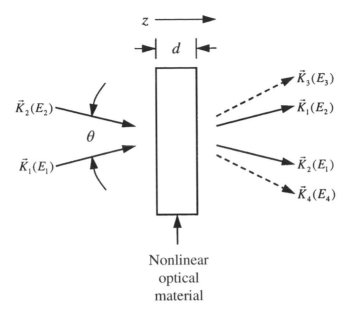

Figure 11.3. Forward wave mixing process involving side-diffracted beams

at an angle θ (within the nonlinear material). On the exit side, there are new, generated waves E_3, E_4, and so forth of the same frequency on the side of the beams E_1 and E_2. These generated waves are due to the spatial mixings of the input waves.

For example, the wave E_3 propagating in the $\vec{k}_3$ direction is due to a nonlinear polarization term P_3 of the form $P_3 \sim E_1 E_2^* E_1$, with an associated wave vector mismatch $\Delta \vec{k}_3 = \vec{k}_3 - (\vec{k}_1 + \vec{k}_1 - \vec{k}_2)$ (see Fig. 11.4).

Similarly, the wave E_4 is due to a nonlinear polarization P_4 of the form $P_4 \sim E_2 E_2^* E_1$, with an associated wave vector mismatch $\Delta \vec{k}_4 = \vec{k}_4 - (2\vec{k}_2 - \vec{k}_1)$ (see Fig. 11.5).

To illustrate the physics, we assume that all waves are polarized in the $\hat{x}$ direction. In plane wave form the total electric field $\vec{E}$ and the nonlinear polarization $\vec{P}$ are given by

$$\vec{E} = \sum_{j=1}^{4} \vec{E}_j = \hat{x} \sum_{j=1}^{4} \frac{1}{2}\left[A_j \exp i(k_j \cdot z - \omega t) + \text{c.c.}\right] \qquad (11.48)$$

and

$$\vec{P} = \sum_{j=1}^{4} \vec{P}_j = \hat{x} \sum_{j=1}^{4} \frac{1}{2}\left[P_j \exp i(k_j \cdot z - \omega t) + \text{c.c.}\right]. \qquad (11.49)$$

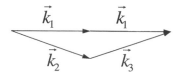

Figure 11.4. Wave vector addition diagram for the process $k_3 = 2k_1 - k_2$.

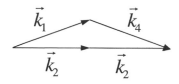

Figure 11.5. Wave vector addition diagram for the process $\vec{k}_3 = 2\vec{k}_1 - \vec{k}_2$.

Furthermore, we impose a simplifying condition, namely, that the intensity of E_1 is much higher than the intensity of E_2; in this case the generated wave E_3 is much more intense than the wave E_4. Under this simplifying assumption, the terms contributing to the generation of waves in the E_3 direction, for example, are $(E_3E_3^*) E_3$, $(E_2E_2^*)E_3$, $(E_1E_1^*)E_3$, $(E_1E_2^*)E_1$, $(E_3E_1^*)E_1$, and $(E_3E_2^*)E_2$.

In writing down those terms, we have put in parentheses products of the electric field amplitude which represent interference among the waves. The terms $E_3E_3^*$, $E_2E_2^*$, and $E_1E_1^*$ are the so-called static or dc components (i.e., the intensity is spatially and temporally nonvarying). On the other hand, the terms $E_3E_1^*$ and $E_3E_2^*$ and $E_1 E_2^*$ are the so-called (intensity) grating components, which are spatially periodic with spatial grating wave vectors given by $(\vec{k}_3 - \vec{k}_1)$ and $(\vec{k}_3 - \vec{k}_2)$, and $(\vec{k}_1 - \vec{k}_2)$ respectively. From this point of view, the term $(E_3E_1^*)E_1$, for example, is associated with the scattering of wave E_1 from the refractive index grating generated by the intensity grating $E_3E_1^*$ into the direction of E_3.

The effects associated with these static and grating terms become clearer when we write down the corresponding coupled Maxwell wave equations. From Equation (11.47), ignoring the transverse Laplacian term, the equations for the amplitudes A of these three waves can be derived to give

$$\frac{dA_1}{dz} = ig(|A_1|^2 + 2|A_2|^2 + 2|A_3|^2)A_1 + ig\left[(A_2A_1^*)A_3 + (A_3A_1^*)A_2\right]\exp i\Delta k_3 z, \quad (11.50)$$

$$\frac{dA_2}{dz} = ig(|A_2|^2 + 2|A_1|^2 + 2|A_3|^2)A_2 + ig[(A_1A_3^*)\exp - i\Delta k_3 z]A_1, \quad (11.51)$$

$$\frac{dA_3}{dz} = ig(|A_3|^2 + 2|A_1|^2 + 2|A_2|^2)A_3 + ig[(A_1A_2^*)\exp - i\Delta k_3 z]A_1, \quad (11.52)$$

where

$$g - \frac{\mu_0 \omega^2 \chi_{xxxx}}{8}, \tag{11.53}$$

$$\Delta k = \left| 2\vec{k}_1 - \vec{k}_2 - \vec{k}_3 \right|. \tag{11.54}$$

For simplicity, we have let all the nonlinear coefficients associated with the various contributing terms $[(E_3 E_3^*)E_3, (E_3 E_1^*)E_1$, etc.] in these equations be equal to χ_{xxxx}. A typical nonlinear polarization term (for P_3^{NL}) associated with $E_1 E_2^* E_1$ in terms of the amplitude A, is given by

$$P_3^{NL} = \tfrac{1}{4} \chi_{xxxx} A_1 A_2^* A_1. \tag{11.55}$$

Because of nonlinear coupling among the waves, Equations (11.50)–(11.52), in general, cannot be solved analytically. However, if one assumes that the pump beam is so strong that its depletion is negligible (i.e., $|A_1|^2 =$const) and that $|A_1|^2 >> |A_2|^2$, $|A_3|^2$ so that only the phase modulation caused by the pump beam (the term $-ig\,|A_1|^2$ on the right-hand side of these equations) is retained, Equations (11.50)–(11.52) may be readily solved to yield

$$A_1(d) = A_1(0)\exp\left[ig\,|A_1(0)|^2\,d\right], \tag{11.56}$$

$$A_2(d) = A_2(0)\exp\left[ig\,|A_1(0)|^2 d\right]\left[\cosh(qd) - i\frac{P}{q}\sinh(qd)\right], \tag{11.57}$$

and

$$A_3(d) = iA_2^*(0)\exp\left[ig\,|A_1(0)|^2 d\right]\left[\frac{p^{*2} + q^{*2}}{q^*}\sinh(q^*d)\right], \tag{11.58}$$

where

$$p = \frac{\Delta k}{2} - g\,|A_1(0)|^2, \tag{11.59}$$

$$q = \left[g\,|A_1(0)|^2 - \frac{\Delta k}{4}\right]^{1/2}. \tag{11.60}$$

From these solutions one can also see that the intensity of the transmitted probe beam acquires a gain factor $G(d)$ given by

$$G(d) = \frac{I_2(d)}{I_2(0)} = \left| \cosh(qd) - \frac{iP}{q} \sinh(qd) \right|^2 . \tag{11.61}$$

Notice that the generated wave $A_3(d)$ in Equation (11.58) is proportional to the complex conjugate of the wave A_2; for this reason, it is sometimes called the forward phase conjugate of A_2. In the following section we will discuss another phase conjugate process.

11.5.2. Optical Phase Conjugation

Figure 11.6 shows a particular degenerate four-wave mixing process involving three input waves E_1, E_2, and E_3, and the generated wave E_4, which is counterpropagating to E_3. The wave E_4 can arise from various combinations of these three input waves.

For example, it could be due to the scattering of E_1 from the (refractive index) grating formed by the interference of E_2 and E_3 [i.e., a nonlinear polarization term of the form $(E_2 E_3^*)E_1$]. Another possibility is the scattering of E_2 from the grating formed by the interference of E_1 and E_3 [i.e., a nonlinear polarization term of the form $(E_1 E_3^*)E_2$].

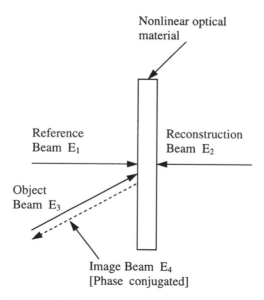

Figure 11.6. Schematic depiction of optical wave front conjugation by degenerate four-wave mixing process.

Both processes carry a wave vector mismatch $\Delta k_4 = k_4 - (k_1 + k_2 - k_3)$. Since $k_2 = -k_1$ and $k_4 = -k_3$, we have $\Delta k_4 = 0$ (i.e., a perfectly phase-matched four-wave mixing process). However, the grating constant associated with $(E_2 E_3^*)$ is much smaller than that associated with $(E_1 E_3^*)$ as a result of the larger wave mixing angle.

If the physical mechanism responsible for the formation of the refractive index grating is independent of the grating period, then these two terms will contribute equally to the generation of the wave E_4. Usually owing to diffusion or other intermolecular physical processes (e.g., in liquid crystals the torque exerted between molecules situated at the intensity minima on molecules at intensity maxima or heat diffusion in the thermal grating), the larger the grating period, the higher will the wave mixing efficiency be. In this case $(E_1 E_3^*)$ is the main contribution term.

Taking into account only the grating term $(E_1 E_3^*)$, the coupled equations for the four interacting waves within the nonlinear medium are given by

$$\frac{dA_1}{dz} = ig'(|A_1|^2 + 2|A_2|^2 + 2|A_3|^2 + 2|A_4|^2) + ig'\{\text{wave mixing terms}\}, \tag{11.62}$$

$$\frac{dA_2}{dz} = -ig'(|A_2|^2 + 2|A_1|^2 + 2|A_3|^2 + 2|A_4|^2) - ig'\{\text{wave mixing terms}\}, \tag{11.63}$$

$$\frac{dA_3}{dz} = ig(|A_2|^2 + 2|A_1|^2 + 2|A_3|^2 + 2|A_4|^2) + ig\{A_1 A_2 A_4^*\}, \tag{11.64}$$

$$\frac{dA_4}{dz} = -ig(|A_4|^2 + 2|A_1|^2 + 2|A_2|^2 + 2|A_3|^2) - ig\{A_1 A_2 A_3^*\}, \tag{11.65}$$

where $g = \mu_0 \omega^2 \chi / 8$, $g' = q/\cos\theta = \mu_0 \omega^2 \chi / 8 \cos\theta$, and χ is the nonlinear susceptibility involved in this process.

As pointed out in the preceding section, the first terms (in square brackets) on the righthand sides of these equations are phase modulation effects caused by the spatially static terms in the optical intensity function. Notice that, because of the presence of these phase modulations, the initially perfect phase-matching condition ($\Delta k = 0$) will be degraded as these waves interact and propagate deeper into the nonlinear medium.

To qualitatively examine the results and implications of optical phase conjugation, we ignore here the phase modulation terms.[7] We also assume that both waves E_1 and E_2 (usually referred to as the pump waves) are very strong compared to the probe wave E_3 and the generated wave E_4 (i.e., $|A_1|$ and $|A_2|$ are constant). We thus have

$$\frac{dA_3}{dz} = igA_1 A_2 A_4^*, \tag{11.66}$$

$$\frac{dA_4}{dz} = -igA_1 A_2 A_3^*.$$

(11.67)

For the configuration depicted in Figure 11.6, the initial conditions are $A_3 = A_3(0)$ at $z=0$ and $A_4 = 0$ at $z=L$, where L is the thickness of the nonlinear material. The solutions for A_3 and A_4 are

$$A_3(z) = A_3(0) \frac{\cos\left[g|A_1 A_2|(z-L)\right]}{\cos\left[g|A_1 A_2|L\right]},$$

(11.68)

$$A_4(z) = -iA_3^*(0) \frac{\sin\left[g|A_1 A_2|(z-L)\right]}{\cos\left[g|A_1 A_2|L\right]}.$$

(11.69)

These equations show the following:

- The transmitted probe beam is amplified by an amplitude gain factor

$$G_a = \frac{A_3(L)}{A_3(0)} = \sec\left[g|A_1 A_2|L\right],$$

(11.70)

  just as in the previous four-wave mixing case.

- The generated wave E_4 is proportional to the complex conjugate of the probe beam E_3. This feature of the generated wave bears a resemblance to the result obtained in the previous section [cf. Eq. (11.58)]. In fact, they are both generated by the same kind of wave mixing process except that a different set of propagation wave vectors is involved. For the same reason, the wave E_4 is termed the optical phase conjugate of E_3. The generated wave E_4 carries a *phase* which is a conjugate of that carried by the input. A consequence of this, which has been well demonstrated in the literature, is that if E_3 passes through a phase-aberrating medium, picking up a phase distortion factor $\exp i \phi_D$, the generated wave E_4 (if allowed to traverse the phase-aberrating medium backward) will pick up a phase factor $\exp i \phi_D$. The net result is therefore the exact cancellation of these phase distortion factors in the signal beams E_4.

Another interesting feature of Equations (11.68) and (11.69) is that if $g|A_1 A_2|L = \pi/2$, both $A_3(z)$ and $A_4(z)$ will assume finite values for a vanishing input $A_3(0)$ (i.e., "oscillations" will occur). To put it another way, imagine that a cavity is formed with its optical propagation axis along $\vec{k}_3$ (or $\vec{k}_4$). Then any coherent noise generated in this direction may be set into oscillation.

An example of *coherent* noise is the scattering noise from the pump beam, which is coherent with respect to, and can therefore interfere with, the pump beam. Numerous reports in the current literature of self-pumped phase conjugators, oscillators, and so forth are based on this process or its variants.[7] In Chapter 12, we describe two recent observations of self-starting phase conjugation using nematic liquid films.

11.5.3.   Nearly Degenerate and Transient Wave Mixing

As pointed out before, laser-induced nonlinear polarizations amount to changes in the dielectric constant. If these dielectric constant changes are spatially coincident with the imparted intensity grating, the materials are said to possess local nonlinearity. In some situations, however, the induced dielectric constant change is spatially shifted from the optical intensity; that is, there is a phase shift ϕ between the dielectric constant and the intensity grating function (see Fig. 11.7).

A phase shift between the intensity and index grating function can arise, for instance, if the frequencies of the two interference waves E_1 and E_2 are not equal (i.e., $\omega_2 = \omega_1 + \Omega$) (see Fig. 11.8). In this case the intensity grating imparted by E_1 and E_2 on the nonlinear medium is moving with time. Because of the finite response time τ of the medium, the induced index grating will be delayed (i.e., a phase shift ϕ will occur between the intensity and the index grating).

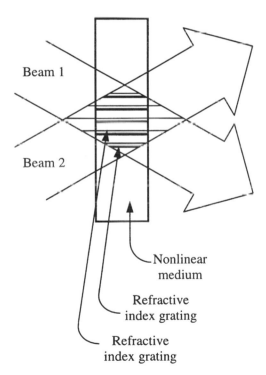

Figure 11.7.   Induced refractive index grating is spatially shifted from the imparted optical intensity grating (e.g., in photorefractive material).

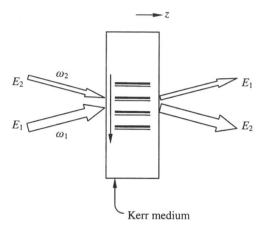

Figure 11.8. Moving intensity grating generated by two coherent incident lasers of different frequency.

The magnitude and sign of the phase shift depend on various material, optical, and geometrical parameters. For the purpose of our present discussion on the consequence of having ϕ in the wave mixing process, we will represent it generally as ϕ. If the total intensity function is given by

$$I_{\text{total}} \propto (E_1 + E_2) \cdot (E_1 + E_2)^* = |E_1|^2 + |E_2|^2 + E_1 E_2^* + E_2 E_1^*, \tag{11.71}$$

then the induced index grating function will be of the form

$$\Delta n \sim \Delta n_0 + n_2 (E_1 E_2^* e^{i\phi} + E_2 E_1^* e^{-i\phi}), \tag{11.72}$$

where Δn_0 is the (spatially) static index change caused by the static intensity $(|E_1|^2 + |E_2|^2)$.

Consider the scattering of wave E_1 or E_2 from the induced index grating; one can associate third-order nonlinear polarization terms of the form $E_1^* E_2 E_1^{i\phi}$, $E_1 E_2^* E_2^{i\phi}$ and so on. If we limit our discussion to only these two waves and follow the procedure used in deriving Equations (11.63) and (11.64), the corresponding coupled two-wave mixing equations are

$$\frac{dA_1}{dz} = ig\big[|A_1|^2 + |A_2|^2\big]A_1 + ig\big(A_2^* A_1\big)e^{i\phi}A_2, \tag{11.73}$$

$$\frac{dA_2}{dz} = ig\big[|A_1|^2 + |A_2|^2\big]A_2 + ig\big(A_1^* A_2\big)e^{i\phi}A_1. \tag{11.74}$$

The effects of the phase shift ϕ on these two-wave couplings are more transparent if these equations are expressed in terms of the intensities I_1 and I_2 and individual phases ϕ_1 and ϕ_2 of the waves. Then we have

$$I_{1,2} = \frac{\varepsilon_0 nc}{2} \langle E^2 \rangle = \frac{\varepsilon_0 nc}{2} \frac{1}{2} A_1 A_1^* = \frac{\varepsilon_0 nc}{4} |A_1|^2 \tag{11.75}$$

and

$$A_1 = |A_1| e^{i\phi_1}, \tag{11.76}$$

$$A_1 = |A_1| e^{i\phi_2}. \tag{11.77}$$

The complex coupled wave equations become two sets of real variable equations. For the intensities we have

$$\frac{dI_1}{dz} = -2g' \sin\phi I_1 I_2, \tag{11.78}$$

$$\frac{dI_2}{dz} = +2g' \sin\phi I_1 I_2. \tag{11.79}$$

For the phases we have

$$\frac{d\phi_1}{dz} = g'[(I_1 + I_2) + I_2 \cos\phi], \tag{11.80}$$

$$\frac{d\phi_2}{dz} = g'[(I_2 + I_1) + I_1 \cos\phi], \tag{11.81}$$

where $g' = 4g/\varepsilon_0 nc$.

From these equations one can readily see that if $\sin\phi$ is nonvanishing, there is a flow of energy between the waves E_1 and E_2. In materials having local nonlinearity (i.e., phase shift $\phi = 0$), side diffraction provides a means for energy exchanges between the two incident beams.

Temporally nonlocal nonlinearities naturally arise if the response of the medium is not instantaneous and generally will be manifested if short laser pulses are used. Because of the delayed reaction of the induced refractive index changes, a *time-dependent* phase shift between the intensity and the index grating function will occur (see Fig. 11.9). There is therefore a time-dependent energy exchange process between the incident beams.

Usually, the process is such that energy flow is from the stronger incident beam to the weaker one. Detailed formalisms, with quantitative numerical solutions, for treating such time-dependent wave mixing processes are given in the work by Khoo and Zhou.[8]

Temporally and spatially shifted grating effects are naturally present in stimulated scattering processes. A frequently studied process is stimulated Brillouin scattering, where an incident optical field E_1 (with a frequency ω_1) generates a coherent backward propagating wave E_2 (with a frequency $\omega_2 = \omega_1 - \omega_s$, where ω_s is the sound frequency of the medium). The physical reason for the transfer of energy from E_1 to E_2 is analogous to what we discussed previously, although the actual dynamics and mechanisms are a lot more complicated. This is discussed in detail in Section 11.6.2.

11.5.4.   Nondegenerate Optical Wave Mixing: Harmonic Generations

Nondegenerate optical wave mixing, where the frequencies of the incident and the generated waves are different, are due to nonlinear polarization of the form given by Equation (11.35). For example, the incident frequencies ω_1, ω_2, and ω_3 can be combined to create new frequencies $\omega_4 = \omega_1 \pm \omega_2 \pm \omega_3$ involving sums or differences. These wave mixing processes are sometimes termed sum-difference frequency generations.

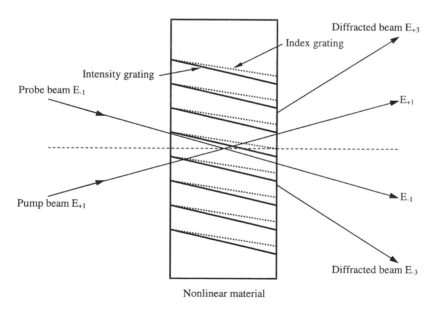

Figure 11.9.   Time-dependent spatially moving index grating generated by two coherent equal-frequency short laser pulses.

In this section we discuss two basically different sum frequency generation processes, namely, second- and third-harmonic generations. These processes are associated with nonlinear polarizations of the following:

Second-harmonic generation:

$$P_i^{2\omega} = \tfrac{1}{2}\left[\chi_{ijk}^{(2)}(-2\omega,\omega,\omega)A_j^{\omega}A_k^{\omega}\right] \tag{11.82}$$

Third-harmonic generation:

$$P_i^{3\omega} = \tfrac{1}{4}\left[\chi_{ijkl}^{(3)}(-3\omega,\omega,\omega,\omega)A_j^{\omega}A_k^{\omega}A_l^{\omega}\right], \tag{11.83}$$

where i, j, and k refer to the crystalline axes, and a sum over $j = 1, 2, 3$ and $k = 1, 2, 3$ in the preceding equation is implicit.

The main fundamental difference between these two processes is that second-harmonic generations involve a polarization that is *even order* in the electric field (i.e., $P \sim EE$), whereas third-harmonic generations involve an *odd-order* polarization ($P \sim EEE$). Consider a centrosymmetric nonlinear medium. If we invert the coordinate system, then Equation (11.82) becomes

$$\tilde{P}_i^{2\omega} = \tfrac{1}{2}\tilde{\chi}_{ijk}^{(2)}\tilde{A}_j^{\omega}\tilde{A}_k^{\omega}. \tag{11.84}$$

Since the polarization and the electric field are vectors, they must change signs (i.e., $\tilde{P}_i^{2\omega} = -P_i^{2\omega}$, $\tilde{A}_i = -\tilde{A}_i$, $\tilde{A}_i \sim -A_i$), whereas $\tilde{\chi}_{ijk}^{(2)} = \chi_{ijk}$ since it is a measure of the material response. Equation (11.84) therefore gives

$$P_i^{2\omega} = -\tfrac{1}{2}\chi_{ijk}^{(2)}A_j^{\omega}A_k^{\omega}. \tag{11.85}$$

Comparing Equations (11.84) and (11.85), the only logical conclusion is that $\chi_{ijk}^{(2)} = 0$ in a medium possessing centrosymmetry. Using similar symmetry considerations, one can show that in centrosymmetric media, all even-power nonlinear susceptibilities are vanishing. Second-, fourth-, and so on harmonic generations are possible therefore only in nonlinear media that are noncentrosymmetric.

The other fundamentally important issue is phase matching, owing to the much larger phase mismatch between the incident fundamental wave at frequency ω and the generated second or third harmonic at 2ω or 3ω respectively. In the case of second-harmonic generation, the phase mismatch Δk in the coupled wave equation [see Eq. (11.59)] is given by

$$\Delta \vec{k} = \vec{k}_{2\omega} - 2\vec{k}_{\omega},$$

with a magnitude

$$\Delta k = \frac{2\pi n(2\omega)}{\lambda/2} - 2\left(\frac{2\pi}{\lambda}\right)n(\omega) = \frac{4\pi}{\lambda}\left[n(2\omega) - n(\omega)\right]. \qquad (11.86)$$

One possibility to achieve phase matching (i.e., $\Delta k = 0$) is to have the fundamental and the harmonic waves propagate as different types of waves (extraordinary and ordinary); that is, utilize the birefringence of the medium. For example, one can have the fundamental wave as an ordinary wave with a refractive index n_o^ω that is independent of the direction of propagation, while the harmonic wave is an extraordinary wave propagating at an angle θ_m with respect to the optical axis; that is, its refractive index is $n_e^{2\omega}(\theta_m)$ such that

$$n_e^{2\omega}(\theta_m) = n_o^\omega. \qquad (11.87)$$

This may be seen in Figure 11.10, which is drawn for a crystal, where $n_e^{2\omega} < n_o^{2\omega}$ (i.e., a negative uniaxial crystal where $n_e < n_o$). From Equations (11.86) and (11.87) we have

$$\frac{1}{\left[n_o^\omega\right]^2} = \frac{1}{\left[n_e^{2\omega}(\theta_m)\right]^2} = \frac{\cos^2\theta_m}{\left(n_o^{2\omega}\right)^2} + \frac{\sin^2\theta_m}{\left(n_e^{2\omega}\right)^2}, \qquad (11.88)$$

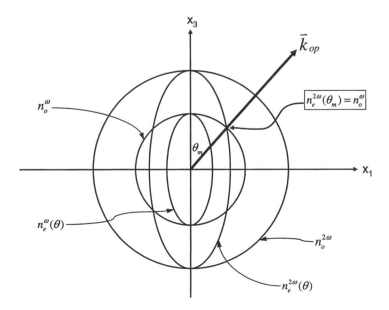

Figure 11.10.   Matching the extraordinary refractive index $n_e(\theta m)$ of the second harmonic (at 2ω) with the ordinary refractive index n_o of the fundamental (at ω).

which gives, for $n_e < n_o$,

$$\sin^2 \theta_m = \frac{\left(n_o^\omega\right)^{-2} - \left(n_o^{2\omega}\right)^{-2}}{\left(n_e^{2\omega}\right)^{-2} - \left(n_o^{2\omega}\right)^{-2}}. \tag{11.89}$$

By drawing a figure similar to Figure 11.10 and going through the preceding analysis for a *positive* ($n_e > n_o$) uniaxial crystal, the reader can readily show that the phase-matching angle for this case is given by [from equating $n_o^{2\omega}$ to $n_e^\omega (\theta)$]

$$\sin^2 \theta_n = \frac{\left(n_o^\omega\right)^{-2} - \left(n_o^{2\omega}\right)^{-2}}{\left(n_o^\omega\right)^{-2} - \left(n_e^\omega\right)^{-2}}. \tag{11.90}$$

11.5.5. Self-Focusing and Self-Phase Modulation

The passage of a laser beam through a nonlinear optical material is inevitably accompanied by an intensity-dependent phase shift on the wave front of the laser, as a result of the intensity-dependent refractive index. Since the intensity of a laser beam is a spatially varying function, the phase shift is also spatially varying. This, together with large amplitude changes, leads to severe distortions on the laser in the form of self-focusing, defocusing, trapping, beam breakups, filamentations, spatial ring formations, and others.

For thick media the problem of calculating the beam profile, by solving Equation (11.45) or (11.47) within the medium, is extremely complicated. Numerical solutions are almost always the rule.[1]

If the material is thin (i.e., in nematic liquid crystal film), the problem of optical propagation is greatly simplified. In its passage through such a nonlinear film, the laser beam is only phase modulated and suffers negligible amplitude change. In other words, the transmitted far-field intensity pattern is basically a diffraction pattern of the incident field, with the nonlinear film playing the role of an intensity-dependent phase screen.

Consider now a *steady-state self-phase modulation* effect (e.g., that induced by a cw laser). If the response of the medium to the laser field is local, the spatial profile of the phase shift follows that of the laser; that is, given the laser transverse profile $I(r)$ (assuming cylindrical symmetry), the nonlinear part of the phase shift is simply $\phi(r) = d\Delta n(r)2\pi/\lambda$, where d is the thickness of the film.

If the incident laser is a Gaussian, that is,

$$I(\text{laser}) = I_0 \exp\left(-\frac{2r^2}{\omega^2}\right), \tag{11.91}$$

the nonlinear (intensity-dependent) phase shift ϕ_{NL} is given by

$$\phi_{NL} = kn_2 dI_0 \exp\left(-\frac{2r^2}{\omega^2}\right), \tag{11.92}$$

where $k = 2\pi/\lambda$ and ω is the laser beam waist. This is the phase shift imparted on the laser upon traversing the nonlinear film.

On the exit side (see Fig. 11.11) at a distance Z from the film, the intensity distribution on the observation plane P is given by the Kirchhoff diffraction integral:[9,10]

$$I(r_1,Z) = \left(\frac{2\pi}{\lambda Z}\right)^2 I_0 \left| \int_0^\infty r\,dr\,J_0\left(\frac{2\pi r r_1}{\lambda Z}\right) \times \exp\left(-\frac{2r^2}{\omega^2}\right) \exp\left[-i(\phi_D + \phi_{NL})\right] \right|^2 , \quad (11.93)$$

where the diffraction phase ϕ_D is given by

$$\phi_D = k\left(\frac{r^2}{2Z} + \frac{r^2}{2R}\right). \quad (11.94)$$

Using the following definitions:

$$y = \frac{r}{\omega}, \quad (11.95a)$$

$$C_1 = \frac{4\pi}{(\lambda Z)^2}\omega^4, \quad (11.95b)$$

$$C_2 = \frac{2\pi}{\lambda}\omega\tan\alpha_0, \quad \alpha = \frac{\lambda}{\pi\omega_0}, \quad (11.95c)$$

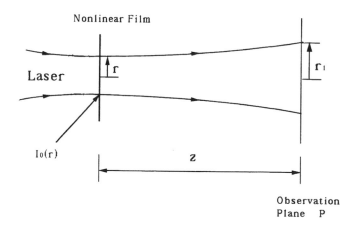

Figure 11.11.   Nonlinear transverse phase shift and diffraction of a laser beam by a thin nonlinear film.

$$C_a = \frac{2\pi}{\lambda} \bar{n}_2 dI_0,$$

(11.95d)

$$C_b = \frac{\pi}{\lambda} \omega^2 \left(\frac{1}{Z} + \frac{1}{R} \right),$$

(11.95e)

and

$$\theta = \frac{r_1}{Z \tan \alpha_0},$$

(11.95f)

we can rewrite Equation (11.93) as

$$I(r_1, Z) = C_1 I_0 \left| \int_0^\infty \exp(-2y^2) y \times \exp\{-i[C_a \exp(-2y^2) + C_b y^2]\} J_0(C_2 \theta y) dy \right|^2.$$

(11.96)

Khoo et al.[9] discuss in detail the various types of intensity distributions and how they evolve as a function of the input intensity. Essentially, the principal deciding factor is the sign of C_b (i.e., the sign of $1/Z + 1/R$).

If $n_2 > 0$, we have the following cases:

1. $(1/Z + 1/R) < 0$ case. The intensity distribution evolves from a Gaussian beam to one where the central data area tends to be dark (low intensity) surrounded by bright rings (see Fig. 11.12).
2. $(1/Z + 1/R) > 0$ case. The intensity distribution evolves as the incident intensity I_0 is increased, from a Gaussian shape to one with a bright central spot and concentric bright and dark rings (see Fig. 11.13).
3. $(1/Z + 1/R) \approx 0$ case. This is the so-called intermediate case, where the intensity distribution at the center area assumes forms that are intermediate between those of cases 1 and 2.

If $n_2 < 0$, the intensity distribution is almost a "mirror" image of the $n_2 > 0$ case; that is, the distribution for cases 1 and 2 are interchanged.

Such intensity redistribution effects, in which the transmitted on-axis laser intensity is reduced owing to the nonlinear self-defocusing effect caused by the laser-induced index change, may be employed for optical limiting applications.[11] A simple setup is shown in Figure 11.14. The aperture situated at the observation plane P will transmit totally a low-power laser, as well as a signal/image. If the laser is intense, the defocusing effect will reduce the transmission through the aperture, thereby protecting the detector or sensor placed behind the aperture.

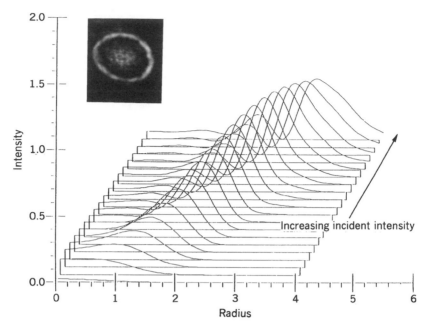

Figure 11.12. Radial intensity as a function of increasing input intensity showing self-phase-modulation-induced redistribution of the axial intensity to the radial portion. Insert is a photograph of the observed pattern on the observation plane P as a result of the passage of a laser beam through a nematic film.

For self-limiting application dealing with a pulsed laser, a dynamical theory is needed. While a dynamical theory is extremely complicated (almost insoluble) for self-focusing effects in long media, such a theory is relatively easier to construct for *thin* nonlinear media.

Consider, for example, a Gaussian–Gaussian incident laser pulse:

$$I(\vec{r},t) = I_0 \exp\left(-\frac{2r^2}{\omega^2}\right)\exp\left(-\frac{t^2}{\tau_p^2}\right) \tag{11.97}$$

(cf. the input pulse shape of Fig. 11.15). The intensity distribution at the observation plane can be deduced from the diffraction theory discussed previously:

$$I(\vec{r}_p,z,t) = \left(\frac{2\pi}{\lambda z}\right)^2 I_0 \exp\left(-\frac{t^2}{\tau_p^2}\right)$$

$$\times \left|\int_0^\infty r\,dr J_0\left(\frac{2\pi r r_p}{\lambda z}\right)\exp\left(-\frac{2r^2}{\omega^2}\right)\times \exp[-i(\Delta\phi_D + \Delta\omega_{\mathrm{NL}})]\right|^2, \tag{11.98}$$

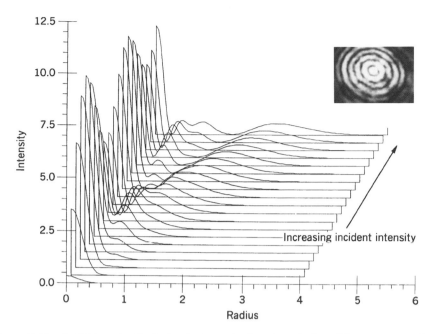

Figure 11.13. Radial intensity distribution for increasing input intensity for a different geometry showing the intensification of the axial laser intensity. Insert is a photograph of the actual pattern using a nematic film.

where the diffractive phase shift is

$$\Delta\phi_D = k\left(\frac{r^2}{2z} + \frac{r^2}{2R}\right),\tag{11.99a}$$

and the nonlinear phase shift is

$$
\begin{aligned}
\Delta\phi_{NL}(\vec{r}_p, z, t) &= kd\,\Delta n(t)\\
&= kd\sum_t\left[\frac{1}{\tau_i}\int_{-\infty}^{t} n_{2i}(t')I(\vec{r}_p, z, t')dt'\right]\\
&= kdI_0\exp\left(-\frac{2r^2}{\omega^2}\right)\sum_i\left[\frac{1}{\tau_i}\int_{-\infty}^{t} n_{2i}(t')\exp\left(-\frac{t'^2}{\tau_p^2}\right)dt'\right],
\end{aligned}\tag{11.99b}
$$

where $n_{2i}(t')$ are the time-dependent nonlinear coefficients associated with all the contributing nonlinear mechanisms.

The final form for $\Delta\phi_{NL}(t)$, as well as the output intensity distribution on the observation plane P, depends on the form of $n_{2i}(t')$ and the response time t_i versus the laser pulse duration t_p. If the response time t_i is longer than t_p, the self-limiting effect is basically an integrated effect, and thus it is more appropriate to discuss the process in terms of energy/area rather than intensity (power/area).

Figure 11.15 shows a theoretical plot of the input and output on-axis intensity profiles for the case of a single nonlinearity with $t_i \gg t_p$. It shows that as the intensity-dependent phase modulation effect is accumulated in time, the output intensity is progressively reduced, finally to a highly diminished value. The energy contained in

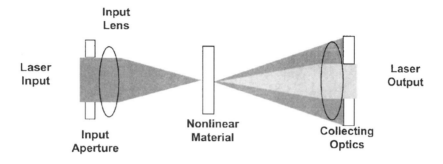

Figure 11.14.   An experimental arrangement for achieving the optical limiting effect using the self-phase modulation or self-focusing/self-defocusing effect. See chapter 12, however, for a fiber array limiting device that does not suffers from small field of view associated with this configuration involving pin-hole.

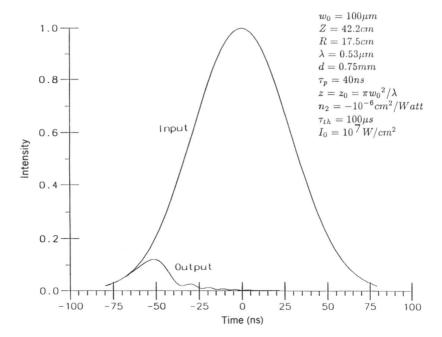

Figure 11.15.   Theoretical plot of the output (energy limited) and input laser pulses.

the transmitted pulse, when plotted against the input energy, will also exhibit a limiting behavior.

11.6. STIMULATED SCATTERINGS

Stimulated scatterings occupy a special place in nonlinear optics. They have been investigated ever since the dawn of nonlinear optics and lasers. Devoting only a section to the discussion of this subject is not likely to do justice to the voluminous work that has been done in this area. The main objective of this section, however, is to outline the fundamental principles involved in these scattering processes in the context of the wave mixing processes discussed previously.

In this section, we will discuss three exemplary types of stimulated scattering processes:

1. Stimulated Raman scattering (SRS)
2. Stimulated Brillouin scattering (SBS)
3. Stimulated orientational (SOS) and thermal (STS) scattering

They correspond to three distinctive types of material excitations. In Raman processes, as in orientational and thermal processes, the excitations are not propagative. On the other hand, Brillouin processes involve propagating sound waves. These differences will be reflected in the coupling and propagation of the incident and generated optical waves in the material.

11.6.1. Stimulated Raman Scatterings

Stimulated Raman scatterings involve the nonlinear interaction of the incident and generated lasers with the vibrations or rotations (or other Raman transitions) of the material. These Raman transition involve two energy levels of the material connected by a two-photon process, as discussed in Chapter 5.

From Equation (5.51) the induced polarization associated with Raman scattering is of the form

$$P_{\text{ind}} \sim q_v E \cdot \varepsilon_0 \left(\frac{\partial \alpha}{\partial q} \right)_{q_0} \left(\frac{N}{V} \right), \tag{11.100}$$

where (N/V) is the density of the molecules. For simplicity, we also neglect the vector nature of the field $\vec{E}$ and the polarization $\vec{P}$. The displacement of the normal coordinate q_v obeys an equation of motion for a forced harmonic oscillator, which is of the form

$$\ddot{q}_v^0 + \omega_R q = F, \tag{11.101}$$

where the force F is related to the optical field molecular interaction energy u (owing to the induced polarization) given by

$$F = \frac{\partial u}{\partial q_v},$$

(11.102)

where

$$u = \frac{1}{2}\vec{P}_{\text{ind}} \cdot \vec{E} = \frac{1}{2}q_v\vec{E} \cdot \vec{E}\varepsilon_0\left(\frac{\partial \alpha}{\partial q}\right)_{q_0},$$

(11.103)

that is,

$$F \propto E^2.$$

(11.104)

From Equation (11.101),

$$q_v \sim E^2.$$

(11.105)

Therefore, the induced polarization, from Equation (11.100), is of the form

$$P_{\text{ind}} \sim E^3.$$

(11.106)

In other words, the nonlinear polarization associated with stimulated Raman scattering is a third-order one.

To delve further into this process, we note from Equations (11.101) and (11.105) that, because of the vibrational frequency $\omega_R = \omega_2 - \omega_1$, only terms in $F \propto E^2$ containing the frequency component $\omega_2 - \omega_1$ (i.e., $E_2 E_1^*$ or $E_1 E_2^*$) contribute to q_v. In other words,

$$q_v \sim \left(E_2 E_1^* + \text{c.c.}\right).$$

(11.107)

The induced third-order nonlinear polarization P_{NL} is therefore of the form

$$P_{\text{NL}} \sim \left(E_2 E_1^* + \text{c.c.}\right)\left(E_1 + E_2 + \text{c.c.}\right).$$

(11.108)

The nonlinear polarization therefore produces scattering at various frequencies $\omega_1, \omega_2, 2\omega_1 - \omega_2$ and $2\omega_2 - \omega_1$.

Focusing our attention on the Stokes waves at frequency ω_2 the nonlinear polarization is given by

$$P \sim \left(E_2 E_1^*\right)E_1. \tag{11.109}$$

Using the formalism given in Section 11.2, we can thus write

$$
\begin{aligned}
P^{\omega_2} &= \frac{1}{2^2}\chi_s\left(-\omega_2,\omega_2,-\omega_1,\omega_1\right)A^{\omega_2}A^{*-\omega_1}A^{\omega_1} \\
&= \frac{1}{2^2}\chi_s\left(-\omega_2,\omega_2,-\omega_1,\omega_1\right)A^{\omega_2}\left|A^{\omega_1}\right|^2,
\end{aligned}
\tag{11.110}
$$

where the subscript s on χ_s signifies that it is for the Stokes wave. χ_s can be deduced from Equations (11.100) – (11.107).

The equation describing the amplitude A_2 of the Stokes wave, from Section 11.5, becomes

$$\frac{\partial A_2}{\partial z} = ig|A_1|^2 A_2, \tag{11.111}$$

where

$$g = \frac{\mu_0 \omega_2^2}{2k_2}\left(\frac{\chi_s}{4}\right). \tag{11.112}$$

Accounting for the loss α (due to random scatterings, absorption, etc.) experienced by E_2 in traversing the medium, Equation (11.111) becomes

$$\frac{\partial A_2}{\partial z} = \left[ig|A_1|^2 - \alpha\right]A_2. \tag{11.113}$$

Since the nonlinear susceptibility χ_s is complex, we may write it as

$$\chi_s = \chi_s' - i\chi_s''. \tag{11.114}$$

Substituting Equation (11.114) in Equation (11.111), we can readily show that the amplitude A_2 will grow exponentially with z, with a *net gain* constant given by

$$\bar{g}_s = \left(\frac{\mu_0 \omega_2^2}{2k_2} \frac{\chi_s''}{4} - \alpha \right). \tag{11.115}$$

In the work by Boyd,[2] the Stokes susceptibility χ_s (sometimes called the Raman susceptibility) is explicitly derived for a multilevel system. However, in practice, it is easier and quantitatively correct to relate the gain constant to the scattering cross section as indicated in Chapter 5. The reader is reminded of the role played by the population difference $N_1 - N_2$ in χ_s, which could reverse sign depending on N_1 and N_2.

From Equation (11.108), as we mentioned earlier, other frequency components besides the Stokes waves can also be generated. An example is the so-called coherent anti-Stokes Raman scattering (CARS), which involves the mixing of the incident laser with its (phase coherent) Stokes waves (see Fig. 11.16). The nonlinear polarization is of the form

$$P_{NL} \sim \tfrac{1}{4} \chi(-\omega_a, \omega_1, \omega_1, -\omega_2) \bar{A}_1 \bar{A}_1 \bar{A}_2^* \tag{11.116}$$

with a phase mismatch

$$\Delta \vec{k} = \left[\vec{k}_a - (2k_1 - k_2) \right]. \tag{11.117}$$

Notice that this is the analog of the degenerate four-wave mixing process discussed in Section 9.5; the anti-Stokes beam assumes the role of the diffracted beam E_3 (cf. Fig. 11.4).

11.6.2. Stimulated Brillouin Scatterings

Brillouin scatterings are caused by acoustic waves in the material. In analogy to Raman scattering, the nonlinear polarizations responsible for the coupled-wave equations are obtainable from consideration of the appropriate material excitations.

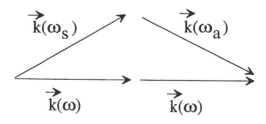

Figure 11.16. Coherent anti-Stokes Raman scattering (CARS) involving the pump beam at ω and its Stokes wave ω_s.

There are, in fact, many distinctly different physical mechanisms that will alter the density ρ of a material, such as pressure p, entropy S, or temperature T. These parameters are in most cases strongly coupled to one another. Accordingly, there are quite a variety of stimulated Brillouin scatterings depending on the underlying physical mechanisms.

In this section we focus our attention on *nonabsorptive materials*, where the density change is due to the spatially and temporally varying strain caused by a corresponding optical density. An example of such an intensity pattern is the one obtained by the interference of two lasers of different frequencies ω_1 and ω_2 (see Fig. 11.8). Analogous to Raman scattering, the frequency difference $\omega_1 - \omega_2$ should match the acoustic frequency.

There is one very important difference between Raman scattering and Brillouin scattering. Raman scattering is associated with *nonpropagating* material oscillations; Brillouin scattering, however, involves the creation of propagating sound waves. Since the wave vector $\vec{k}_s$ of the sound wave carries a momentum $\hbar \vec{k}_s$, the resulting directions of the generated acoustic wave (at $\vec{k}_s$ and ω_s) and the frequency downshifted optical wave ($\vec{k}_2, \omega_2 = \omega_1 - \omega_s$) must obey the wave vector matching conditions. The equation governing the induced density change $\Delta\rho$ is, in SI units, given by,[2,12]

$$\left(\frac{\partial^2}{\partial t^2} + 2\Gamma_B \frac{\partial}{\partial t} - v_s^2 \nabla^2 \right) \Delta\rho = \nabla \cdot \left(-\frac{\gamma_e}{2} \nabla (E \cdot E) \right). \qquad (11.118)$$

If we write $\Delta\rho$ in terms of a slowly varying envelope $\bar{\rho}$ as

$$\Delta\rho = \frac{1}{2} \left[\bar{\rho}(\vec{r}_s) \exp\left[i(\vec{k}_s \cdot \vec{r} - \omega_s t) \right] + \text{c.c} \right], \qquad (11.119)$$

then ∇ in Equation (11.118) is given by $\nabla \equiv \partial/\partial r_s$. In Equation (11.118) v_s^2 is the sound velocity and Γ_B is the damping constant.

Writing now the optical fields in terms of their amplitudes A_1 and A_2 in plane wave form

$$E_1 = \frac{1}{2} \hat{x}_1 \left[A_1 \exp\left[i(\vec{k}_1 \cdot \vec{r} - \omega_1 t) \right] + \text{c.c.} \right], \qquad (11.120)$$

$$E_2 = \frac{1}{2} \hat{x}_2 \left[A_2 \exp\left[i(\vec{k}_2 \cdot \vec{r} - \omega_2 t) \right] + \text{c.c.} \right], \qquad (11.121)$$

and substituting Equations (11.119)−(11.121) into Equation (11.118) and using the slowly varying envelope approximations

$$\left| \frac{d^2\bar{\rho}}{dr_s^2} \right| \ll \left| k_s^2 \rho \right|, \left| k_s \frac{d\bar{\rho}}{dr_s} \right|, \qquad (11.122)$$

Equation (11.118) becomes

$$\left(\frac{\partial}{\partial r_s} + \frac{\Gamma_B}{v_s}\right)\bar{\rho} = \frac{ik_s\gamma_e(\hat{x}_1 \cdot \hat{x}_2)}{4v_s^2}A_1A_2^*e^{i\Delta kr_s},\qquad(11.123)$$

where $\Delta k = |k_1 - k_2 - k_3|$.

This is the material excitation equation. Then the coupled optical wave equations for A_1 and A_2 can be obtained by identifying the nonlinear polarization associated with the density change $\Delta\rho$. This is given by

$$P_{NL} = \Delta\varepsilon E = \left(\frac{\partial\varepsilon}{\partial\rho}\cdot\Delta\rho\right)E.$$
$$(11.124)$$

Since $\Delta\rho$ oscillates as $e^{-i(\omega_1 - \omega_2)t}$ we have

$$\frac{P_{NL}^{\omega_1}}{2} = \frac{\partial\varepsilon}{\partial\rho}\left(\frac{\bar{\rho}}{2}\right)\left(\frac{A_2}{2}\right)$$
$$(11.125)$$

and

$$\frac{P_{NL}^{\omega_2}}{2} = \frac{\partial\varepsilon}{\partial\rho}\left(\frac{\bar{\rho}}{2}\right)^*\left(\frac{A_1}{2}\right).$$

The nonlinear coupled-wave equations for A_1 and A_2 are, from Equation (11.47),

$$2ik_1\frac{\partial A_1}{\partial r_s} = \frac{-\mu_0\omega_1^2}{2}\frac{\partial\varepsilon}{\partial\rho}\bar{\rho}A_2e^{-i\Delta kr_s},\qquad(11.126)$$

$$2ik_2\frac{\partial A_2}{\partial r_s} = \frac{-\mu_0\omega_2^2}{2}\frac{\partial\varepsilon}{\partial\rho}\bar{\rho}^*A_1e^{i\Delta kr_s},\qquad(11.127)$$

where we have replaced z by r_s as the reference coordinate for propagation. We have also ignored the $\nabla^2 A$ term with our plane wave approximation. With a little rewriting, Equations (11.126) and (11.127) become

$$\frac{\partial A_1}{\partial r_s} = -\alpha_1 A_1 + ig_1\bar{\rho}^*A_2e^{-i\Delta kr_s},\qquad(11.128)$$

$$\frac{\partial A_2}{\partial r_s} = -\alpha_2 A_2 + ig_2 \bar{\rho}^* A_1 e^{+i\Delta k r_s},$$ (11.129)

and Equation (11.123) becomes

$$\frac{\partial \bar{\rho}}{\partial r_s} = \frac{-\Gamma_B}{v_s} \bar{\rho} + ig_s A_1 A_2^* e^{i\Delta k r_s}.$$ (11.130)

In Equations (11.128) and (11.129) we have included the phenomenological loss terms $-\alpha_1 A_1$ and $-\alpha_2 A_2$ respectively, to account for the losses (e.g., absorptions or random scatterings) experienced by A_1 and A_2 in traversing the medium.

The coupling constants g_1, g_2, and g_s are given by

$$g_1 = \frac{\mu_0 \omega_1^2}{4k_1} \left(\frac{\partial \varepsilon}{\partial \rho} \right),$$ (11.131)

$$g_2 = \frac{\mu_0 \omega_2^2}{4k_2} \left(\frac{\partial \varepsilon}{\partial \rho} \right),$$ (11.132)

$$g_s = \frac{k_s \gamma_e}{4v_s^2}.$$ (11.133)

Because of the nonlinear couplings among A_1, A_2, and ρ, solutions to Equations (11.128)–(11.130) are quite complicated. A good glimpse into stimulated Brillouin scattering may be obtained if we solve for the density wave amplitude $\bar{\rho}$ in a perturbative manner.

Ignoring the r_s dependence of A_1 and A_2, Equation (11.130) can be integrated to give

$$\bar{\rho} = \left(\frac{g_s A_1 A_2^*}{\Delta k - i\Gamma_B / v_s} \right) e^{i\Delta k r_s}.$$ (11.134)

Substituting the complex conjugates of Equation (11.134) into Equation (11.129), we get

$$\frac{\partial A_2}{\partial r_s} = -\alpha_2 A_2 + \frac{g_2 g_s |A_1|^2 (\Gamma_B / v_s) A_2}{(\Delta k)^2 + (\Gamma_B / v_s)^2} + \frac{i\Delta k g_2 g_s |A_1|^2 A_2}{(\Delta k)^2 + (\Gamma_B / v_s)^2}.$$ (11.135)

If $|A_1|^2 =$ (i.e., the pump beam is so strong that its depletion is negligible), the solution for the generated wave A_2 at frequency $\omega_2 = \omega_1 - \omega_s$ is an exponential function, with a gain constant

$$G_{sB} = \frac{g_2 g_s |A_1|^2 (\Gamma_B/v_s)}{(\Delta k)^2 + (\Gamma_B/v_s)^2} - \alpha_2 = g_{sB} - \alpha. \tag{11.136}$$

In other words, starting from some weak initial (spontaneous) Brillouin scattering, greatly enhanced stimulated emission at ω_2 will be produced as it propagates along with the pump wave at ω; a strong sound wave at ω_s will also be generated. As shown in Figure 11.17, the directions of propagation of these waves are related by the phase-matched condition ($\Delta k = 0$, which gives $k_1 = k_2 + k_s$ or $k_1 - k_2 = k_s$).

In Brillouin scattering, owing to the large difference between the acoustic and optical frequencies ($\omega_s \ll \omega_1, \omega_2$), usually $\omega_2 \approx \omega_1$ and therefore $k_2 \approx k_1$. From Figure 11.17, $k_2 \approx k_1$ means that

$$k_s \approx 2k_1 \sin\left(\frac{\theta}{2}\right), \tag{11.137}$$

which is maximal at $\sin(\theta/2) = 1$ (i.e., $\theta = 180°$).

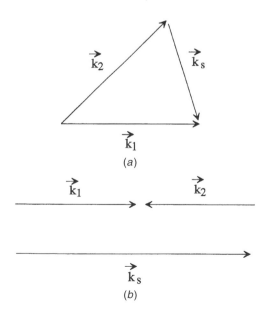

Figure 11.17. (a) Wave vector phase matching of the pump wave $\vec{k}_1$, the generated wave $\vec{k}_2$ and sound wave $\vec{k}_s$. (b) Direction of incident and stimulated Brillouin waves. Optimal gain occurs if the generated wave is counterpropagating to the incident pump wave.

Since g_s in G_{sB} is proportional to k_s, we therefore conclude that Brillouin gain is maximal if the generated wave is backward propagating. The maximum value for g_{sB}, (for $\Delta k = 0$, $\theta = 180°$) is given by

$$g_{sB} = \frac{g_s g_2 |A_1|^2}{(\Gamma_B/v_s)} \qquad (11.138)$$

$$= \frac{\gamma_e \mu_0 \omega_{12}^2 k_1}{16 v_s \Gamma_B k_2} \left(\frac{\partial \varepsilon}{\partial \rho}\right) |A_1|^2. \qquad (11.139)$$

As one may expect, the gain is governed mainly by the electrostrictive coefficient γ_e, $(\partial \varepsilon/\partial \rho)$ and the incident laser intensity $|A_1|^2$. Note that since $\gamma_e = \rho_0 (\partial \varepsilon/\partial \rho)$, the factor $\gamma_e (\partial \varepsilon/\partial \rho)$ in Equation (11.139) can be written as $\rho_0 (\partial \varepsilon/\partial \rho)^2$ or (γ_e^2/ρ_0).

11.6.3. Stimulated Orientational Scattering in Liquid Crystals

In liquid crystals, laser induced director axis reorientations lead to the possibility of another type of stimulated scattering process,[13-15] in which the spontaneous noise originates from scattering of the incident light by director axis fluctuations, as depicted schematically in Figure 11.18. Just as in the case of stimulated Brillouin scattering, the incident pump beam interferes coherently with its scattered noise, forming moving gratings along the direction of propagation.

Consider, for example, the case of an extraordinarily polarized laser incident on a planar nematic film with the optical electric field $E_e - e_x E_e e^{i(k_e z - \omega_1 t)}$ parallel to the director axis. Due to thermal fluctuations of the director axis, an ordinary polarized wave (noise) component $E_o = e_y E_o^{noise} e^{i(k_o z - \omega_2 t)}$ is generated at a spectrum of frequencies ω_2's. Here e_x and e_y are unit vectors along the polarization directions of the extraordinary and ordinary waves (e and o waves), E_e and E_o are their amplitudes, and $k_e = (\omega/c)n_e$ and $k_o = (\omega/c)n_o$ are the wave vectors of the e and o waves, respectively.

The difference in the wave vectors of the e and o waves results in modulation of the polarization state of the light in the NLC. Consequently, the torque acting on the NLC director is also modulated, leading to modulation of director axis reorientation, cf. Figure 11.18 with a grating wave vector

$$q = \frac{2\pi}{\lambda}(n_e - n_o). \qquad (11.140)$$

This grating will mediate stimulated scattering between the ordinary and extraordinary wave components. In analogy to stimulated Brillouin scattering, the growth of the scattered wave component is dependent on the intensity of the incident laser, and

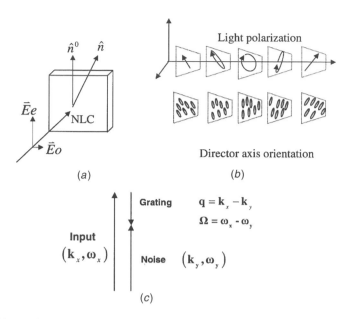

Figure 11.18.   (a) Scattering of a polarized laser by director axis fluctuations in a NLC; (b) Optical polarization grating formed by coherent pump beam and the scattered orthogonal polarized component and the resulting director axis reorientation grating; (c) Wave vector matching of the input pump, the scattered noise and the orientation grating for maximum energy exchange.

will become stimulated once the intensity reaches a threshold at which the gain exceeds the loss. The major difference here is the threshold intensity and the interaction length required for the process to initiate. Compared to other nonlinear optical materials which require cm's interaction length with MW (Megawatt) laser power, in nematic liquid crystals, the process can be realized with microns thick sample using mW (milliwatt) laser power—thus enabling the application of the process for polarization conversion in microphotonic devices.

To analyze the process quantitatively, the starting point is the Euler–Lagrange equation for the free energy f of the system of NLC film under an applied field:[15]

$$\frac{\partial}{\partial x_j}\frac{\delta f}{(\partial U_m/\partial x_j)} - \frac{\delta f}{\delta U_m} - \frac{\delta R}{\delta U_m} = 0. \tag{11.141}$$

Taking $\theta(z,t)$ as the independent variable, the equation governing the reorientation angle $\theta(z,t)$ becomes

$$\eta\frac{\partial\theta(z,t)}{\partial t} - K_2\frac{\partial^2\theta(z,t)}{\partial z^2} = \frac{1}{2}\varepsilon_a\big[\cos[2\theta(z,t)](E_xE_y^* + E_x^*E_y) \\ + \sin[2\theta(z,t)](|E_y|^2 - |E_x|^2)\big]. \tag{11.142}$$

The orientation angle follows the interference grating between the input e wave (E_x) and its scattered noise component o wave (E_y), and may be expressed as $\theta(z,t) = \theta(z,t)e^{i(\vec{q}\cdot\vec{r}-\Omega t)} + $ c.c. Using the slowly varying envelop approximation, that is, $\partial^2\theta(z,t)/\partial z^2$ and $[q\ \partial\theta(z,t)/\partial z]$ are negligible compared with $q^2\theta\ (z,t)$, the equations of electric field and director reorientation become

$$\frac{dE_x(z)}{dz} = \frac{i\varepsilon_a}{2k_x}\left[-2\,|\theta(z)|^2\,E_x(z)\frac{\omega_x^2}{c^2} + E_y(z)\frac{\omega_y^2}{c^2}\theta^*(z)\right] - \frac{\alpha}{2}E_x, \quad (11.143)$$

$$\frac{dE_y(z)}{dz} = \frac{i\varepsilon_a}{2k_y}\left[2\,|\theta(z)|^2\,E_y(z)\frac{\omega_y^2}{c^2} + E_x(z)\frac{\omega_x^2}{c^2}\theta(z)\right] - \frac{\alpha}{2}E_y, \quad (11.144)$$

$$\left[K_2q^2 - i\eta\Omega + \frac{1}{2}\varepsilon_a(|E_y|^2 - |E_x|^2)\right]\theta(z,t) = \frac{1}{2}\varepsilon_aE_xE_y^*. \quad (11.145)$$

(1) Steady-State Small Signal Regime. In the small signal ($E_y \ll E_x$, $\theta \ll 1$) regime, usually the terms containing $|\theta|^2$ on the right-hand side (RHS) of Equations (11.143) and (11.144) and the third term on the left-hand side (LHS) of Equation (11.145) are ignored. Ignoring losses, in this limit, the above equations give

$$\theta(z) \sim E_xE_y^*,$$

$$\frac{dE_x(z)}{dz} \sim \frac{i\varepsilon_a}{2k_x}\left[E_y(z)\frac{\omega_y^2}{c^2}\theta^*(z)\right] \sim g\,|E_y|^2\,E_x(z),$$

$$\frac{dE_y(z)}{dz} \sim \frac{i\varepsilon_a}{2k_x}\left[E_y(z)\frac{\omega_y^2}{c^2}\theta^*(z)\right] \sim -\,g\,|E_x|^2\,E_y(z). \quad (11.146)$$

These equations show that the noise component E_x will grow with an exponential constant $g|E_y|^2$, that is, proportional to the intensity of the pump beam. If we denote the intensity of the o-wave noise beam by I_o, the solution of the above equation yields[13–15]

$$I_o(z=L) = I_o^{(\text{noise})}\,e^{GI_ez}, \quad (11.147)$$

where

$$G = f(\Omega)\frac{\pi\varepsilon_a^2}{2cn_e\lambda K_2q^2}, \quad f(\Omega) = \frac{2\Omega/\Gamma}{1+\Omega^2/\Gamma^2}, \quad \Gamma = \frac{K_2q^2}{\eta}, \quad (11.148)$$

where $\varepsilon_a = n_e^2 - n_o^2$ is the optical anisotropy of the NLC, K_2 is the elastic constant of the NLC, $\Omega = \omega_1 - \omega_2$ is the frequency difference between the pump and signal waves, and η is the orientational viscosity. The gain coefficient is maximal, $f(\Omega_m) = 1$, for an optimal frequency shift of the scattered wave from the frequency of the incident beam, $\Omega_m = \Gamma$.

The gain maximum coefficient G_m given in Equation (11.148) can be rewritten as

$$G_m\left(\text{cm} \cdot \text{W}^{-1}\right) \cong \frac{(n_e + n_o)^2 \lambda}{8\pi c n_e K_2}. \qquad (11.149)$$

Equation (11.149) shows that G is independent of the optical anisotropy $\Delta\varepsilon$ of the NLC, and determined mainly by the elastic constant K_2 of the NLC. This is due to the fact that, in Equation (11.148), we have the factor $q^2 \sim (n_e - n_o)^2$ in the denominator that cancels the factor $\varepsilon_a^2 = (n_e^2 - n_o^2)^2 \simeq (n_e - n_o)^2 (n_e + n_o)^2$ in the numerator. G_m is therefore dependent only on the elastic constant of NLCs. This inverse dependence of the gain on the elastic constant K suggests that *for nematic liquid crystals with lower elastic constant, the effective gain constant will be proportionately larger.*

For visible wavelength $\lambda \approx 0.5$ μm, and using the NLC E7 parameters: $n_o \approx 1.54$, $n_e \approx 1.75$, the grating constant $\Lambda \approx 2.5$ μm. Using these values and the values for the elastic constant $K_2 \approx 3 \times 10^{-7}$ dyne and the dielectric anisotropy $\varepsilon_a = \varepsilon n_e^2 - n_o^2 \approx 0.6$, the intensity gain factor is estimated to be $G_m \sim 1 \times 10^{-2} \text{cm}^2/\text{W}$. For an interaction length (NLC film thickness) $d = 200$ μm, this gives a gain coefficient $G_m d \sim 2 \times 10^{-4} \text{ cm}^2/\text{W}$ for a film thickness $d = 200$ μm. On the other hand, the scattering loss coefficient of NLCs is about 4.6 cm^{-1}, or a total loss of ~ 0.1. In other words, the gain coefficient will be >1 for an input intensity $>10^3 \text{ W/cm}^2$. This threshold value is indeed in good agreement with results reported previously.[13]

Also, *since n_e and n_o remain essentially constant throughout the visible–near IR regime*, one would expect the nonlinear wave mixing process to occur with similar efficiency throughout the visible–infrared regime $(0.5 - 12$ μm$)$, especially in the hitherto underexploited infrared regime. An important point pertaining to near-IR spectral regime is that, besides being transparent and possessing a large birefringence, the orientational fluctuation induced scattering loss in NLC scales inversely as the laser wavelength (cf. Chapter 5), that is λ^{-n} $(n \geq 2)$. Therefore, the scattering loss of liquid crystals in the near-IR wavelength will be significantly less than in the visible wavelength. For 1.55 μm laser, the scattering loss will be an order of magnitude smaller than the visible ~ 0.5 μm region. This means longer interaction, higher process efficiency, and therefore lower optical power threshold requirement.

(2) Steady-State Pump-Depletion Limit. In order to account for severe pump depletion associated with high conversion efficiency, the small signal theory as outlined above in Equations (11.146)–(11.149) is not adequate. Complete quantitative solutions of the coupled dynamical Equations (11.143)–(11.145) for the e and o

waves, and the laser induced director axis reorientation as they interact in the bulk of the NLC film are required. Time independent and dynamical solutions have been presented in the work by Khoo and Diaz.[15] In the work by Khoo and Liang,[13] extensions of the theory to describe self-starting optical phase conjugation based on the stimulated orientational and thermal scatterings are also presented.

Figure 11.19 shows the computed output e and o waves for visible (0.5 μm) and near IR (1.55 μm) laser as a function of the input laser power for the case where the input laser is an e wave (E_x). The liquid crystal (E7 from EM Chemicals) parameters are $n_e = 1.75$ and $n_o = 1.54$, elastic constant K_2 is 3.0×10^{-12} N, and the viscosity coefficient η is 0.07 P. For these parameters, we calculated that the optimum gain coefficient G_m ($\Omega\Gamma = 1$) corresponds to a frequency difference ($\omega_x - \omega_y = 31$ Hz). The sample thickness is 400 μm. For these values of n_e and n_o, the grating spacing is λ (1.55 μm)/0.21 = 7.38 μm.

In general, the o wave (originated as scattered noise) grows exponentially to the maximum value while the input e wave increases linearly at low input power, and begins to show sign of depletion as soon as the o wave power is substantial, eventually dropping to a value much lower than the generated o wave. The thresholds are 60 and 15 mW for input laser beam size spot diameters of 80 and 40 μm, respectively. With such nonlinear (quadratic) dependence of the threshold on the spot diameter, clearly it is possible to realize a sub-mW threshold using more tightly focused beams.

The calculation[14] also shows that with the 1.55 μm laser, a complete conversion is achieved at an input power level on the order of 80 mW, whereas the visible (0.5 μm)

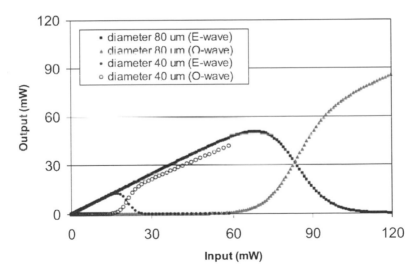

Figure 11.19. Calculated output e- and o-wave as functions of input e-wave power. Parameters used: 400-micron thick planar aligned nematic liquid crystal sample; $n_e = 1.75$, $n_o = 1.54$; elastic constant $K_2 = 3.0 \times 10^{-12}$ N.

laser requires a much higher input laser power of 140 mW. This is expected from the dependence of the gain constant on wavelength.

11.6.4. Stimulated Thermal Scattering (STS)

Analogous to SBS and SOS discussed above, the thermal index change dn/dT in liquid crystals could mediate energy transfer between an incident (pump) beam and its scattered noise.[16] Since the creation of thermal grating requires an absorption of light energy, the process generally requires that the pump beam and its scattered noise be copolarized, unlike the SOS process discussed in the preceding section. By considering the mixing of two copolarized light, the incident E_L and its scattered coherent noise E_S, an analogous set of equations can be derived. In the small signal limit, again we have[13,16]

$$E_L(z) = E_L(0)e^{-(\alpha/2)z}, \tag{11.150a}$$

$$\frac{dE_S}{dz} = \gamma_T e^{\alpha z} E_S - \frac{\alpha}{2} E_S, \tag{11.150b}$$

$$
\gamma_T = i\frac{n_0\omega^2}{k_{Sz}c^2}\left\{ \frac{\alpha \frac{dn}{dT}\left(\frac{n^\circ c}{8\pi}\right)|E_L|^2}{\rho c_p Dq^2} \cdot \frac{1}{1 - i\Omega\tau_T} \right\}
$$

$$
= i\frac{\omega\alpha\left(\frac{dn}{dT}\right)I_L}{c\rho c_p Dq^2 \cos\frac{\Theta}{2}} \cdot \frac{1 + i\Omega\tau_T}{1 + (\Omega\tau_T)^2} = g_T + i\kappa_T. \tag{11.151}
$$

[It is important to note that in the notation of Ref. 13, the frequency difference Ω is defined as $\Omega = \omega_S - \omega_L$. So a lower frequency noise than the pump will have $\Omega < 0$.]

$$
g_T = \text{Re}\{\gamma_T\} = \frac{\omega\alpha\left(\frac{dn}{dT}\right)I_L}{c\rho c_p Dq^2 \cos\frac{\Theta}{2}} \cdot \frac{-2\Omega\tau_T}{1 + (\Omega\tau_T)^2}. \tag{11.152}
$$

Just as in the case of SOS, the STS gain are maximum for a lower frequency noise at $\Omega\pi = -1$. In terms of the optical intensity, these maximum gain can be wrtitten and $g_T = G_T I_o$.

For a SBS-like wave mixing geometry as depicted in Figure 11.17, we have in STS,

$$q = 2k \sin\left(\frac{\Theta}{2}\right),$$

and therefore

$$G_T = \frac{\omega \dfrac{dn}{dT} \alpha}{2c\rho c_p D q^2 \cos \dfrac{\Theta}{2}} = \frac{\lambda \dfrac{dn}{dT} \alpha}{16\pi \rho c_p D} \cdot \frac{1}{\sin^2 \dfrac{\Theta}{2} \cos \dfrac{\Theta}{2}}, \qquad (11.153)$$

where Θ is the crossing angle between E_1 and E_2.

For thermal effect, the parameters of the liquid crystal are $\rho \approx 1$ gm cm^{-3}, $c_p \approx 2$ J gm^{-1} °K^{-1}, $D \approx 2 \times 10^{-3}$ cm^2 s^{-1}, $dn/dT \approx 10^{-3}$/°K, and the absorption is about $\alpha_a \approx 20$ cm^{-1} (this value can be adjusted by a dopant). Using the experimental parameters,[13] $\lambda \approx 0.5$ μm, $\Theta = 1°$, we have $\cos(\Theta/2) \approx 1$ and $\sin^2(\Theta/2) \approx \Theta^2/4 \approx (1.3 \times 10^4)^{-1}$. From these parameters, we get the intensity gain factor $2G_T = 1.3 \times 10^{-1}$ cm/W which is an order of magnitude larger than $2G_m$. Consequently, the threshold optical intensity for observing STS is also about an order of magnitude lower. Again, this is in good agreement with the experimental observation of STS.[16]

REFERENCES

1. Shen, Y. R. 1984. *Principles of Nonlinear Optics*. New York: Wiley.

2. Boyd, R. W. 1992. *Nonlinear Optics*. San Diego: Academic Press.

3. Flytzanis, C. 1975. In *Quantum Electronics* H. Rabin and C. L. Tang, (eds.) Vol. 1, Part A, p. 30. New York: Academic Press.

4. Jackson, J. D. 1963. *Classical Electrodynamics*, p. 119. New York: Wiley.

5. Bloembergen, N. 1965. *Nonlinear Optics*, p. 69. New York: Benjamin.

6. Khoo, I. C., and S.T. Wu. 1993. *Optics and Nonlinear Optics of Liquid Crystals*. Singapore: World Scientific.

7. See, for example, Khoo, I. C., and W. Wang. 1991. Effects of side diffractions and phase modulations on phase conjugations in a Kerr medium. *IEEE J. Quantum Electron.* QE-27:1310, and reference therein.

8. Khoo, I. C., and P. Zhou. 1990. Transient multiwave mixing in a nonlinear medium. *Phys. Rev. A.* 41(3):1544; Odulov, S. G., and M. S. Soskin. 1979. Dynamic self-diffraction of coherent light beams. *Usp. Fiz. Nauk.* 129:113; 1979. *Sov. Phys. Usp.* 22(9):742.

9. Khoo, I. C., J. Y. Hou, T. H. Liu, P. Y. Yan, R. R. Michael, and G. M. Finn. 1987. Transverse self-phase modulation and bistability in the transmission of a laser beam through a nonlinear thin film. *J. Opt. Soc. Am. B.* 4:886.

10. Khoo, I. C., P. Y. Yan, T. H. Liu, S. Shepard, and J. Y. Hou. 1984. Theory and experiment on optical transverse intensity bistability in the transmission through a nonlinear thin (nematic liquid crystal) film. *Phys. Rev. A.* 29:2756.

11. Khoo, I. C., M. V. Wood, B. D. Guenther, Min-Yi Shih, P. H. Chen, Zhaogen Chen and Xumu Zhang. 1998. Liquid crystal film and nonlinear optical liquid cored fiber array for ps-cw frequency agile laser optical limiting application. *Optics Express,* 2(12):471–482. See also Chapter 12.

12. Bartra, I. P., R. H. Enns, and D. Pohl. 1971. Stimulated thermal scattering of light. *Phys. Status Solid.* 48:11.

13. Khoo, I. C., and Y. Liang. 2000. Stimulated orientational and thermal scatterings and self-starting optical phase conjugation with nematic liquid crystals. *Phys. Rev. E.* 62:6722–6733.

14. Khoo, I. C., and J. Ding. 2002. All-optical cw laser polarization conversion at 1.55 micron by two beam coupling in nematic liquid crystal film. *Appl. Phys. Lett.* 81:2496–2498.

15. Khoo, I. C., and A. Diaz. 2003. Nonlinear dynamics in laser polarization conversion by stimulated scattering in nematic liquid crystal films. *Phys. Rev. E.* 68:042701-1–4.

16. Khoo, I. C., H. Li, and Y. Liang. 1993. Self-starting optical phase conjugation in dyed nematic liquid crystals with a stimulated thermal-scattering effect. *Opt. Lett.* 18:1490.

12

Nonlinear Optical Phenomena Observed in Liquid Crystals

Liquid crystals possess wonderful light-scattering abilities, linear or nonlinear. As a result, studies of their nonlinear optical responses have been vigorously pursued in various contexts. As in the case of electro-optics, it would require a treatise to summarize all the work done to date, as almost all conceivable nonlinear optical phenomena have been observed in liquid crystals in all their mesophases. Some of these phenomena were studied for their novelty; others have been developed into diagnostic tools or practical devices. In this chapter, we limit our attention here to only *exemplary* studies which are fundamentally interesting and/or practically important.

12.1. SELF-FOCUSING, SELF-PHASE MODULATION, AND SELF-GUIDING

12.1.1. Self-Focusing and Self-Phase Modulation and cw Optical Limiting with Nematic Liquid Crystals

In earlier studies of self-focusing effects in liquid crystals, the main emphasis was on the understanding of fundamental phenomena in nonlinear optics or liquid crystals such as laser-induced ordering and self-focusing and the associated beam breakups in an extended interaction region.[1] In the nematic phase, the fact that the liquid crystal film is thin but highly nonlinear allows one to employ the so-called nonlinear diffraction theory and the external self-focusing and self-phase modulation effects discussed in Chapter 11.

The discoveries of photorefractivity[2] and supraoptical nonlinearity[3] in nematic liquid crystals have ushered in the era of external self focusing with laser power as low as nanowatts in thin optical media. Figure 12.1 shows a typical optical limiting setup using external self-defocusing effect. A linearly polarized laser beam is focused by a 15 cm focal length input lens to a spot diameter of 0.1 mm onto a 25 μm thick liquid crystal film placed just behind the focal plane of the input lens. The nematic film is tilted to enhance the nonlinear refractive index change experienced

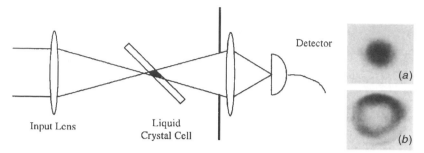

Figure 12.1. Experimental set up for optical limiting action using external self-defocusing effect. Insert are photographs of the transmitted laser intensity distribution at input power (a) below 70 nanoWatt and (b) above 100 nanoWatt.

by the extraordinary incident ray.[4] An aperture of 5 mm diameter is placed at 40 cm behind the sample to monitor the central region of the transmitted beam. Above an input power of ~70, it is observed that the central region of the transmitted beam becomes progressively darkened, and the beam divergence increases dramatically, as depicted by the two photographs in Figure 12.1. This change in the central beam power as a function of the input power is plotted in Figure 12.2, which exhibits a typical limiting behavior.

The optical limiting action involved in the setup as depicted in Figure 12.1 is based on the Gaussian intensity profile of the input laser, which induces a corresponding Gaussian profile in the director axis reorientation, and therefore a transverse spatial phase shift (similar to a lens) on the wave front of the laser. In the far-field diffraction zone, this phase shift is manifested in an increased divergence (defocusing) and formation of spatial rings, and so on, as discussed in the previous chapter.

A limitation common to such self-defocusing or self-phase modulation effect on a Gaussian beam is that the nonlinearity in the central region could reach a saturation point as the laser intensity increases. In that case, the reorientation profile and the laser self-induced defocusing effect will diminish, thereby restricting the dynamic range of the limiting operation.

An alternative way[5] of to achieve optical limiting with the same methyl-red-doped nematic film is depicted in Figure 12.3. Since the input/output polarizers and the liquid crystal cells are thin planar structures without the need for an aperture or pinhole situated at a distance from the "device," and the underlying limiting mechanism holds at very high incident laser power, this polarizers+LC cell setup is a practical configuration that will give very high dynamic range sensor protection capability and eye-safe clamped transmission.

In the experiment,[5] the liquid crystal used is methyl-red-doped (1% by weight) 5CB or E7. The cell is made by sandwiching the lightly doped liquid crystal between two rubbed polyvinyl alcohol (PVA) coated glasses. The director axis of the nomadic liquid crystal lies on the plane of the cell windows, which are placed such that the director axis of the nomadic film rotates by 90° from the incident to the exit windows (see Fig. 12.3).

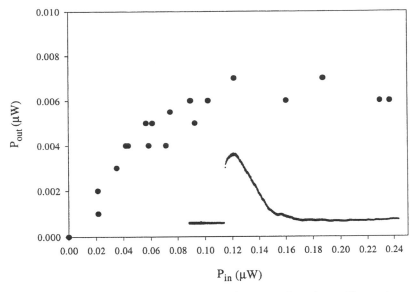

Figure 12.2.   Plot of detected output power versus input laser power. Insert is a oscilloscope trace of the transmitted on-axis laser power for a step-on input laser.

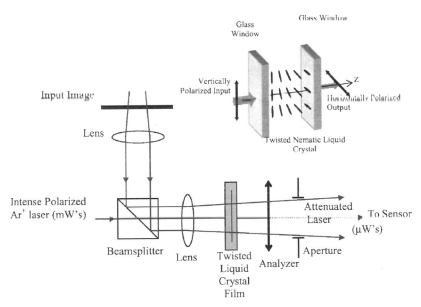

Figure 12.3.   Experimental set up for optical limiting action using laser induced nematic liquid crystal axis realignment effect. Upper diagram shows the geometry of the twisted nematic cell. The focused spot diameter of the Argon laser on the film is 150 microns. Sample thickness: 25 microns. Dye concentration: 0.5 %.

A polarized argon laser at 488 nm is used to simulate a "threat" laser over a background low-light-level "scenery," which consists of an image of a resolution chart illuminated by a low-power He-He laser. The polarization of the low-power argon laser follows adiabatically the director axis, and thus is rotated similarly. At high laser power, two mechanisms come into play:

1. The induced space-charge field in the direction normal to the plane of the window will cause the director axis to tilt toward the normal to the cell window. This changes the polarization state of the laser upon its exit from the sample. Consequently, the analyzer at the exit end will cause the output laser power to continuously decrease towards vanishing value as the incident laser power is increased.
2. Another mechanism is simply due to the finite absorption in the doped sample. The index and overall director orientation will also change as the cell temperature increases towards the nomadic→isotropic transition temperature T_{ic}. It is also likely that the *trans–cis* isomerism occurring in the azo-dye dopant causes degradation of the order parameter, leading to randomization of the director axis reorientation and transition to the isotropic phase. Therefore, as the incident laser power is increased, the exit polarization begins to take up more orthogonal component, resulting in attenuation of the transmission. In the isotropic liquid phase, the polarization of the exit laser will be completely orthogonal to the analyzer, and thus the transmitted laser power is totally attenuated (to the level defined by the crossed polarizers).

It was found that both reorientation and thermal effect in such 90° twisted nematic film, in combination with the analyzer at the output end, are highly effective in reducing the transmission of the incident laser as its power increases. Since the clamping action gets better for higher input laser power, the system thus possesses an extremely large dynamic range.

Figure 12.4 shows the photographs of the transmitted resolution chart with the threat laser embedded. The threat argon laser gives rise to the bright spot. At high input power, the argon laser is suppressed, and the spot practically disappears, leaving the background scenery intact and safe to view by eye. At the onset of the limiting action, self-defocusing effect due to the Gaussian beam intensity distribution also contributes significantly to the optical limiting action. The dramatic self-limiting action of the incoming laser beam is captured in the series of photographs of the exiting laser as the incident power is raised (see Fig. 12.5).

The clamped output depends on the sample initial alignment qualities. In general, it is in the range of a few µWs, for input laser as high as 140 mW,[5] that is, a dynamic range exceeding 10^5. The clamped transmission is about 3×10^{-5}. The low power "linear" transmission of the sample is 10% (0.1), due mostly to interface losses through the polarizer and uncoated cell windows, and so this clamped transmission corresponds to an attenuation factor (low-power transmission/high-power transmission) of over 3000.

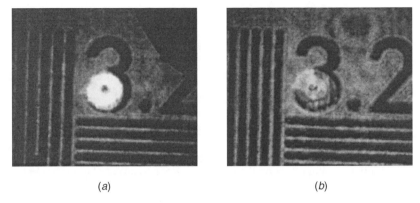

(a)   (b)

Figure 12.4. Photograph of the transmitted laser beam with the liquid crystal film (a) at low laser intensity, showing no limiting effect. Bright spot is due to the laser. The photo is deliberately overexposed to highlight the weak laser spot. (b) Above the limiting threshold, showing greatly attenuated laser spot.

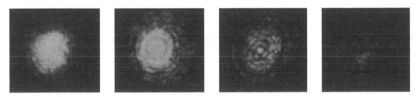

Figure 12.5. Photographs of the transmission through the LC film as incident laser intensity is increased — in order from left to right.

Using step function millisecond laser pulses, the response time for the laser pulse height to switch to e^{-2} of the input level is on the order of 40 ms. With the optimization of sample configuration and alignment properties, and dopant and nematic liquid crystals choice, these limiting performance characteristics can obviously be improved considerably. In particular, using (faster) laser-induced order parameter and phase change effects in azo-dye-doped liquid crystals discussed previously, one can envision designing devices capable of microseconds or faster limiting responses. Since the threshold for limiting action is rather low, this study has clearly demonstrated the feasibility of constructing practical all-optical limiters for cw or long-pulse laser or other intense light sources (e.g., sun glares, flashes, welding torches, etc.).

It is important to note that since the thermal/density and order parameter changes could be induced in microseconds or tens of nanoseconds (cf. preceding chapters), these director-axis-reorientation or order-parameter-change mediated limiting actions will work well for sensor protection in these time scales. For shorter laser pulses, for example, nanosecond and picosecond or subpicosecond laser pulses, the response will not be able to build up sufficiently in time to provide the necessary attenuation effect. In those time regimes, electronic optical nonlinear mechanisms, in particular, nonlinear photonic absorptions, have to be employed. This is discussed in

Section 12.6, where other nonlinear device/concepts that are effective for the short time scale will be introduced. It is interesting to note that if one combines the nematic films with the nonlinear (image transmitting) fiber array for nanosecond/picosecond limiting application[4–6] (cf. Sec. 12.6), it is possible to fashion optical sensor protection devices capable of covering an extremely wide temporal range (from ps laser pulses to cw light sources).

12.1.2. Self-Guiding, Spatial Soliton, and Pattern Formation

For an interaction geometry in which light propagates through a longer interaction length, the self-focusing effect on the beam gives rise to a spatial confinement of the beam size opposite to the diverging effect of diffraction, leading to a self-guiding beam. Such self-guiding beam propagation processes have been studied as long ago as the 1960s, and have in recent years received intense renewed interest mostly in the context of spatial solitons.[7,8] These efforts result from a new understanding of the phenomena combined with the possibility of developing low threshold light-controlled readdressing and switching devices[7,8] with the help of ultranonlinear optical materials.

In this context, nematic liquid crystals offer themselves as attractive candidates as the "test beds" for theories, concepts, and prototype device demonstrations[9,10] as the millisecond response speed allow convenient dynamical observations, and actual practical solutions where there are less stringent requirements on the speed. An exemplary experimental observation of self-guiding effect in NLC is discussed here.

Figure 12.6 depicts the experimental setup used in the *first* demonstration of spatial soliton in an aligned nematic liquid crystal utilizing the purely dielectric field induced director axis reorientation nonlinearity.[9] The liquid crystal cell consists of planar aligned liquid crystal between E7 and two ITO coated windows which allow application of ac voltage. If the applied voltage is above the Freedericksz transition for the geometry, it will create a pretilt θ_0 of the director axis, making it easier for the

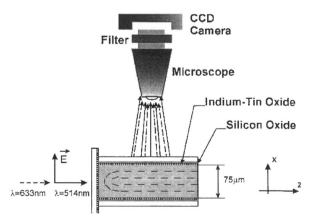

Figure 12.6. Experimental set up used in the *first* demonstration of spatial soliton in a aligned nematic liquid crystal (after Ref. 9).

optical field to further reorient the axis to an angle θ—an interaction geometry that is analogous to the one employed for the first observation of light-induced molecular reorientation above the Freedericksz transition.[11]

Consider the case of a linearly polarized laser that is incident from the side window. If we denote its optical electric field by $\vec{A}=A \cdot \hat{x}$ *and its direction of propagation is* along z in a bulk NLC, the evolution of the amplitude of this optical field is described by a Schrödinger-like equation in a slightly inhomogeneous medium:[9]

$$2ik\frac{\partial A}{\partial z}+\left[\frac{\partial^2 A}{\partial x^2}+\frac{\partial^2 A}{\partial y^2}\right]+k_0^2\Delta\varepsilon(\sin^2\theta-\sin^2\theta_0)A=0, \tag{12.1}$$

where $k\approx k_0\sqrt{n^2_\perp+\Delta\varepsilon}\,\sin\theta_0$ is the wave vector and $k_0=2\pi/\lambda$, with λ the wavelength. On the other hand, the director axis reorientation is described by a typical torque balance equation for this interaction geometry.

Figure 12.7 shows a numerical simulation for an input beam power of 3.9 mW and a beam waist of 3 μm (the laser used is the 514.5 nm line of an Ar[+] laser). In the absence of the external bias ($\theta_0=0$), the linearly polarized electric field of the input light is perpendicular to the director axis. Since its intensity is below the optical Freedericksz threshold value (cf. Chapter 8), it cannot create director axis reorientation; that is, there is no self-action effect. The focused beam thus diffracts freely in x and y as it propagates along z (Fig. 12.7a). When the external voltage above the Freedericksz threshold is applied, the director axis is reoriented (for the numerical simulation, the angle is assumed to be 45°). In this "above Freedericksz" condition,[11] the optical field will reorient the director axis without threshold, and initiate the self-guiding process leading to the formation of a spatial soliton—a beam that maintains its beam waist over many Rayleigh lengths (see Fig. 12.7b).

Such fundamental soliton characteristics as well as other more sophisticated phenomena, including beam steering, beam combining, and incoherent light processing,

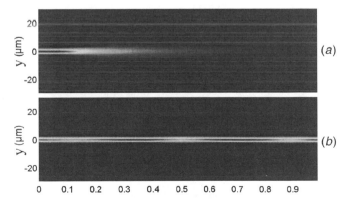

Figure 12.7. Computer simulations of the propagation in the (y-z) plane of a beam launched with a 3mm waist and 3.9mW power. (a) no applied bias voltage, linear diffraction when $\theta_0=0$. (b) solitary wave when director axis is pretilted.

have been experimentally verified.[9,10] These and other work using photorefractive crystals[12,13] have clearly demonstrated the possibility of employing such solitons for beam/signal processing using spatial solitons.

Such beam propagation phenomena have also been studied in the context of (self-started) pattern formations.[14,15]. Due to the combined effect of feedback and spatial phase modulation[16] a single incident beam could break up into an array of beams and exhibit many dynamical patterns and bistabilities/instabilities as various LC and optical parameters are varied. In Section 12.2, we will also discuss a special case where a linearly polarized incident beam self-converts dynamically into an orthogonally polarized beam and exhibits various oscillatory and chaotic behaviors without feedback.

12.2. OPTICAL WAVE MIXING

Laser-induced dynamic gratings and optical wave mixings in nonlinear optical media have been studied under various contexts. For practical applications, a great deal of interest is centered on liquid crystals[17-25] and photorefractive crystals[26-29] because of the low-power requirements associated with these highly nonlinear optical materials. Various two-beam coupling effects involving energy exchanges from a strong pump beam to a weak probe, such as self-starting optical phase conjugations,[21-24,26-28] beam and image amplification,[20,21,25] polarization rotation, and stimulated scatterings,[29-31] have been observed.

Most studies in photorefractive crystals and polymers employ visible and near-infrared (around 1 μm) lasers, primarily owing to the fact that the underlying photorefractive effects[19,26] require resonant excitation of photocharges and space-charge fields. In the near-infrared regime, in particular, the communication wavelength channel (1.55 μm), there have been recent studies on the electro-optical and nonlinear optical responses[32-34] of liquid crystals.

From many perspectives, liquid crystals are particularly suited for applications in this regime. First, they are nonabsorptive and possess large dielectric anisotropy ($\Delta\varepsilon = \varepsilon_e - \varepsilon_o \sim 1$) over the entire near-UV to infrared spectrum and beyond (400 nm–20 μm; cf. Chapter 3). Secondly, the scattering loss α in this regime is an order of magnitude smaller than in the visible region, since the principal mechanism for scattering loss are director axis fluctuations [$\alpha \sim 1/\lambda^n$ ($n > 2$)].

Accordingly, unlike most nonlinear optical materials, liquid crystals enable optical wave mixings to occur with high efficiency in the visible as well as infrared regime. In this section, we discuss two exemplary wave mixing processes in liquid crystals: stimulated orientational scatterings and optical phase conjugation processes.

12.2.1. Stimulated Orientational Scattering and
Polarization Self-Switching: Steady State

Earlier observation[35] of transient SOS effects was conducted in the spirit of stimulated Brillouin scattering and involved fairly high power pulsed ruby laser. However, the more interesting and practically useful one is the steady-state case analyzed previously. Typically, for the orientation grating constant $\lambda_q = 2\pi/|q| = \lambda/\Delta n \sim 3\mu m$, the corresponding orientational relaxation time constant τ is on the order of 10 ms. Using a laser pulse duration much longer than this will allow one to observe the steady-state version of the o-e ray stimulated scattering effects at relatively low laser power, as first demonstrated in the work by Khoo and Liang.[31] As a result, there have been several studies[30,36–39] conducted with low-power cw lasers in the visible as well as infrared spectral regions.

As an example of the simplicity and compactness that the phenomenon can be realized in practice, consider the experimental setup used in the work by Khoo and Ding[37] as shown in Figure 12.8. The linearly polarized erbium-doped fiber laser ($\lambda = 1.55$ μm) is focused at normal angle onto the planar aligned nematic film as an e or o wave, that is, with the optical electric field parallel or perpendicular, respectively, to the director axis of the planar aligned liquid crystal (director axis lying in the plane of the sample). A polarizing beam splitter at the exit end is used to separate the e and o waves. The absorption of E7 at this wavelength (1.55 μm) is negligibly small. The transmission of the e wave through the 400 μm thick cell is about 85%, with about 8% contribution from air–glass interface reflection loss. It was observed that in order to mediate energy exchange efficiently, the divergence of the input laser should match the "cone" of its scattered noise. This corresponds to focusing the laser to a beam waist on the order of tens of micrometers or less. Under these conditions, and as illustrated in the photoinserts in Figure 12.9, the noise builds up to an intense

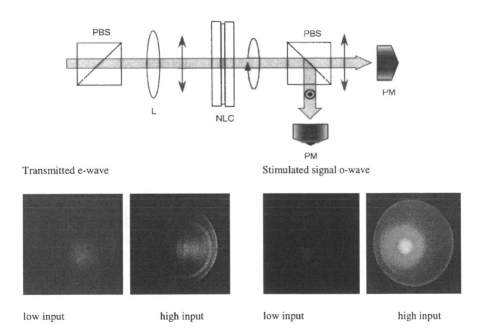

Figure 12.8. Experimental set up for observing stimulated orientational scattering in a nematic liquid crystal. Photos show the transmitted pump (e-wave) and stimulated (o-wave) at low and high input laser power.

beam, while the input intense beam diminishes when the incident laser power exceeds the stimulated scattering threshold.

The observed pump and signal beam powers at the exit end are shown in Figure 12.9, and their overall dependence on the input laser power is in very good agreement with theory (cf. Chapter 11). The o-wave "noise" grows exponentially from diffuse side scatterings to an intense beam, reaching the maximum value at ~48 mW input. On the other hand, at low input power, the input e wave initially increases linearly with the input, and begins to drop dramatically above the stimulated scattering threshold, eventually dropping to an almost diminishing value. Because of the presence of multifrequency components in the director axis fluctuations, considerable oscillatory behavior in the e- and o-wave outputs are observed. At high intensity, we also observe considerable side scattering and defocusing of the output beams; effects are likely cause for the drop-off in the exit beam power after reaching the maximum.

Since the two-beam coupling effect depends only on the frequency difference between the pump beam and its scattered noise, the conversion process should hold whether it is from e to o waves, or o to e waves. By repeating the same experiments with an o-wave input (rotating the planar nematic sample by 90°), similar stimulated e-wave scatterings have been observed at similar laser power levels (see Figs. 12.10a and 12.10b).

These observations with near-infrared (1.55 mm) laser in light of other studies involving visible laser has clearly illustrated the point made previously about the "nonresonant" nature of the two-beam coupling processes. Unlike other materials where resonant (and therefore very wavelength selective or restricted) interaction between the lasers and the material is needed to produce the necessary two-beam coupling effect, orientation-mediated wave mixing processes in nematic can be realized with laser of any wavelength outside the absorption "bands" of liquid crystals. In view of the great arsenal of available liquid crystals, this latter condition amounts to very little restriction. Therefore, one can envision application of

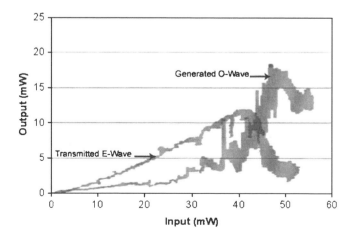

Figure 12.9.   Recorded transmitted pump and signal as a function of input laser power.

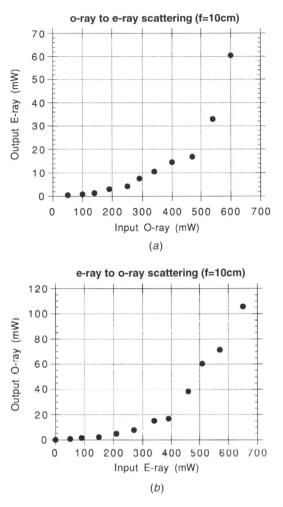

Figure 12.10. Stimulated *o*- to *e*- wave scattering. (a) Observed *o*-wave power as a function of the input *e*-wave power showing a typical stimulated scattering dependence above a certain threshold. (b) Observed *e*-wave power as a functrion of the input *o*-wave power showing a typical stimulated scattering dependence above a certain threshold.

orientation outside the visible–near-IR regime, for example, the 2–5 μm spectral regimes, where highly nonlinear optical materials are scarce.

For fundamental studies, these polarization self-switching processes could also be integrated into the spatial soliton formation discussed in the previous section, and thus a host of novel optical phenomena await the inquisitive.

12.2.2. Stimulated Orientational Scattering: Nonlinear Dynamics

Self-organization and collective phenomena have been studied extensively in various material and optical systems including liquid crystals.[14,15,40–47] These studies provide fundamental insights into how seemingly incoherent individual molecules or noises would self-select or self-organize by mutual interactions into a crystalline or coherent form/signal, with or without an external driving field. In these contexts, the multifrequency wave mixing and "self-selection" stimulated scatterings in nematic liquid crystals discussed above offer a natural setting to explore and gain further understanding of nonlinear dynamical system. It is interesting to note that although the driving or pump beam is a continuous-wave constant-intensity one, the output waves are found to exhibit various dynamical and oscillatory phenomena, as a result of the underlying complex nonlinear nature of the field-driven director axis reorientation process.

In a recent study,[39] the coupled electromagnetic waves and director axis reorientation equations, cf. Chapter 11, for the stimulated orientation scattering process are solved by treating all the molecular and optical parameters involved as spatial-temporal variables. For example, the director axis reorientation angle θ is represented as $\theta(z,t) \equiv \sum_n A_n(t)\sin(n\pi z/L)$. In that case, the torque balance equation becomes

$$\frac{dA_n}{dt} = \left[i\Omega - \frac{K_2}{\eta}\left[q^2 + \left(\frac{n\pi}{L}\right)^2 \right] \right]A_n + \frac{\varepsilon_a}{\eta L}\int_0^L \theta_s \sin\left(\frac{n\pi z}{L}\right)\left(|E_y|^2 - |E_x|^2 \right)dz$$

$$+ \frac{\varepsilon_a}{\eta L}\int_0^L \sin\left(\frac{n\pi z}{L}\right)\left(E_x E_y^* - \frac{1}{2}\theta_s^2 E_x^* E_y - |\theta_s|^2 E_x E_y^* \right)dz. \qquad (12.2)$$

The propagation equations for the fields and the evolution equations for the director expansion coefficients were solved numerically to analyze experimental observations. Here we summarize some of the major findings that may serve as a guide for further study.

Figure 12.11a shows the computed "orbits" of the director axis in the phase space defined by $A_1 - A_3$ in the intensity regime when these oscillations occur. These intensity oscillations observed just above the stimulated scattering threshold are reminiscent of director motion above the optically induced Freedericksz threshold in previous studies[45–47] in nematic liquid crystals. According to the theoretical simulation, see Figure 12.11b, for a particular laser wavelength and liquid crystalline parameters (elastic constant, viscosity, and index difference $n_e - n_o$), these oscillations will occur preferentially close to the characteristic frequency component $\Omega_{max} = \Gamma$. Notice how even though the oscillation of the director axis is extremely small, there is an effective periodic transfer of energy (*while the total energy remains constant*) between the e and o waves through the sample (see Fig. 12.11b). This is indeed quantitatively verified in the experiments see Figures 12.11c and 12.11d.[39]

Detailed measurements of the Fourier spectra of the oscillations at various fixed input power above the lasing threshold, using both visible and infrared lasers, have also confirmed the theoretical predictions. Using the liquid crystalline (E7) parameters: $n_e = 1.75$, $n_o = 1.54$, elastic constant $K_2 = 3.0 \times 10^{-12}$ N, and viscosity coefficient $\eta \sim 0.1$ P, the optimum gain coefficient is obtained for a frequency shift $W_{max} = \Gamma \sim 32/2\pi$ rad/s (or 32 Hz) at a laser wavelength of 532 nm. The corresponding polarization grating spacing is $0.532/(n_e - n_o) = 2.5$ μm. If the laser wavelength

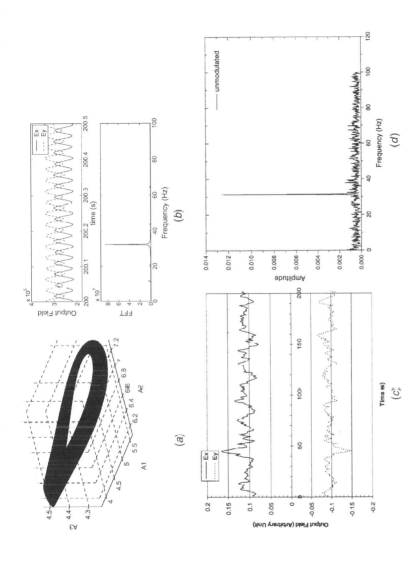

Figure 12.11. (a) Phase diagram of the director angle coefficients A_i (in mrads) for an input e-wave of 3.8×10^5 V/m. Laser wavelength: 532 nm; sample thickness: 200 μm; laser spot diameter: 40 μm. (b) Output field dynamics and Fourier transform for an input e-wave of 3.8×10^5 V/m. The behavior is quasi-periodic with a well-defined oscillation frequency around 32 Hz. (c) Simultaneous oscilloscope traces of the output pump and generated waves showing the oscillatory energy exchanges between the two beams. Pump laser power: 60 mW; laser wavelength: 532 nm; sample thickness: 250 μm; laser spot diameter: 40 μm. (d) Measured Fourier spectrum of the output e-wave showing a sharp spike at the characteristic frequency $\Gamma = 32$ Hz. Pump laser power: 60 mW; laser wavelength: 532 nm; sample thickness: 250 μm; laser spot diameter: 40 μm.

is 1.55 μm, the characteristic frequency Γ is ~4 Hz. These theoretical estimates are very close to the experimentally observed characteristic frequencies for the corresponding laser wavelengths used.

Furthermore, applying an ac control field of appropriate frequency will enhance the stimulated scattering efficiency. This is perhaps the first instance where an applied ac field could influence laser-induced stimulated scattering in a material. It is simply due to the rather low optical threshold field (~1000 V/cm) for stimulated scattering in nematic liquid crystals, in contrast to other nonlinear materials[48] that require orders of magnitude higher threshold field. The enhancement effect and appearance of frequency harmonics under the action of a modulating ac field provide another means of control on the stimulated scattering processes/devices, besides a new ground for further fundamental pursuits similar to previous studies en route to chaos and control of nonlinear dynamics.[42,45] The driven coupled oscillator model as depicted in Figure 12.11 and the build up of noise to a coherent signal also resemble the coupled-oscillator-neural-network model, which shows how arbitrary, randomly assembled oscillators can establish a desired "coherent" configuration.[49] In this regard, nematic liquid crystals remain an interesting and low-power "user friendly" nonlinear material for practical simulations and understanding of fundamental processes in many disciplines.

12.2.3. Optical Phase Conjugation with Orientation and Thermal Gratings

Optical phase conjugations using liquid crystals were first conducted in the work by Khoo and Zhuang[50] using coherent visible lasers and later by Leith et al.[51] using partially spatially coherent laser. By reducing the spatial coherence of the laser with rotating ground glass, these researchers have demonstrated a high-resolution, low-noise phase conjugation imaging and aberration correction technique (see Fig. 12.12), which is complementary to the multiple-exposure method devised by Huignard et al.[52] Since liquid crystals are inherently noisy due to their large

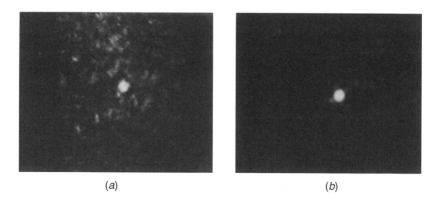

(a)   (b)

Figure 12.12. (a) Photograph of the phase-conjugated reconstructed image of the laser beam with a spatially coherent light (i.e., no use of rotating ground glass); the coherent noise produces serious interference and degradation of the signal. (b) Photograph of the reconstructed signal with spatially incoherent light showing elimination of coherent noise to give a well-defined image.

birefringence and orientational fluctuations, these methods of reducing coherent noise effects are clearly important for phase conjugations or other real-time non-linear optical holographic imaging processes involving liquid crystal films.

Subsequently, optical phase conjugation and related wave mixing studies have also been conducted with infrared (10.6 μm) lasers[53–55] in an effort to extend the successful applications of optical phase conjugation (primarily performed with inorganic photorefractive crystals[56]) from the visible to the infrared spectrum. One major drawback of using liquid crystals for infrared (around 10 μm region) optical wave mixing applications is the generally large absorption losses in this spectral regime. Typically, they range from 40 to 100 cm^{-1} and impose a severe limitation on the interaction length (i.e., less than 200 μm) and the choice of thermal index change as the wave mixing mechanism.

On the other hand, in the mid-infrared regime (1–5 μm) nematics are less absorptive; typically the absorption contrast is about 10 cm^{-1} or less, and thus longer interaction lengths are possible. Since there are few competitive materials (in terms of the large nonlinearity and low laser power threshold requirement) in these spectral regimes, nematic liquid crystals will be the natural choice for phase conjugation related applications.

Stimulated scattering and phase conjugation effects have also been briefly observed in a smectic-A liquid crystal film,[57] where the nonlinearity involved is the laser-induced density change and may be called a stimulated Brillouin scattering process. These density effects are also responsible for the interference effects observed in the diffraction from nanosecond laser pulse induced gratings in smectic liquid crystal film. To date, however, there has not been much activity on optical wave mixing studies in the smectic-A phase. In the isotropic phase optical phase conjugation effects have been reported by Fekete et al.,[58] Madden et al.,[59] and Khoo et al.[60] Recently, optical phase conjugation with OALCSLM has demonstrated remarkable aberration correction capabilities for large optical systems.[61]

12.2.4. Self-Starting Optical Phase Conjugation

Self-starting optical phase conjugation (SSOPC), in which a single incident laser beam generates its phase-conjugated replica via some optical wave mixing effect in a nonlinear optical material, is a fundamentally interesting and practically useful process. Usually, the signal originates as some coherently scattered noise from the pump laser beam (e.g., owing to scatters in a crystal, spontaneous Brillouin scattering, etc.). This noise signal interacts with the pump beam and grows into a strong coherent signal. This phenomenon is commonly observed in stimulated Brillouin scattering involving high-power pulsed laser[48,62] and in photorefractive materials with low-power cw lasers.[56]

Studies of similar SSOPC processes have been conducted using stimulated orientational and thermal scatterings in nematic liquid crystals.[31,36,63] Again, owing to the extraordinarily large optical nonlinearities of liquid crystals, the thresholds needed for initiating the self-starting phase conjugation oscillations are typically in the mW range, compared to the mW power requirements in typical SBS liquid.

A typical experimental setup used is depicted in Figure 12.13. The liquid crystal used is E7 (from EM Chemical), which has a nematic–isotropic phase transition temperature of 63°C. The material has negligible absorption (<0.01 cm^{-1}) at the argon laser wavelength (5145 Å line) used. A 100 μm thick planar aligned nematic liquid crystal film is made by sandwiching the E7 between two rubbed polymer-coated glass slides. The experiment is conducted at room temperature without the use of a temperature cell.

A linearly polarized cw argon laser operating at the 5145 Å line is electrically chopped to yield pulses of variable millisecond duration. It is focused onto the nematic liquid crystal sample with its electric field polarization vector $\vec{E}$ parallel to the director axis $\vec{n}_o$ (i.e., an e wave). The transmitted beam is reflected, focused back on the sample, where it intersects the incident beam at a crossing angle in air of 3°. The polarization of the reflected beam is rotated so that it is orthogonal to the polarization direction of the incident beam (i.e., an o wave).

The insert in Figure 12.13 depicts the wave vector matching condition for the case where the incident wave $E_x^{(i)}$ is an e wave. The scattered o-wave component $E_y^{(i)}$ is coherent with respect to, and interferes with, $E_x^{(i)}$ to produce an orientational grating with a wave vector $\vec{q}$. Energy is transferred from $E_x^{(i)}$ to $E_y^{(i)}$ via the stimulated orientational scattering effect mentioned previously. Similarly, the reflected o wave $E_y^{(r)}$ interacts with its scattered e-wave component $E_x^{(r)}$ with an orientational grating $\vec{q}$ which matches that produced by the incident wave. These processes thus reinforce one another, leading to coherent signal output; the scattered "noise" signals $E_y^{(i)}$ and $E_x^{(r)}$ will grow into a coherent beam when the laser power exceeds the threshold for SOS.

The SSOPC signal is photographed at an observation plane located about 5 m away. The onset dynamics of the SSOPC signal is monitored by a photodiode and recorded on a storage oscilloscope. The threshold power for SSOPC is similar to that required for the forward SOS effect in the same sample. When the power of the incident beam is small (<300 mW), only speckle noise appears on the observation plane. Above an input pump power of 300 mW, a well-defined beam of the phase-conjugated signal becomes clearly visible, as shown in the insert in Figure 12.13. The phase-conjugated signal spatial quality is similar to the input laser beam and has approximately the same divergence, in spite of the aberrations imposed by all the optics in front of the nematic cells. At an input power of 1 W, the maximum efficiency of the SSOPC effect is about 5%.

The dynamics of the self-starting phase conjugation and the forward stimulated o-e scattering process are shown in Figures 12.14a and 12.14b, respectively. Below the threshold, only background noises, due mainly to random orientational fluctuation-induced cross-polarization scattering, are detected. The coherent signals are detected as monotonically increasing signals with build up times on the order of 45 ms (input power of 300 mW) and 30 ms (input power of 560 mW) for the e-o self-starting phase conjugation (Fig. 12.14a), and 7ms for the e-o stimulated scattering process (Fig. 12.14b). The latter is in agreement with the theoretical estimate given previously.

As in most self-starting optical phase conjugation originating from speckle noise,[64] the onset dynamics of the signal consist of two regimes: an onset time τ^* and

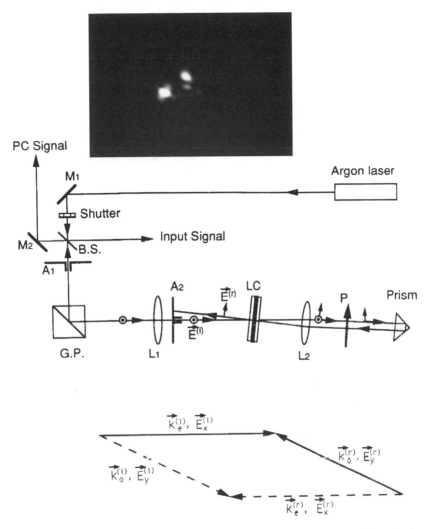

Figure 12.13. Experimental setup for *e–o* phase conjugation based on SOS. Photograph shows the observed signal (double images are due to glass slides surfaces). Also shown is the *e–o* wave vector phase-matching condition. Spot diameter of the laser at the location of the sample is about 50 μm. Crossing angle is 3°.

a buildup time τ, see Figure 12.15. The length of the onset time is dependent on the noise characteristics and amplitude, whereas the buildup time is dependent on the material response time. For a grating constant of about 9 μm (wave mixing angle of 3°), the orientational grating response time is given by $\tau_r \approx \gamma / K_2 q^2$. Using typical values of $\gamma \approx 1.2$ P, $K_2 = 5 \times 10^{-7}$ dyne, and $q = 2\pi/10$ μm, we get $\tau_r = 50$ ms, which is in good agreement with the experimental observation (see Fig. 12.14a). For a larger crossing angle (i.e., a smaller grating constant), the SOS response time will be considerably reduced by the q^{-2} dependence but at the expense of higher threshold power

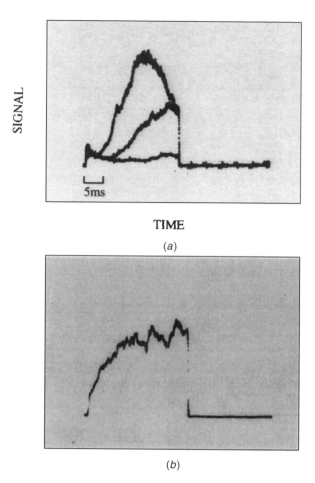

Figure 12.14.   (a) Observed buildup dynamics of the *e-o* phase conjugation signal for three input powers: 300 mW (lowest trace showing noise only), 400 mW (signal appears), and 560 mW (upper trace, signal appears earlier). Time scale is 5 ms/div. Pulse duration is 25 ms. (b) Oscilloscope trace of the time evolution of the stimulated *o*-wave signal for an input *e*-wave square pulse power of 500 mWatt. Input pulse duration – 25 ms. Arrow indicates the background noise level.

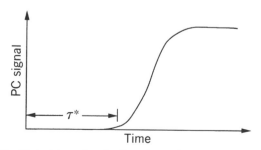

Figure 12.15.   Typical dynamics of self-starting optical phase conjugation process.

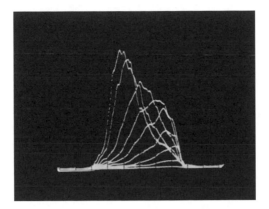

Figure 12.16. Osiclloscope traces of the time evolution of SSOPC signal as a function of input power ranging from 200 mW (lowest curve) to 800 mW (uppermost curve), showing the shortening of the onset and build-up times. Time scale: 0.5 ms/div.

(cf. Chapter 11). On the other hand, the onset time is dependent on the noise characteristics and amplitudes. As shown in Figure 12.14a, the onset times are measured to be about 15 and 5 ms for input powers of 300 and 560 mW, respectively.

As pointed out previously, thermal effects can be faster than the orientation counterpart. Figure 12.16 shows the oscilloscope traces of the thermal SSOPC signal as a function of the input pump power. Both the onset time and buildup time are shortened as the input power is increased. The total time it takes for the signal to build up to maximum can be as short as 0.5 ms, at a pump power of about 800 mW. This shortening of the buildup and onset times has also been observed as the sample temperature is increased toward T_c.[36] A principal drawback of thermal grating mediated SSOPC in liquid crystal is that the efficiency is highly dependent on the proximity of the phase transition temperature T_c and the requirement of very stable temperature control. On the other hand, laser-induced nematic axis reorientation effects do not require such proximity to the phase temperature.

12.3. LIQUID CRYSTALS FOR ALL-OPTICAL IMAGE PROCESSING

12.3.1. Liquid Crystals as All-Optical Information Processing Materials

As a flat thin film structure, nematic liquid crystals which possess large intensity-dependent refractive index changing coefficient n_2 are particularly suited for all-optical image processing applications. Since their discoveries in 1994 and 1998, respectively, photorefractivity[17] and supranonlinear[3] optical dye-doped liquid crystals have received intense interest because of their possible applications in optical information processing, communication, and storage applications. A good review of work done till 1999 on photorefractivity and dye-doped liquid crystals based beam/image processing and optical storage may be found in the work by Khoo et al.[65]

In Table 12.1, we list the *effective nonlinear index coefficients* n_2 of these NLC materials along with other well-known classes of nonlinear optical materials1.[66-70] Note that $\chi^{(3)}$ is the equivalent third-order nonlinear susceptibility; α is the absorption constant, and τ is the response time. We also insert the corresponding data for optically addressed liquid crystal spatial light modulator (LC-OASLM) (see Chapter 6) by estimating its index changing efficiency under a given illumination optical intensity.

For practical applications, the absorption constant and response times of the material have to be taken into account by defining a figure of merit termed the *switching efficiency* $\chi^{(3)}/\alpha\tau$. The typical values for these materials are listed in Table 12.2. One must note that the values could vary greatly within the same material, depending on the optical interaction geometry, so they are meant as a quick guide only. Also, for some applications, the required response times could not be "compensated" by the magnitude of the nonlinearity, for example, in picosecond laser pulse switching, so that slower response materials are not suitable, even if the switching efficiency meets the requirement. On the other hand, for whole-image processing, where an entire

Table 12.1. Refractive Index Coefficients of Some Nonlinear Optical Materials

Materials	Order of Magnitude of n_2 (cm^2/W)
Nematic liquid crystal	
Purely optically induced	10^{-4}
Thermal and order parameter change	10^{-4}
Excited dopant (dye molecule) assisted	10^{-3}
Photorefractive -C60 doped	10^{-3}
Methyl-red doped	10
Azo-benzene LC (BMAB) doped NLC	>2
C60/nanotube doped film	>20
OASLM-LC [estimated]	~10
GaAs bulk (Ref. 66)	10^{-5}
GaAs multiple quantum well (MQW) (Ref. 67)	10^{-3}
Photorefractive crystals/polymers (Ref. 68 and 69)	$\sim 10^{-4}$
Bacteriorhodopsin (Ref. 70)	10^{-3}

Table 12.2. Switching Efficiency $\chi^{(3)}/\alpha\tau$ of Various Materials

Materials	$\chi^{(3)}/\alpha\tau$ (10^{-10} m^{3}V^{-2}s^{-1})
GaAs bulk (Ref. 1)	30
GaAs MQW (Ref. 2)	300
Bacteriorhodopsin (Ref. 5)	0.05
Photorefractive crystals/polymers (Refs. 3 and 4)	10^{-1}
Methyl-red doped LC film	>100
C60/nanotube doped LC film	>100
OALCSLM estimated	>100

[Note: $n_2 = 0.105 \times \chi^{(3)}$cgs/$n_o^2$ (cm^2/W); $\chi^{(3)}$ (in m^2/V^2) = 1.39×10^{-8} $\chi^{(3)}$cgs (in esu). For MRNLC, $\alpha = 150$ cm^{-1}, $\tau = 10$ ms, $\chi^{(3)} = 3.13 \times 10^{-6}$ (m^2/V^2), so $\chi^{(3)}/\alpha\tau = 209$ (10^{-10} m^{3}V^{-2}s^{-1}).]

image is processed in a "parallel" fashion, a frame speed of 1 KHz is quite sufficient, making liquid crystal films a promising all-optical image processing material.

We close this section with a further remark on dye- or other dopant-enhanced nematic liquid crystals as a material choice for optical processing applications (e.g., optical wave front conjugation and aberration correction[61] versus their electronic counterpart—LCSLM. Because of their supranonlinearity, the optical intensity/power requirement is comparable to the intensity level required for the operation of LCSLMs. Moreover, the resolution capability of these commercial spatial light modulators is also limited. On the other hand, the entire bulk of the nematic liquid crystal film can be used as a holographic recording medium. In Table 12.3, we compare and contrast the performance characteristics of a commercial liquid crystal spatial light modulator (LCSLM) with what could be expected of a dye-doped liquid crystal (DDLC) film, for example. Clearly the development of DDLC to replace LCSLM is an attractive endeavor, since there exist, many dyes which will cover the entire visible–infrared spectrum, whereas the photosensitive semiconductor layer in LCSLM has a limited spectral response bandwidth.

12.3.2. All-Optical Image Processing

Some of these specialized nonlinear all-optical processes have been discussed in the preceding sections and the literature quoted therein. Since these nematic liquid crystals possess extremely high nonlinear index change coefficient, they require no external bias, and function at much lower optical powers and shorter exposure times than other conventional methods.[2–8] Here we discuss more exemplary cases.

Using the setup as depicted in Figure 12.17, Khoo et al.[71] has demonstrated incoherent to coherent, wavelength conversion and contrast inversion. The image bearing optical beam, at a wavelength of 488 nm, creates a spatial phase shift on the nematic film, which is sensed by a coherent He-Ne laser. Visible coherent images can be created with input incoherent beam intensities as low as 90 μW/cm^2. This is comparable to the sensitivity of liquid crystal spatial light modulators.[61]

Table 12.3. Comparison of the Performance Characteristics of Commercial Liquid Crystal Spatial Light Modulator (LCSLM) with What Could be Expected of Dye-Doped Liquid Crystal (DDLC) Film

	LCSLM	DDLC
Speed	FLC:$>$1 kHz	$>$30 Hz
	NLC: 30 Hz	
Sensitivity	40 μW/cm^2	40 μW/cm^2
Resolution	30 lp/mm	$>$200 lp/mm
Optical	FLC:$<$10%	
Efficiency	NLC: 15%	$\geq$30%
Aperture	25 mm	$>$25 mm

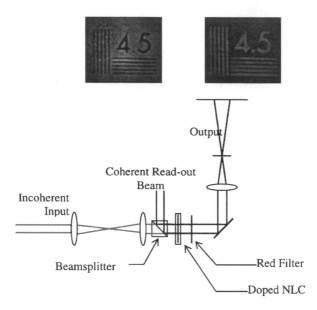

Figure 12.17. Experimental set up for incoherent to coherent image conversion. Insert are photographs showing the contrast inversion (bright–dark) operation. Note that this set up could also be used for wavelength conversion and incoherent-coherent image conversion.

Using methyl-red-doped nematic liquid crystal (MRNLC) film, Shih et al.[72] has also demonstrated other image processing functions such as edge enhancement and image addition/subtraction, by using the MRNLC film set at the focal plane of a 4-f optical system[72–74] (see Fig. 12.18) to filter out appropriate spatial frequency components. The filtering action is based on the optical intensity limiting effect discussed earlier, and is further explained in the following section. In the study, a 10 µm thick 90° twisted MRNLC film (0.5 wt% methyl-red-doped 5CB pentyl-cyano-biphenyl) is used. When a linearly polarized light passes through this twisted film, the plane of polarization of the incident light follows the twist of the NLC directors and emerges with its polarization axis rotated by 90°. This twisted PNLC film is located at the Fourier plane and oriented so that the LC surface alignment direction D1 is parallel to the transmission axis P1 of the polarizer, as shown in Figure 12.18. The analyzer is oriented so that its P2 is parallel to the LC alignment direction D2.

In the Fourier plane, generally, lower spatial frequency components are associated with higher optical intensity while higher spatial frequency components (emanating from edges of the object) have lower intensity. The low-intensity higher-order components which carry information about the edges within the image pass through the whole optical system without perturbing the nematic alignment, that is, its polarization will follow the twisted nematic and gets rotated by 90°. However, the higher intensities of the lower-order spatial frequency components will induce a reorientation

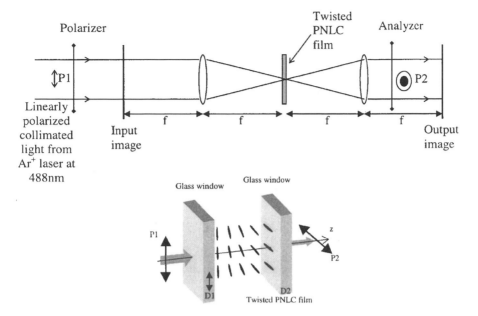

Figure 12.18. Experimental set up. Insert shows the planar twisted nematic liquid crystal (NLC) film. The NLC sample is situated near the Fourier plane.

of the NLC molecules and give rise to a change of their polarization. As a result, the analyzer will partially or totally block these lower-order components. The output image therefore exhibits an edge-enhancement effect.

Using a different setup based on the Weigert effect,[74] all optical image addition and subtraction operations have also been demonstrated with the use of MRNLC films.[72] The response time for these photoinduced molecular reorientational effects within the NLC films are in millisecond regime at an optical power of $\sim$1 mW.

12.3.3. Intelligent Optical Processing

With the advent of micro- and nanotechnologies, imaging techniques/devices/systems have progressed to the point whereby sensors, processors, and inter-and intrapixel communication and output can all be integrated on a single chip or pixel element to enable what has been termed "smart" or intelligent pixel.

Among the various material systems that make up smart pixels, liquid crystals are particularly interesting because of their transparency, broadband electro-optical response, and other unique material properties. Liquid crystal on silicon (LCOS) devices, liquid crystal spatial light modulators (LCSLMS), incorporating neural net functions enabled by the backplane electronics and inter- and intrapixel electronic connections, such as optoelectronic neuron array, LC/silicon retina, and winner take-all circuit, have been developed previously.[75,76] In these devices, the photosensitivity needed for sensing the input comes from the photoconductive layer processed into the liquid crystal device, and require complex and expensive backplane electronic drive to operate.

A recent study[77] has shown that these supranonlinear liquid crystals could perform similar neural-net-like signal in an all-optical and electronic-free manner. Although the response speed of the liquid crystal is much slower that those electronic pixels, the actual processing speed is quite fast, since an entire image can be simultaneously processed, rather than sequentially by converting the input to digital bit trains for digital image processing. The process thus circumvents the complex problem of integrating a large number of electronic connections and optical-electronic converters.

Although the demonstration was a simple image edge enhancement process, such neural-net-like operation can be generalized to more complex objects and image processing operations such as aberration correction by optical phase conjugation and associative memory by self-starting phase conjugation oscillators (see Fig. 12.19), as demonstrated previously[78,79] using inorganic photorefractive crystals.

Another interesting and potentially useful property of nematic liquid crystal is that it is capable of nonlocal photorefractive response, as in C60 or carbon nanotube–doped NLC. The nonlocal response allows one to simulate neural net operation of smart pixels where "connections" to nearest neighbors are made.[75,76,80] In photorefractive materials, the space-charge field distribution is spatially shifted from the incident optical intensity function as the fields originate from a gradient function

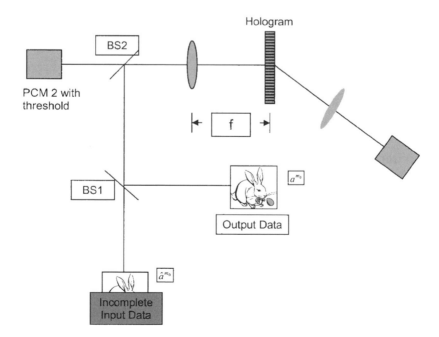

Figure 12.19. Schematic depiction of experimental set up used to demonstrate associative memory with self-starting optical phase conjugation, where a complete image is reconstructed from partial input data (78).

of the optically induced space-charge distribution. In other words, the material responds to the *gradients of the input intensity distribution*, which is basically the same as the operation used in obtaining edge enhancement in image processing.

12.4. HARMONIC GENERATIONS AND SUM-FREQUENCY SPECTROSCOPY

Harmonic generations occupy a special place in the field of nonlinear optics; not only were they the first observed nonlinear optical effect, they are also the most widely researched and used in current lasers and electro-optic technology. While most commercially available materials for harmonic generations are inorganic crystals, organic materials have received considerable interest owing to their inherently large molecular nonlinear susceptibilities and more recently, their chirality.[81,82]

Liquid crystal molecules are well known to be highly anisotropic and noncentrosymmetric and possess large second-order nonlinear molecular polarizability. For a typical liquid crystal such as 8CB (4′-*n*-octyl-4-cyanobipenyl), the polarizability is measured to be on the order of 25×10^{-32} esu,[81] which is much larger than the molecular polarizability of KDP. However, when these molecules assemble themselves in the liquid crystalline phases, they tend to assume configurations where this centroasymmetry is reduced to a vanishing value (e.g., by having the polar direction of molecules or molecular layers lined up in opposite ways). As a result, although the individual molecular polarizability of the liquid crystals is quite large, liquid crystals have not been shown to be efficient harmonic generators. Accordingly, harmonic generations have largely been developed/employed as surface spectroscopic tools,[83] rather than as the means for new laser sources.

The centrosymmetry may, however, be broken by the application of an applied dc electric field;[84] such symmetry can also be broken on a surface either as a freely suspended film or by a surface alignment modification technique that induces the flexoelectric effect.[85] Sukhov and Timashev have shown that the centrosymmetry can also be broken optically.[86] As explained in Chapter 11, the main obstacle in getting efficient harmonic generation is the phase matching of the fundamental and second-harmonic wave vectors.

In the work by Saha and Wong,[84] phase matching is achieved by the birefringent dispersion method discussed in Chapter 11, which is very commonly used for second-harmonic generations in nonlinear crystals. The fundamental wave propagates as the extraordinary ray and the harmonic wave as the ordinary ray. A dc electric field of 15 kV/cm is applied perpendicularly to the director axis of a planar nematic (5CB, *p-n*-pentyl-*p′*-cyanobiphenyl) sample. The observed harmonic signal as a function of the angle of deviation from the phase-matched direction is shown to be in good agreement with theory. From this experiment it appears that if the right geometry is chosen and if one uses the less "lossy" smectic phase, larger interaction lengths and perhaps higher generation efficiency could be attained.

Another possibility to break the centrosymmetry of liquid crystal is to make use of the flexoelectric effect.[85,86] There have been research earlier on second-harmonic generation in smectic and ferroelectric liquid crystals,[87] and recent work on chiral smectic-A liquid crystals.[88] Basically, if a nematic liquid crystal undergoes orientational deformations of the splay or bend type, a spontaneous polarization, the so-called flexoelectric effect, will occur, resulting in a second-order nonlinear susceptibility. In the work by Sukhov and Timashev,[86] a spatially periodic orientational distortion with a wave vector $\vec{q}$ is created by stimulated orientational scattering of an o wave $\vec{E}_R$ into an e-polarized wave $\vec{E}_s$, where $\vec{E}_R$ is derived from a ruby laser. A weakly focused Nd:YAG laser fundamental beam at 1.06 μm is then incident on the sample, propagating along the $\vec{K}_H$ direction; its second harmonic is generated in the same direction with a wave vector $\vec{K}_N$. Phase matching is achieved by matching the orientational distortion grating vector $\vec{q}$ with the phase mismatch $\vec{K}_H - 2\vec{K}_N$ [i.e., $\vec{q}(\alpha_0) = \vec{K}_H - 2\vec{K}_N$, where α is the (phase matched) angle made by $\vec{K}_R$ (the wave vector of the incident ruby laser) with the nematic director axis]. The experimental results obtained by Sukhov and Timashev are in good agreement with their theoretical model.

12.5. OPTICAL SWITCHING

Just as the linear electro-optical effects can be utilized in various electrically controlled switching processes, optically induced refractive index changes can be applied in opto-optical switching. These switching processes may be in the form of mixings of several waves or self-actions. In these processes the fundamental nonlinear optical parameter involved is the intensity-dependent *phase shift*, which involves a combination of the nonlinear index change and the optical path length. Such intensity-dependent shifts are also responsible for bistabilities, differential gain, transistor action, and so forth in nonlinear Fabry-Perot processes.[89]

Optical switching can also be mediated by the *refractive index change alone*. One good example is the nonlinear interference switching process.[90–92] Consider Figure 12.20. If medium 2 possesses an intensity-dependent index, then the transmission/reflection at the interface will also be intensity dependent. If medium 1 is of a higher index than medium 2, a total internally reflected (TIR) optical field in the medium may switch over to the transmission state if medium 2 possesses a positive optical nonlinearity (i.e., its refractive index is an increasing function of the optical intensity). On the other hand, if medium 2 possesses a negative nonlinearity, the reverse switching process is possible (i.e., from a transmissive to a total reflective state). With a proper choice of "antireflection" coating material of refractive index n_1, the device could be highly transmissive at low incident light intensity and could switch to the TIR state at higher incident laser power.

Nematic liquid crystals possess both positive and negative signs of nonlinearities associated with the ordinary and extraordinary thermal index changes, respectively. Furthermore, both index gradients are extraordinarily large near the phase transition temperature. This translates into much smaller optical power (or fluence) needed to turn on the switching processes.

A principal concern in these liquid-crystal-based switching devices is the response time. However, unlike electro-optical effects which involve the molecular reorientations of liquid crystals in their nematic phase and are slow, nonlinear liquid

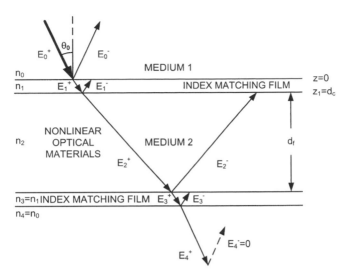

Figure 12.20. Schematic depiction of total internal reflection ↔ transmission switching by a dielectric (medium 1) cladded nonlinear material (medium 2). Two optional thin films of index , on both sides of the nonlinear material, are used for maximizing the transmission of the device in the transmission state. Also shown are the optical electric fields for the reflected, the transmitted, and the incident light.

crystal optical switching devices can actually respond quite quickly. This is because the amount of refractive index change required to turn on the device can be created in a very short time by a sufficiently intense laser pulse. An example has been discussed in the work by Khoo et al.[93] where the self-defocusing and the resulting switching to a low-transmission state can be achieved in nanoseconds, using the laser-induced index change.

Likewise, in the transmission to a TIR switching process as depicted in Figure 12.21, the required index change and therefore the switching to a low-transmission state can be achieved in the nanosecond time scale, using the thermal or density effects of nematic or isotropic liquid crystals. For Q-switched infrared laser pulses (e.g., CO_2 lasers), which are typically in the microsecond regime, nematic films of thickness on the order of a few wavelengths will serve as very good switches[91] as thermal time constants are typically also in the microsecond regime (cf. Chapter 9).

The experiments reported in the work by Khoo et al.[94] have confirmed these observations. Figure 12.21 shows an experimental setup for observing the transmission TIR switching process. In the experiment involving CO_2 laser pulses, the liquid crystal used is E7 ($n_e = 1.75$ and , sandwiched between ZnSe prisms ($n = 2.64$) which are transparent at both the visible and infrared spectral regimes. The sample is homeotropically aligned and is 83 μm thick. The incident laser is p polarized, and thus it "sees' the extraordinary refractive index of the liquid crystal. Note that dn_e/dT is negative, thus the system could undergo transmission → TIR switching.

The incident angle is set 2° away from the TIR ($\theta - \theta_{TIR} = -2°$). It is estimated that, owing to the reflection loss at the air–prism and prism–liquid crystal interfaces, especially near the TIR state, only 20% of the incident laser is effectively incident on the liquid crystal. This fact emphasizes the importance of the antireflection coating[96] design proposed specifically for a ZnSe prism–liquid crystal transmission-TIR cell.

Figure 12.22 shows five oscilloscope traces of the transmitted light corresponding to input powers of 0.46 (lowest curve), 1.05, 2.06, 2.72, and 3.05 W (uppermost curve). At high input power (2.06 W), the transmission of the later portion of the pulse is diminished. The decrease becomes more rapid as the input power is raised. At an input power of 3 W, a "switch-off" time of about 10 ms is measured. This decrease in the response time with increasing input power is in good agreement with the quantitative theory given in the work by Khoo et al.[93,95]

Since the operating temperature is 22°C (room temperature), which is far from the nematic–isotropic transition of E7 ($T_c = 60°C$), this long switch-off time is understandable. If the temperature of the sample is raised to near T_c, the required temperature rise for the required refractive index change will be smaller (because of a much larger thermal refractive index gradient near T_c); the switching time will be shorter. Correspondingly, the required laser energy fluence (in J/cm^2) will also decrease. It is estimated in the work by Lindquist et al.[94] that the threshold fluence required for switching microsecond CO_2 lasers can be as small as 0.25 J/cm^2.

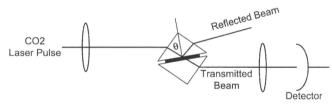

Figure 12.21. Experimental setup for observing the transmission TIR switching process with infrared pulsed or cw lasers.

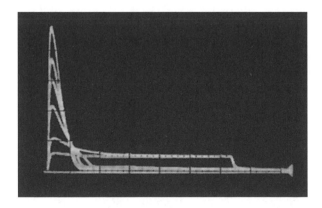

Figure 12.22. Oscilloscope traces of the transmitted infrared laser pulse. Incident pulse width is 130 ms. The traces correspond to increasing input power of 0.46, 1.45, 2.06, 2.72, and 3.05 W.

An alternative material that can be used for transmission-TIR switching is an isotropic liquid crystal. The nonlinearity involved is the density decrease caused by laser heating (i.e., a negative refractive index change). Accordingly, a 25 μm thick sample which consists of an isotropic liquid crystal (TM74A from EM Chemicals) sandwiched between two ZnSe prisms ($n=2.402$) is made.[94] Q-switched CO_2 laser pulses, of total pulse duration on the order of 10 μs as shown in Figure 12.24a, are focused (at a spot diameter of 250 μm) on the sample and directed at incidence angles just below the TIR condition.

Figures 12.23a–12.23c shows the corresponding transmitted CO_2 laser pulse. At $\theta-\theta_c=10°$, the transmitted pulse resembles the input pulse (see Fig. 12.24a). As we approach θ_c, as shown in Figures 12.24b and 12.24c, the later parts of the laser pulses are observed to be greatly attenuated, as a result of the (negative) index induced by the front part of the pulse. The switching threshold fluence is measured to be about 0.25 J/cm^2.

To reduce the transmission-TIR switching/limiting threshold, one possibility is to combine it with a self-defocusing effect (SDE), as shown in Figure 12.24. The process now has the added advantage that the defocusing effect, in conjunction with the aperture placed in front of the detector, would further lower the transmission of the high-power harmful radiation. Note that the limiting process (by SDE) will continue even as the liquid crystal turns into liquid (when the cell effectively functions as a total reflector), in contrast to self-limiting using SDE alone. Greater dynamic ranges as well as lower switching thresholds are therefore expected of such so called hybrid cells.

Study[96] has also shown that prior to transmission-TIR switching, the incident laser will suffer an even more severe intensity-dependent phase shift, the

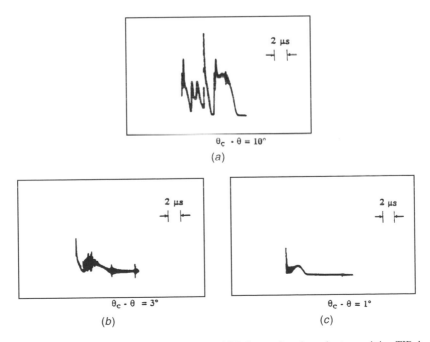

Figure 12.23. Oscilloscope traces of the transmitted CO_2 laser pulses through a transmittion-TIR device for various incident angles near the TIR condition (a) $\theta-\theta_c=10°$; (b) $\theta-\theta_c=3°$; (c) $\theta-\theta_c=1°$.

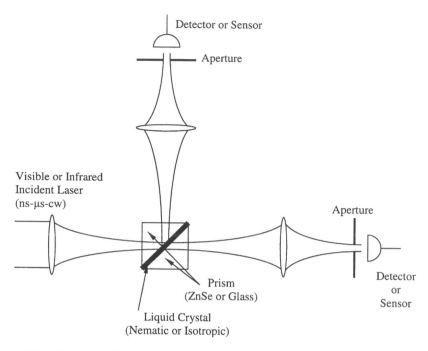

Figure 12.24. Schematic of TIR-transmission switching/limiting device. Note that with the lens system, it can also give rise to a self-defocusing limiting effect in either the transmission of the reflection mode.

so-called tunneling phase shift (TPS), associated with the increasing optical path length of the laser through the liquid crystal film. These TPS effects will further contribute to defocusing of the laser and lower the switching threshold for the self-limiting/switching configuration depicted in Figure 12.24.

From the point of view of waveguiding, the switching of the TIR state to the transmissive state as discussed previously may be viewed as an intensity-dependent optical propagation mode change; that is, the system switches from a guided (total reflection) to a lossy mode (transmission loss). In more well-defined guided wave geometries, such as fibers and planar waveguides (Ref. 97, see also Section 12.1.2), these intensity-dependent guided wave switching phenomena, with nematic liquid crystals as the nonlinear claddings, have also been observed.

12.6. NONLINEAR ABSORPTION AND OPTICAL LIMITING OF SHORT LASER PULSES IN ISOTROPIC PHASE LIQUID CRYSTALS AND LIQUIDS

12.6.1. Introduction

Lasers are now extensively used in various devices and systems. In general, because of the sensitivity of the human eye and sensors, direct intentional or accidental exposures to laser radiation will cause temporary or severe permanent damages to the sensor (eye). For known laser sources, fixed-wavelength filters offer effective means of protection. However, these filters would be totally ineffective against frequency agile lasers (see Table 12.4), the wavelengths of which can range from the UV to far infrared. Next generation electro-optical devices, tunable filters where the transmission window can be electronically tuned away from the incident laser wavelength, are effective against cw lasers but cannot respond fast enough to short laser pulses. Agile frequency pulsed lasers, therefore, pose the most challenging task for eye and sensor protection. Typical laser pulse duration ranges from picoseconds to nanoseconds, with pulse energy ranging from microjoules (μJ) to Joules (J). On the other hand, the typical damage threshold of optical sensors is $\sim$J/cm^2.[98] Typically, these sensors are placed at the focal plane of a focusing optical system/device with a gain of $\sim 10^6$. Assuming an entrance pupil area of 1 cm^2, the so-called maximum permissible exposure (MPE) value at the entrance to the device or optical system is $\sim$1 μJ (1×10^{-6} J), that is, a fraction of the available energy from pulsed laser will damage the sensor irreparably.

In order to protect the sensor, an ideal optical limiting material/device should be one that is transparent at low input light intensity level, and becomes (in a time shorter than the laser pulse duration) increasingly "opaque" as the incident light intensity increases, in such a manner that the energy of the transmitted laser pulse is clamped below the MPE value. Furthermore, the ability to increasingly attenuate the incident light should be maintained for a large range of incident laser pulse energies, from less than a microjoule to over tens of millijoules, that is, an optical density (O.D.) of >4.

Passive all-optical switching, also known as nonlinear self-action effects, offers a practical solution to such challenges. In these processes, the incident light interacting with the nonlinear optical material creates the complex phase shift necessary to change its polarization state, transmission, and other spatio-optical characteristics.

Table 12.4.   cw and Pulsed Laser

Laser Type	Wavelength (μm)	Wavelength Region
Nd:YAG	1.064	
Nd:Glass	1.06	
Nd:YAG*2	0.532	
Ruby	0.694	
Excimer+Raman	0.48	
Ar	0.488–0.514 (Lines)	Visible $\sim$ Near IR
Kr	0.407–0.8 (Lines)	
Diodes	0.7–0.9	
Alexandrite	0.7–0.8 (Continuous)	
Dyes	0.4–0.7 (Continuous)	
Ti:Sapphire	0.65–1.1 (Continuous)	
OPO	0.4–2 (Continuous)	
CO$_2$*2	4.5–5.6 (Lines)	
CO	4.6–5.2 (Lines)	

Since earlier studie,[99] were made tremendous advances have been made recently as a result of the rapid development of highly nonlinear materials capable of multiphotonic absorption processes, and as a result of better understanding of the detailed molecular photonics and dynamics as well as more novel practical device configurations.[100–110] In particular, materials that possess reverse saturable absorption (RSA), two-photon absorption (TPA), and excited state absorption (ESA) properties have been shown to be quite effective for optical limiting applications. In the following sections, we review the fundamentals of these nonlinear absorbers in conjunction with a novel imaging/limiting optical device—nonlinear fiber array[104–108] as schematically depicted in Figure 12.25.

12.6.2. Nonlinear Fiber Array

The fiber array is constructed by filling a capillary array (refractive index of 1.53) with the nonlinear liquid (refractive index ~ 1.61). These liquid filled capillaries thus act as waveguiding cores and the entire structure functions as an *imaging faceplate*, transmitting the image brought to the front focal plane to the exit plane. In this configuration, the scenery is imaged onto an area covering many fibers, whereas the laser, originating at some point source and impinging on the fiber array from a particular direction (spatial frequency), will be imaged (focused) onto a single fiber (a pixel on the imaging plane) (see Fig. 12.25). Thus the fiber array is equivalent to an ensemble of several 100,000 focusing optics+pinhole, each targeted to attenuate intense light sources arriving at the fiber array, that is, a specific spatial frequency component. It may be rightly called an *intensity-dependent spatial frequency filter.*

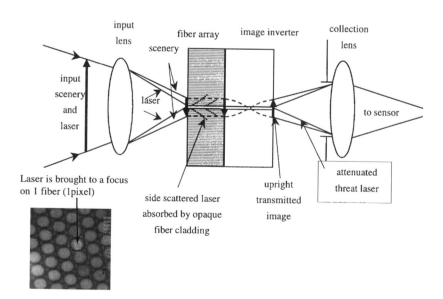

Figure 12.25. Schematic depiction of the nonlinear fiber array placed in the focal plane of an imaging system for optical limiting application.

If the core material is a nonlinear absorber, for example, a TPA material, qualitatively speaking, the transmitted laser intensity I_{out} is related to the input intensity I_n by $I_{out} \sim I_{in} \exp{-\beta(IL)}$, where (IL) is the integrated value of the laser over the interaction length L and β is the two-photon absorption constant, for example. For a given incident laser power P, the IL value for a "free-space" focusing optics (see Fig. 12.26) in a bulk nonlinear liquid is given by

$$(IL)_{bulk} \sim nP/\lambda \tan^{-1}(L/z_0),$$

where $z_0 = \pi n \omega_0^2/\lambda$. On the other hand, in fiber, we have

$$(IL)_{fiber} \sim PL/\pi a^2.$$

With the confined/guided-wave propagation geometry, the fiber array will provide a much larger (IL) value. For example, if $L\sim5$ mm, $l=0.532$ µm, $a=\omega_o\sim15$ µm, and $n=1.5$, we have $(IL)_{fiber}/(IL)_{bulk}\sim7$. For F1 optics used in many sensor systems, the focused beam radius $\omega_o\sim5$ µm, and the corresponding fiber advantage factor will be much larger. This translates to greatly enhanced clamping ability of the fiber array as confirmed by quantitative numerical simulations and experimental studies.[104–108]

12.6.3. RSA Materials (C60 Doped ILC)

Optical limiting by materials exhibiting RSA has been widely studied.[99,101] These materials, for example, C60 molecules, possess a molecular energy level scheme as depicted in Figure 12.27, in which the excited state absorption cross sections σ_1 (of the singlet state) and/or σ_T of the triplet excited state are greater than the ground state cross section σ_o.

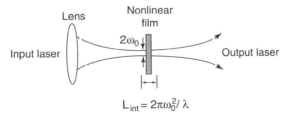

$$L_{int} = 2\pi\omega_0^2/\lambda$$

For $\omega_{int} = 5$ µm, $\lambda = 532$ nm, $L_{int} = 0.3$ mm

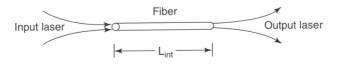

Figure 12.26. Illustration of the extended interaction length provided by fiber waveguide geometry.

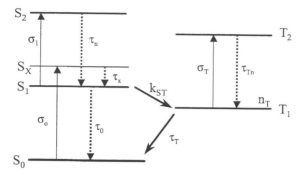

Figure 12.27.   Molecular energy levels of a RSA materials, for example, C60 doped ILC fiber core liquid.

Molecules are excited from the ground state S_0 by laser radiation to a vibrational substate S_x of the first electronic excited singlet state S_1 (or of higher-lying singlet states) with an absorption cross section σ_o. From this state, the molecules decay rapidly ($\tau_x \sim$ picoseconds) to the singlet state S_1. This level relaxes either to the ground state (with rate $1/\tau_o$) or to the triplet state T_1 with intersystem crossing rate k_{ST}. Absorption of laser radiation may further excite the molecules in levels S_1 or T_1 to higher energy states S_2 and T_2, with absorption cross sections σ_1 and σ_T, respectively. These excited states decay rapidly to states S_1 and T_1. The effectiveness of transitions to the triplet state may be limited by a small intersystem crossing rate.

The limiting capability of the RSA mechanism is determined mainly by the cross section ratio $=\sigma_1/\sigma_o$ (or σ_T/σ_1) and the population of the molecular states as they change in time. The cross-section ratio is wavelength dependent, and materials that exhibit RSA at one frequency may not show it at other wavelengths. For example, fullerenes C_{60} and C_{70} show the RSA effect at 308 and 534 nm, but not at 335 nm.[110] For a pure three-level molecular system exhibiting RSA, the transmission function ranges from $T_{lin} = \exp(-\sigma_o N_o L)$ for low-intensity incident beams to $T_{sat} = \exp(-\sigma_1 N_o L)$ for high degrees of excitation. For phthalocyanines and fullerenes, transitions to both excited singlet and triplet states create an effective excited state cross section.

The theoretical simulation of a pulse propagating through the fiber array requires formulating the rate equations and the equation of propagation along the fiber axis. For the molecular energy level scheme, as shown in Figure 12.27, we have

$$\frac{dS_0(t,z)}{dt} = -\frac{\sigma_0}{h\nu}IS_0 + \frac{1}{\tau_1}S_1 + \frac{1}{\tau_T}T_1,$$

$$\frac{dS_1(t,z)}{dt} = \frac{\sigma_0}{h\nu}IS_0 - \frac{1}{\tau_1}S_1 - \frac{1}{\tau_{ST}}S_1 + \frac{S_2}{\tau_2} - \frac{\sigma_1}{h\nu}IS_1,$$

$$\frac{dS_2(t,z)}{dt} = -\frac{S_2}{\tau_2} + \frac{\sigma_1}{h\nu}IS_1,$$

$$\frac{dT_1(t,z)}{dt} = \frac{S_1}{\tau_{ST}} - \frac{\sigma_T}{h\nu}IT_1 + \frac{T_2}{\tau_3} - \frac{1}{\tau_T}T_1,$$

$$\frac{dT_2(t,z)}{dt} = \frac{\sigma_T}{h\nu}IT_1 - \frac{T_2}{\tau_3}.$$

Assuming a multimode laser intensity distribution in the fiber, the nonlinear absorption of the laser intensity is described by

$$\frac{dI(t,z)}{dz} = -I\alpha_{nl},$$

with α_{nl} the nonlinear optical absorption coefficient:

$$\alpha_{nl} = \sigma_0 S_0 + \sigma_1 S_1 + \sigma_T T_1.$$

In the numerical solution of these coupled equations, the shape of the incident laser pulse is assumed to be Gaussian, with FWHM pulse duration t_p. The range of t_p covers picoseconds to nanoseconds. Note that because the intersystem crossing times are much greater than picoseconds, the excited state contribution usually becomes significant for nanosecond laser pulses.

For illustration purposes, we have computed the steady-state nonlinear transmission of the C_{60}-ILC filled fiber array. The results for a 3-mm long fiber are shown in Figures 12.28a and 12.28b for a 3 mm long fiber. The material parameters used are experimental values obtained for the singlet state of C_{60} at a wavelength of 532 nm.[110] These nonlinear transmission curves at low and high input laser energies are plotted in Figures 12.28a and 12.28b, respectively. At C60 concentration of $S_T = 7.22 \times 10^{17}$ cm^{-3} or $S_T - 1.9 \times 10^{18}$ cm^{-3}, the respective linear transmission are 50% and 15%. As expected, the higher concentration offers a greater dynamic range at the expense of linear transmission.

These simulations, and more realistic simulations involving pulsed lasers,[108] show that such RSA material cored fiber arrays are capable of clamping the transmission of the laser pulse energy to below the sensor/eye MPE level for incident laser energies up to a certain point, corresponding to the *minimum transmission value*, as indicated in Figure 12.28. Beyond that, the absorbing molecular levels will be depopulated by the laser, and the transmission will begin to increase, that is, the limiting action begins to deteriorate.

The dynamic range (input intensity range for which the material is able to limit effectively) depends on various material and optical parameters. For the molecular energy levels as depicted in Figure 12.27, one of the critical parameters is the decay rate τ_n and τ_{Tn}, which determine how fast the excited levels S_1 and T_1 can repopulate after the molecule has been excited to levels S_2 and T_2. Shorter decay times translate into a greater availability of molecules in the excited level which will aid in the nonlinear absorption (limiting) process. The dynamic range also

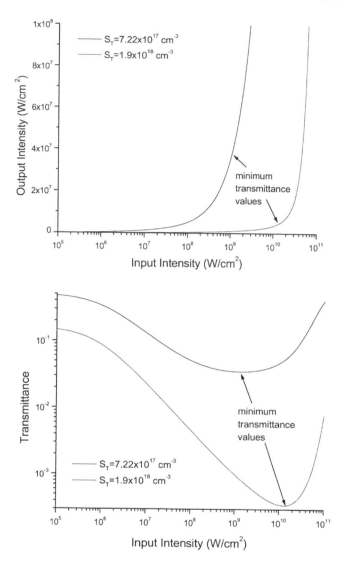

Figure 12.28. (a) Limiting curve and (b) nonlinear transmission of C60-ILC cored fiber array (3mm length and 20 μm core diameter). The parameters used are wavelength $\lambda = 532$ nm; cross sections $\sigma_0 = 3.2 \times 10^{-18}$ cm^2, $\sigma_1 = 1.6 \times 10^{-17}$ cm^2, $\sigma_T = 1.4 \times 10^{-17}$ cm^2; relaxation times $\tau_0 = 32.5$ ns, τ_n and $\tau_{Tn} \sim 1$ ps; and intersystem crossing times $\tau_{ST} = 1.35$ ns and $\tau_T = 40$ ms.

depends on the absorption cross-section ratio $R = \sigma_1/\sigma_o$ (or σ_T/σ_1), which will help determine the rate at which transitions from the ground state to upper levels take place.

For large dynamic range operation, one special advantage of C60-ILC (liquid) cored fiber array is that in the high input intensity regime, the doped ILC will exhibit nonlinear scattering and defocusing effect. These processes, together with

the light blocking action[104–108] of the opaque fiber cladding, provide further attenuation of the laser transmission. Also, for laser energies above about 1 mJ, bubbles will be created, which will cause very wide-angle scattering of the laser as
well and greatly lower the laser transmission. The bubbles stay in place for a few
minutes, or occasionally float away in a few seconds. The photos in Figure 12.29
depict the exit end view of the fiber array following this sequence of events
(before and after the laser pulses). Studies have shown that under these incident
laser energy levels ($\sim$ a few mJ), the fiber array does not suffer catastrophic damage and still allow acceptable viewing while providing below MPE level clamping
on the laser transmission.

12.6.4.   Optical Limiting by TPA Materials (L34 Fiber Core Liquid)

Two-photon absorbers[103,107–109] (TPA) are particularly desirable because they are
transparent at low light levels, and become increasingly absorptive at higher laser
intensities. The limiting behavior of TPA materials can be further enhanced by the
presence of higher-order nonlinear processes such as excited-state absorption.
Among all the neat organic liquids investigated so far, the so-called L34 discovered and reported in the works by Khoo et al.[107,108] has been shown to possess
both large TPA and higher ESA cross sections, and a molecular energy level diagram shown in Figure 12.30 that is ideally suited for optical limiting application.
The ground singlet state is connected to a two-photon excited singlet state, which
is in turn connected by single-photon absorption to high-lying singlet states, or
via intersystem crossing, to some excited triplet state that could also undergo
single-photon absorptions.

Femto- and picosecond nonlinear transmission (z scan) measurements give the
intrinsic TPA coefficient $\beta \sim 4.7$ cm/GW, whereas nanosecond studies show that β
is intensity dependent, and ranges from 4.7 at low intensity to over 170 cm/GW at
high intensity. The latter is attributed to excited-state contribution, which is confirmed in picosecond dynamical pump-probe studies.[104–108]

If we consider a multimode nanosecond laser pulse propagating through a
fiber core made of L34, at each position and time, the rate equations for the

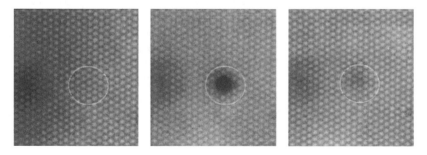

Figure 12.29.   Photographs of the exit plane of the fiber array before, right after, and a few minutes after the
high energy laser pulse. During live imagery, we observed a large bubble float out of the sample test region.

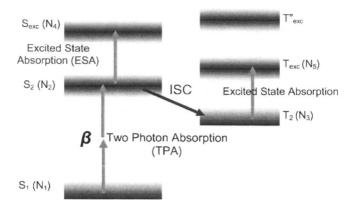

Figure 12.30. Molecular energy level scheme of the neat liquid L34.The symbol N's denote population densities of the corresponding level.

molecular level population densities $N(z,t)$'s are given by (using the energy level scheme of L34)

$$\frac{\partial N_1}{\partial t} = -\left(\frac{\beta I^2}{2h\nu\, N_{0A}} + \frac{\alpha_g I}{h\nu\, N_{0A}}\right) N_1,$$

$$\frac{\partial N_2}{\partial t} = \frac{\beta I^2}{2h\nu\, N_{0A}} N_1 - \frac{1}{\tau_2} N_2 - \alpha_{exc} I N_2 + \frac{\alpha''_{exc} I}{h\nu\, N_{0A}}(N_4 - N_2) + \frac{1}{\tau_4} N_4,$$

$$\frac{\partial N_3}{\partial t} = -\frac{\alpha''_{exc} I}{h\nu\, N_{0A}}(N_3 - N_5) - \frac{1}{\tau_3} N_3 + \frac{1}{\tau_2} N_2 + \frac{1}{\tau_5} N_5,$$

where α_g accounts for the linear absorption or scattering loss of the ground state liquids (α_g of L34, for example, is $\sim$0.1 cm^{-1}), β is the intrinsic two-photon absorption constant, I is the laser intensity, and α_{exc} and α''_{exc} are excited-state absorption constants. In these equations, we have included the possibility of stimulated downward transitions between levels connected by single-photon transitions (e.g., between S_2 and S_{exc} and T_2 and T_{exc}). This would be the case if the laser pulse width (e.g., femtoseconds) is shorter than the relaxation time constants of the upper levels. For the nanosecond time scale, one can ignore those stimulated downward transition terms.

For typical fiber array core diameter and refractive index difference between the core and cladding, the waveguiding process is of a multimode nature,[108] and allows an assumption of uniform radial intensity distribution; that is, both the level populations N's and the laser intensity I are functions of the propagation length z and time t only. In this regime, the laser intensity I is described by an equation of the form

$$\frac{dI}{dz} = -\left(\alpha_g \frac{N_1}{N_{0A}} I + \beta \frac{N_1}{N_{0A}} I^2 + \alpha_{exc} \frac{N_2}{N_{0A}} I + \alpha''_{exc} \frac{N_3}{N_{0A}} I + \cdots\right). \quad (12.3)$$

As an example, we calculated and compared the limiting action and the dynamic range of fiber core made up of materials that possess *only* two-photon absorption (TPA) and a material which possesses both TPA and ESA with varying degrees of population recycling rates (i.e., τ_5 values).

Figures 12.31a–12.31c show the detailed spatial and temporal evolutions of a nanosecond laser pulse as it propagates through the fiber core made of different types of TPA materials. At very low input energies (~ 1 μJ), the excited states become populated. Level N_2 is usually sparingly populated (even at high energies of 200 μJ), but it is useful as a transition level between the two-photon absorption and the excited-state absorption scenarios. Most of the limiting takes place in the entrance region of the fiber, and although the ground level may become depleted, the limiting action may still go on due to continuous transitions between excited states N_5 and N_3. The excited-state transition processes account for an intensity-dependent effective two-photon absorption coefficient that is greater than the intrinsic value.[108]

If no transition is allowed between the high-lying electronic states N_3 and N_5, the limiting action although slightly better than a pure two-photon transition model, does not compare with the complete model which "clamps" the output at a very low level (see Fig. 12.32). Accordingly, materials possessing both TPA and ESA properties

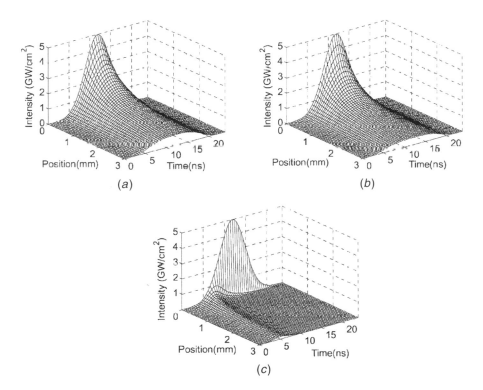

Figure 12.31.   Temporal propagation of a 120 μJ pulse through a 3mm fiber array for (a) a pure TPA; (b) TPA + ESA (no transition N_3 to N_5 allowed); and (c) the complete model of Figure 12.30 with $\tau_2 = 0.1$ns and $\tau_3 = 1$ ps. Laser Pulse FWHM is 7ns. Figure (c) shows that the laser pulse is effectively "limited" with a "clamped" transmission below the MPE level.

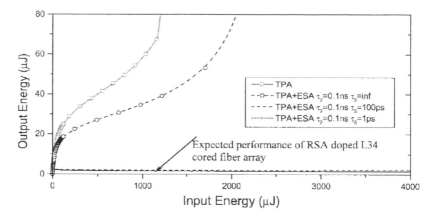

Figure 12.32. Limiting Curves for liquid cored 3mm fiber array (wavelength $\lambda = 532$ nm, pulse width $\tau_p = 7$ns) with core materials that possess TPA or TPA+RSA properties. The parameters used are $\alpha_g = 0.136$ cm^{-1}, $\alpha_{exc} = 10 \times 10^4$ cm^{-1}, $\beta = 4.11$cm/GW. Fiber core diameter: 20 µm; fiber length: 5 mm. L34 will have large dynamic range as indicated.

have a greater dynamic range, being more effective for optical limiting applications. Furthermore, the linear transmission is very high ($\sim$90% for the 3 mm fiber array described). Such transparency allow further doping with RSA materials, and the resulting nonlinear liquid that possess RSA+TPA+ESA will be expected to have a low threshold but very large dynamic range.

12.7.   CONCLUSION

In conclusion, liquid crystals are complex and are wonderful electro-optical and non-linear optical materials. Many novel, interesting, and useful materials (with negative, zero or positive refractive indices), processes, phenomena, and devices are awaiting our further exploration.[111]

REFERENCES

1. Wong, G. K. L., and Y. R. Shen. 1974. Transient self-focusing in a nematic liquid crystal in the isotropic phase. *Phys. Rev. Lett.* 32:527.

2. Khoo, I. C., H. Li, and Y. Liang. 1974. Observation of orientational photorefractive effects in nematic liquid crystals. *Opt. Lett.* 19:1723.

3. Khoo, I.C., S. Slussarenko, B. D. Guenther, and W. V. Wood. 1998. Optically induced space charge fields, dc voltage, and extraordinarily large nonlinearity in dye-doped nematic liquid crystals. *Opt. Lett.* 23:253–255.

4. Khoo, I. C., M. V. Wood, B. D. Guenther, Min-Yi Shih, P. H. Chen, Zhaogen Chen, and Xumu Zhang. 1998. Liquid crystal film and nonlinear optical liquid cored fiber array for ps-cw frequency agile laser optical limiting application. *Opt. Express.* 2(12):471–82.

5. Khoo, I. C., M. V. Wood, M. Y. Shih, and P. H. Chen. 1990. Extremely nonlinear photosensitive liquid crystals for image sensing and sensor protection. *Opt. Express.* 4(11):431–442.

6. Khoo, I. C., J. Ding, A. Diaz, Y. Zhang, and K. Chen. 2002. Recent studies of optical limiting, image processing and near-infrared nonlinear optics with nematic liquid crystals. *Mol. Cryst. Liq. Cryst.* 375:33–44; see also Ref. 108.

7. For a review, see for example, Boardman, A. D. (ed.) *Soliton Driven Photonic.* 2001. NATO ASI 976507. Dordrecht: Kluwer Academic.

8. Stegeman, G. I., D. N. Christodoulides, and M. Segev. 2001. *IEEE J. Sel. Top. Quantum Electron.* 6:1419.

9. Peccianti, M., A. Derossi, G. Assanto, A. de Luca, C. Umeton, and I. C. Khoo. 2000. *Appl. Phys. Lett.* 77:7.

10. Conti, C., M. Peccianti, and G. Assanto. 2005. Spatial solitons and modulational instability in the presence of large birefringence: The case of highly nonlocal liquid crystal. *Phys. Rev. E.* 72:066614, and references therein.

11. Khoo, I. C., and Shu-Lu Zhuang. 1980. Nonlinear optical amplification in a nematic liquid crystal above the Freederick's transition. *Appl. Phys. Lett.* 37:3; Herman, R. M., and R. J. Serinko. 1979. Nonlinear-optical processes in nematic liquid crystals near Freedericksz transitions. *Phys. Rev. A.* 19:1757.

12. Mitchell, M., Z. Chen, M. Shih, and M. Segev. 1996. *Phys. Rev. Lett.* 77: 490.

13. Christodoulides, D. N., T. Coskun, M. Mitchell, and M. Segev. 1997. *Phys. Rev. Lett.* 78:646; 80:2310.

14. D'Alessandro, G., and W. J. Firth. 1991. Spontaneous hexagon formation in a nonlinear optical medium with feedback mirror. *Phys. Rev. Lett.* 66:2597.

15. See, for example, Arecchi, F. Tito, Stefano Boccaletti, and Pier Luigi Ramazza. 1999. *Phys. Rep.* 318:1–83 and references therein.

16. Khoo, I. C., J. Y. Hou, T. H. Liu, P. Y. Yan, R. R. Michael, and G. M. Finn. 1987. Transverse self-phase modulation and bistability in the transmission of a laser beam through a nonlinear thin film. *J. Opt. Soc. Am. B.* 4:886.

17. Khoo, I. C., H. Li, and Y. Liang. 1994. Observation of orientational photorefractive effects in nematic liquid crystals. *Opt. Lett.,* 19:1723.

18. Pagliusi, P., R. Macdonald, S. Busch, G. Cipparrone, and M. Kreuzer. 2001. Nonlocal dynamic gratings and energy transfer by optical two-beam coupling in a nematic liquid crystal owing to highly sensitive photoelectric reorientation. *J. Opt. Soc. Am. B.* 18:1632–1638.

19. Khoo, I. C. 1996. Orientational photorefractive effects in nematic liquid crystal films. *IEEE J. Quantum Electron.* 32:525–534.

20. Fuh, Andy Y.-G., C. C. Liao, K. C. Hsu, C. L. Lu, and C. Y. Tsue. 2001. Dynamic studies of holographic gratings in dye-doped liquid-crystal films. *Opt. Lett.* 26:1767–1769.

21. Ono, H., and N. Kawatsuki. 1997. Orientational photorefractive gratings observed in polymer dispersed liquid crystals doped with fullerene. *Jpn. J. Appl. Phys.* 36:6444 – 6448.

22. Antipov, O. L. 1993. Mechanism of self-pumped phase-conjugation by near-forward stimulated scattering of heterogeneous laser-beams in nematic liquid-crystal. *Opt. Commun.* 103:499; Antipov, O. L., N. A. Dvoryaninov, and V. Sheshkauskas. 1991. Parametric generation and phase conjugation of intersecting laser-beams in a layer of a nematic liquid-crystal containing a dye. *JETP. Lett.* 53:610.

23. Sanchez, F., P. H. Kayoun, and J. P. Huignard. 1988. 2-wave mixing with gain in liquid-crystal at 10.6 μm wavelength. *J. Appl. Phys.* 64:26.

24. Brignon, A., I. Bongrand, B. Loiseaux, and J.-P. Huignard. 1997. Signal-beam amplification by two-wave mixing in a liquid-crystal light valve. *Opt. Lett.* 22:1855–1857.

25. Bartkiewicz, S., A. Miniewicz, F. Kajzar, and M. Zagorska. 1998. Observation of high gain in a liquid-crystal panel with photoconducting polymeric layers. *Appl. Opt.* 37:6871–6877.

26. See, for example, Gunter, P., and J. P. Huignard, eds. 1989. *Photorefractive Materials and their Applications.* Vols. 1 and 2. Berlin: Springer-Verlag. See also Ref. 56.

27. White, J. O., M. Cronin-Golomb, B. Fischer, and A. Yariv. 1982. Coherent oscillation by self-induced gratings in the photorefractive crystal $BaTiO_3$. *Appl. Phys. Lett.* 40:450.

28. Cronin-Golomb, M., B. Fisher, J. O. White, and A. Yariv. 1983. Passive phase conjugate mirror based on self-induced oscillation in an optical ring cavity. *Appl. Phys. Lett.* 42:919.

29. Khoo, I. C., Yu Liang, and Hong Li. 1992. Coherent-beam amplification and polarization switching in a birefringent medium. *IEEE J. Quantum Electron.* 28:1816–1824. proposed the theory. See also Eichler, H. J., G. Hilliger, G. R. Macdonald, and P. Meindl. 1997. Optical vector wave mixing processes in nonlinear birefringent nematic liquid crystals. *Phys. Rev. Lett.* 78:4753–4756 for experimental confirmation.

30. Etchegoin, P., and R. T. Phillips. 1997. Stimulated orientational scattering and third-order nonlinear optical processes in nematic liquid crystals. *Phys. Rev.E.* 55:5603–5612.

31. Khoo, I. C. and Y. Liang, 1995. Observation of stimulated orientational scattering and cross-polarized self-starting phase-conjugation in a nematic liquid-crystal film. *Opt. Lett.* 20:130–132.

32. Kaczmarek, Malgosia, Min-Yi Shih, Roger S. Cidney, and I. C. Khoo. 2002. *IEEE J. Quantum Electron.* 38:451–457.

33. Khoo, I. C., M. Kaczmarek, M. Y. Shih, M. V. Wood, A. Diaz, J. Ding, and Y. Zhan. 2002. *Mol. Cryst. Liq. Cryst.* 374:315–324.

34. Collings, N., M. Bouvier, B. Züger, and J. Grupp. 1998. *Mol. Cryst. Liq. Cryst.* 320:277–285.

35. Ya. Zeldovich, B., S. K. Merzlikin, N. F. Pilipetskii, and A. V. Sukhov, 1985. Observation of stimulated forward orientational light scattering in a planar nematic liquid crystal. *JETP Lett.* 41:515.

36. Khoo, I. C., and Y. Liang. 2000. Stimulated orientational and thermal scatterings and self-starting optical phase conjugation with nematic liquid crystals. *Phys. Rev. E.* 62:6722–6733; see also Khoo, I. C., H. Li, and Yu Liang. 1993. Self-starting optical phase conjugation in dyed nematic liquid crystal with stimulated thermal scattering effect. *Opt. Lett.* 18:3.

37. Khoo, I. C., and J. Ding. 2002. All-optical cw laser polarization conversion at 1.55 micron by two beam coupling in nematic liquid crystal film. *Appl. Phys. Lett.* 81:2496–2498.

38. Khoo, I.C., and A. Diaz. 2003. Nonlinear dynamics in laser polarization conversion by stimulated scattering in nematic liquid crystal films. *Phys. Rev. E.* 68:042701-1 4.

39. Khoo, Iam Choon , Jianwu Ding, and Andres Diaz. 2005. Dynamics of cross-polarization stimulated orientation scattering in nematic liquid crystal film. *J. Opt. Soc. Am. B.* 22:844–851.

40. Peccianti, M., K.A. Brzdkiewicz, and G. Assanto. 2002. *Opt. Lett.* 27:1460.

41. Sandfuchs, O., F. Kaiser, and M. R. Belic. 2001. *Phys. Rev. A.* 64:063809.

42. Simpson, T. B., J. M. Liu, A. Gavrielides, V. Kovanis, and P. M. Alsing. 1995. *Phys. Rev. A.* 51:4181.

43. Odoulov, S. G., B. I. Sturman, S. Shamonina, and K. H. Ringhofer. *Opt. Lett.* 21:854–856.

44. Heurich, J., H. Pu, M. G. Moore, and P. Meystre. 2001. *Phys. Rev. A.* 63:033605.

45. Carbone, V., G. Cipparrone, C. Versace, C. Umeton, and R. Bartolino. 1996. *Phys. Rev. E.* 54:6948–6951; Cipparrone, G., V. Carbone, C. Versace, C. Umeton, R. Bartolino, and F. Simoni. 1993. *ibid.* 47:3741; Russo, G., V. Carbone, and G. Cipparrone. 2000. *ibid.* 62:5036; Carbone, V., G. Cipparrone, and G. Russo. 2001. *ibid.* 63:51701.

46. Demeter, G., and L. Kramer. 2001. *Phys. Rev. E.* 64:20701.

47. Demeter, G. *Phys. Rev. E.* 61:6678.

48. See, for example, Ya. Zeldovich, B., N. F. Pilipetsky, and V. V. Shkunov. 1985. *Principles of Phase Conjugation.* Vol. 42, Springer Series in Optical Sciences. Berlin: Springer-Verlag.

49. See, for example, Hoppensteadt, F. C., and E. M. Izhikevich. 1999. *Phys. Rev. Lett.* 82:2983–2986.

50. Khoo, I. C., and S. L. Zhuang. 1981. Wave front conjugation in nematic liquid crystal films. *IEEE J. Quantum. Electron.* QE-18:246.

51. Leith, E. N., Hsuen Chen, Y. S. Cheng, G. J. Swanson, and I. C. Khoo. 1984. Coherence reduction in phase conjugation imaging. In *Proceedings of the 5th Rochester Conference on Coherence and Quantum Optics.* E. Wolf and L. Mandel (eds.). 1155. London: Plenum.

52. Huignard, J. P., J. P. Herriau, P. Aubourg, and E. Spitz. 1979. Phase-conjugate wavefront generation via real-time holography in Bi12SiO2 crystals. *Opt. Lett.* 4:21.

53. Sanchez, F., P. H. Kayoun, and J. P. Huignard. 1988. Two-wave mixing with gain in liquid crystal at 10.6 μm wavelength. *J. Appl. Phys.* 64.26.

54. Richard, L., J. Maurin, and J. P. Huignard. 1986. Phase conjugation with gain at CO2 laser line from thermally induced grating in nematic liquid crystals. *Opt. Commun.* 57:365.

55. Khoo, I. C., P. Y. Yan, G. M. Finn, T. H. Liu, and R. R. Michael. 1988. Low power (10.6 μm) laser beam amplification via thermal grating mediated degenerate four wave mixing in a nematic liquid crystal film. *J. Opt. Soc. Am. B.* 5:202.

56. Solymar, L., D. J. Webb, and A. Grunnet-Jepsen. 1996. *The Physics and Applications of Photorefractive Materials.* Oxford: Clarendon. and references therein; See also Ref. 26.

57. Khoo, I. C., R. R. Michael, and P. Y. Yan. 1987. Stimultaneous occurrence of phase conjugation and pulse compression in stimulated scatterings in liquid crystal mesophases. *IEEE J. Quantum Electron.* QE-23:1344.

58. Fekete, D., J. Au Yeung, and A. Yariv. 1980. Phase conjugate reflection by degenerate four wave mixing in a nematic crystal in the isotropic phase. *Opt. Lett.* 5:51.

59. Madden, P. A., F. C. Saunders, and A. M. Scott. 1986. Degenerate four-wave mixing in the isotropic phase of liquid crystals: The influence of molecular structure. *IEEE J. Quantum Electron.* QE-22:1287.

60. Khoo, I. C., H. Li, P. G. Lopresti, and Yu. Liang. 1994. Observation of optical limiting and back scattering of nanosecond laser pulses in liquid crystal fibers. *Opt. Lett.* 19:530.

61. Gruneisen, M. T., and J. M. Wilkes. 1997. Compensated imaging by real-time hologra-phy with optically addressed spatial light modulators. In *Spatial Light Modulators*. G. B. Burdge and S. C. Esener (eds.). OSA TOPS Vol. 14.

62. Boyd, R. W. 1992. *Nonlinear Optics*. San Diego: Academic Press.

63. Antipov, O. L., and D. A. Dvoryaninov. 1991. Parametric generation phase conjugation of intersecting laser beams in a layer of nematic liquid crystal containing a dye. *JETP Lett*. 53:611.

64. See, for example, Khoo, I. C., N. Beldyugina, H. Li, A. V. Mamaev, and V. V. Shkunov. 1993. Onset dynamics of self-pumped phase conjugation from speckled noise. *Opt. Lett*. 18:473, and references therein.

65. Khoo, I. C., Min-Yi Shih, M. V. Wood, B. D. Guenther, P. H. Chen, F. Simoni, S. Slussarenko, O. Francescangeli, and L. Lucchetti. 1999. Dye-doped photorefractive liq-uid crystals for dynamic and storage holographic grating formation and spatial light modulation. IEEE Proceedings Special Issue on Photorefractive Optics: Materials, Devices and Applications. *IEEE Proc*. 87(11):1897–1911.

66. Oudar, J. L. 1986. In *Nonlinear Optics Materials and Devices*, C. Flytzanis and J. L. Oudar (eds.). Berlin: Springer. p. 91.

67. Chemla, D. S. 1986. In *Nonlinear Optics Materials and Devices*, C. Flytzanis and J.L. Oudar (eds.). Berlin: Springer. p. 65.

68. Moerner, W. E., and S. M. Silence. 1994. Polymeric photorefractive materials. *Chem. Rev*. 94:127–155; see also Ref. 1.

69. Gunter, P., and J. P. Huignard (eds.). 1989. Photorefractive materials and their applica-tions, Vols. I and II. Berlin: Springer-Verlag.

70. Thoma, R., N. Hampp, C. Brauchle, and D. Oesterhelt. 1991. Bacteriorhodopsin films as spatial light modulators for nonlinear-optical filtering. *Opt. Lett*. 16:651.

71. Khoo, I. C., M. V. Wood, M. Y. Shih, and P. H. Chen. 1999. Extremely nonlinear photosen-sitive liquid crystals for image sensing and sensor protection. *Opt. Express*. 4(11):431–442; Shih, M. Y., I. C. Khoo, A. Shishido, M. V. Wood, and P. H. Chen. 2000. All-optical image processing with a supra nonlinear dye-doped liquid crystal film. *Opt. Lett*. 25:978–980.

72. Shih, M. Y., A. Shishido, and I. C. Khoo. 2001. All-optical image processing by means of photosensitive nonlinear liquid crystal film: edge enhancement and image addition/sub-traction. *Opt. Lett*. 26:1140–1142.

73. Gorecki, C., and B. Trolard. 1998. Optoelectronic implementation of adaptive image pro-cessing using hybrid modulations of Epson liquid crystal television: applications to smoothing and edge enhancement. *Opt. Eng*. 37:924–930.

74. See, for example, Jonathan, J.M.C., and M. May. 1980. Application of the Weigert effect to optical processing in partially coherent light. *Opt. Eng*. 19:828–833.

75. Johnson, K. M., D. J. Mcknight, and I. Underwood. 1993. *IEEE JQE*. 29:699–714.

76. Mears, R. J., W. A. Crossland, M. P. Dames, J. R. Collington, M. C. Parker, S. T. war, T. M. Wilkinson, and A. B. Davey. 1996. *IEEE JSTQE*. 2:35–46.

77. Khoo, I. C., K. Chen, and A. Diaz. 2003. All-optical neural-net like image processing with photosensitive nonlinear nematic film. *Opt. Lett*. 28:2372–2374.

78. Owechko, Y. *IEEE J. Quantum Electron*. 25:619–634, and references therein.

79. Athale, R. A., H. H. Szu, and C. B. Friedlander. 1986. *Opt. Lett*. 11:482–484.

80. Krishnamoorthy, A. V., and D. A. B. Miller. 1996. *IEEE JSTQE*. 2:55–76; Vorontsov, M. A. *J. Opt. Soc. Am. A*. 16:1623–1637.

81. Shen, Y. R. 1989. Studies of liquid crystal monolayers and films by optical second harmonic generation. *Liq. Cryst.* 5:635; see also Shen, Y. R. 1989. Surface properties probed by second-harmonic and sum-frequency generation. *Nature (London)*. 337:519.

82. Belkin, M. A., S. H. Han, X. Wei, and Y. R. Shen. 2001. Sum-frequency generation in chiral liquids near electronic resonance. *Phys. Rev. Lett.* 87:113001; see also Zhuang, X., P. B. Miranda, D. Kim, and Y. R. Shen. 1999. Mapping molecular orientation and conformation at interfaces by surface nonlinear optics. *Phys. Rev. B*. 59:12632–12640.

83. Oh-e, M., H. Yokoyama, S. Yorozuya, K. Akagi, M. A. Belkin, and Y. R. Shen. 2005. Chirality probed by sum-frequency vibrational spectroscopy for helically structured conjugated liquid crystalline polymers. *Mol. Cryst. Liq. Cryst.* 436:1027–1035.

84. Saha, S. K., and G. K. Wong. 1979. Phase-matched electric-field-induced second-harmonic generation in a nematic liquid crystal. *Opt. Commun.* 30:119; see also Ou-Yang, Z.-C., and Y.-Z. Xie. 1986. Theory of second-harmonic generation in liquid crystals. *Phys. Rev. A*. 32:1189.

85. Gu, S.-J., S. K. Saha, and G. K. Wong. 1986. Flexoelectric induced second-harmonic generation in a nematic liquid crystal. *Mol. Cryst. Liq. Cryst.* 69:287.

86. Sukhov, A. V., and R. V. Timashev. 1990. Optically induced deviation from central symmetry; lattices of quadratic nonlinear susceptibility in a nematic liquid crystal. *JETP Lett.* 51(7):415; see also Baranova, N. B., and B. Ya. Zeldovich. 1982. *Dokl. Akad. Nauk SSSR.* 263:325; *Sov. Phys. –Dokl.* 27:222.

87. Liu, J. Y., M. G. Robison, K. M. Johnson, D. M. Wabba, M. B. Ros, N. A. Clark, R. Shao, and D. Doroski. 1990. Second harmonic generation in ferroelectric liquid crystals. *Opt. Lett.* 15:267; see also Taguchi, A., Y. Oucji, H. Takezoe, and A. Fukuda. 1989. *Jpn. J. Appl. Phys.* 28:997; Barnik, M. I., L. M. Blinov, A. M. Dorozhkin, and N. M. Shtykov. 1981. Generation of the second optical harmonic induced by an electric field in nematic and smectic liquid crystals. *Sov. Phys. JETP.* 54:935.

88. Han, S. H., M. A. Belkin, and Y. R. Shen. 2004. Optically active second-harmonic generation from a uniaxial fluid medium. *Opt. Lett.* 29:1527–1529.

89. Khoo, I. C., J. Y. Hou, R. Normandin, and V. C. Y. So. 1983. Theory and experiment on optical bistability in a Fabry-Perot interferometer with an intracavity nematic liquid-crystal film. *Phys. Rev. A*. 27:3251.

90. Kaplan, A. E., 1977. Theory of hysteresis reflection and refraction of light by a boundary of nonlinear medium. *Sov. Phys. JETP.* 45:896.

91. Khoo, I. C., and P. Zhou. 1989. Dynamics of switching total internal reflection to transmission in a dielectric cladded nonlinear film. *J. Opt. Soc. Am. B*. 6:884.

92. Khoo, I. C., W. Wang, F. Simoni, G. Cipparrone, and D. Duca. 1991. Experimental studies of the dynamics and parametric dependences of total-internal-reflection to transmission switching and limiting effects. *J. Opt. Soc. Am. B*. 8:1464.

93. Khoo, I. C., R. R. Michael, and P. Y. Yan. 1987. Theory and experiments on optically induced nematic axis reorientation and nonlinear effects in the nanosecond regime. *IEEE J. Quantum Electron.* QE-23:267.

94. Lindquist, R. G., P. G. Lopresti, and I. C. Khoo. 1992. Infrared and visible laser induce thermal and density nonlinearity in nematic and isotropic liquid crystals. *Proc. SPIE.* 1692:148–158.

95. Khoo, I. C., Ping Zhou, R. R. Michael, R. G. Lindquist, and R. J. Mansfield. 1989. Optical switching by a dielectric cladded nematic film. *IEEE J. Quantum Electron.* QE-25:1755.

96. Khoo, I. C., and P. Zhou. 1992. Nonlinear interface tunneling phase shift. *Opt. Lett.* 17:1325. See also P. Zhou and I. C. Khoo. 1993. Anti-reflection coating for a nonlinear transmission to total reflection switch *Int. J. Nonlinear Opt. Phys.* 2(3). p. 437.

97. Vach, H., C. T. Seaton, G. I. Stegeman, and I. C. Khoo. 1984. Observation of intensity-dependent guided waves. *Opt. Lett.* 9:238; see also Goldburt, E. S., and P. St. J. Russell. 1985. Nonlinear single-mode fiber coupler using liquid crystals, *Appl. Phys. Lett.* 46:338, and Sec. 12.1.2 on self-guiding solitons.

98. ANSI Standard Z136.1. American national standard for the safe use of lasers. 2000. New York : American National Standards Institute, Inc.

99. Tutt, L. W., and T. F. Boggess. 1993. Review of optical limiting mechanisms and devices using organics, fullerenes, semiconductors and other materials. *Prog. Quantum Electron.* 17:299–338.

100. Spangler, C. W. 1999. Recent development in the design of organic materials for optical power limiting. *J. Mater. Chem.* 9:2013–2020.

101. Sutherland, R. L., M. C. Brant, D. M. Brandelik, P. A. Fleitz, D. G. McLean, and T. Pottenger. 1993. Nonlinear absorption study of a C60 - toluene solution. *Opt. Lett.*, 18:858–860.

102. Miles, P. A., 1994. Bottleneck optical limiters: The optimal use of excited-state absorbers. *Appl. Opt.* 33:6965–6979, see also Xia, T., D. J. Hagan, A. Dogariu, A. A. Said, and E. W. Van Stryland. 1997. Optimization of optical limiting devices based on excited-state absorption. *Appl. Opt.*, 36:4110–22.

103. Ehrlich, J. E., X. L. Wu, I. Y. S. Lee, Z. Y. Hu, H. Rockel, S. R. Marder, and J. W. Perry. 1997. Two-photon absorption and broadband optical limiting with bis-donor stilbenes. *Opt. Lett.* 22:1843–1845.

104. Khoo, I. C., P. H. Chen, M. V. Wood, and M.-Y. Shih. 1999. Molecular photonics of a highly nonlinear organic fiber core liquid for picosecond-nanosecond optical limiting application. *Chem. Phys.* 245:517–31.

105. Khoo. I. C., 1996. US Patent 5,589,101, Liquid crystal fiber array for optical limiting of laser pulses and for eye/sensor protection. Issued 31 December 1996.

106. Khoo, I. C., M. V. Wood, B. D. Guenther, M.-Y. Shih, and P. H. Chen. 1998. Nonlinear absorption and optical limiting of laser pulses in a liquid-cored fiber array. *J. Opt. Soc. Am. B.* 15:1533–40.

107. Khoo, I.-C., A. Diaz, M. V. Wood, and P. H. Chen. 2001. Passive optical limiting of picosecond-nanosecond laser pulses using highly nonlinear organic liquid cored fiber array. *IEEE J. Sel. Top. Quantum Electron.* 7:760–768.

108. Khoo, I.C., Andres Diaz, and J. Ding. 2004. Nonlinear-absorbing fiber array for large dynamic range optical limiting application against intense short laser pulses. *J. Opt. Soc. Am. B.* 21:1234–1240.

109. He, G. S., T.-C. Lin, P. N. Prasad, C.-C. Cho, and L.-J. Yu. 2003. Optical power limiting and stabilization using a two-photon absorbing neat liquid crystal in isotropic phase. *Appl. Phys. Lett.* 82:4717–19.

110. Barroso, J., A. Costela, I. Garcia-Moreno, and J. L. Saiz. 1998. Wavelength dependence of the nonlinear absorption of C-60- and C-70-toluene solutions. *J. Phys. Chem. A.* 102:2527–2532.

111. See, for example, Khoo, I. C., D. H. Werner, X. Liang, A. Diaz and B. Weiner. 2006. Nano-sphere dispersed liquid crystals for funable negative-zero-positive index of refraction in the optical and Terahertz regimes. *Opt. Lett.* 31:2592–2594.

Index